Lecture Notes in Statistics 137
Edited by P. Bickel, P. Diggle, S. Fienberg, K. Krickeberg,
I. Olkin, N. Wermuth, S. Zeger

Springer
*New York
Berlin
Heidelberg
Barcelona
Budapest
Hong Kong
London
Milan
Paris
Singapore
Tokyo*

V. Seshadri

The Inverse Gaussian Distribution
Statistical Theory and Applications

 Springer

V. Seshadri
Department of Mathematics and Statistics
McGill University
Montreal, PQ
Canada H3A 2K6

```
Library of Congress Cataloging-in-Publication Data

Seshadri, V.
   The inverse Gaussian distribution : statistical theory and
applications / V. Seshadri.
      p.   cm. -- (Lecture notes in statistics ; 137)
   Includes bibliographical references and index.
   ISBN 0-387-98618-9 (softcover : alk. paper)
   1. Inverse Gaussian distribution.  I. Title.  II. Series: Lecture
notes in statistics (Springer-Verlag) ; v. 137.
QA276.7.S472  1998
519.2'4--dc21                                                 98-38589
```

Printed on acid-free paper.

© 1999 Springer-Verlag New York, Inc.
All rights reserved. This work may not be translated or copied in whole or in part without the written permission of the publisher (Springer-Verlag New York, Inc., 175 Fifth Avenue, New York, NY 10010, USA), except for brief excerpts in connection with reviews or scholarly analysis. Use in connection with any form of information storage and retrieval, electronic adaptation, computer software, or by similar or dissimilar methodology now known or hereafter developed is forbidden.
The use of general descriptive names, trade names, trademarks, etc., in this publication, even if the former are not especially identified, is not to be taken as a s.gn that such names, as understood by the Trade Marks and Merchandise Marks Act, may accordingly be used freely by anyone.

Camera ready copy provided by the author.
Printed and bound by Braun-Brumfield, Ann Arbor, MI.
Printed in the United States of America.

9 8 7 6 5 4 3 2 1

ISBN 0-387-98618-9 Springer-Verlag New York Berlin Heidelberg   SPIN 10690409

PREFACE

This book is written in the hope that it will serve as a companion volume to my first monograph. The first monograph was largely devoted to the probabilistic aspects of the inverse Gaussian law and therefore ignored the statistical issues and related data analyses.

Ever since the appearance of the book by Chhikara and Folks, a considerable number of publications in both theory and applications of the inverse Gaussian law have emerged thereby justifying the need for a comprehensive treatment of the issues involved. This book is divided into two sections and fills up the gap updating the material found in the book of Chhikara and Folks. Part I contains seven chapters and covers distribution theory, estimation, significance tests, goodness-of-fit, sequential analysis and compound laws and mixtures. The first part forms the backbone of the theory and wherever possible I have provided illustrative examples for easy assimilation of the theory.

The second part is devoted to a wide range of applications from various disciplines. The applied statistician will find numerous instances of examples which pertain to a first passage time situation. It is indeed remarkable that in the fields of life testing, ecology, entomology, health sciences, traffic intensity and management science the inverse Gaussian law plays a dominant role. Real life examples from actuarial science and ecology came to my attention after this project was completed and I found it impossible to include them.

I began this project during my sabbatic year spent in the Department of Mathematics at the University of Western Australia. I am most grateful to Tony Pakes and his colleagues for providing me an ideal environment for research. During the course of the writing I have profitted from some correspondence I had with Satish Iyengar, Qiming Liao and Dr.M.W.Levine. The manuscript was first typed by Dr.Ivano Pinneri in Perth. Upon my return to McGill additional typing was done by Heather MacAuliffe and Sithamparapillai Ambikkumar. Dan Bododea helped me with the cumbersome task of preparing the figures. Subsequently I have enjoyed the hospitality of the Mathematics Department at San Diego State University through the kind efforts of Dr. Anantha. The project was completed in the congenial atmosphere of their computing laboratory. I am very appreciative of the hospitality afforded to me there. Nikolaus Kleiner and Ahilan Anantha used their expertise with TEX to format and prepare some complicated tables. I am very grateful to them for their assistance. I thank Raffaella Bruno who spent several hours proof-reading and eradicating many spelling errors. A special word of thanks is due to John Kimmel for his guidance and generosity. I wish to thank my wife and family for their constant encouragement and the patience they have shown during my long hours of absence from home. Financial support from the Natural Sciences and Engineering Research Council of Canada is gratefully appreciated.

I would like to dedicate this work to Etienne Halphen and Ken Tweedie without whose pioneering efforts this distribution would not have attained the popularity it enjoys today. Etienne Halphen had the foresight to discover the generalized inverse Gaussian law but his life was cut short by tragedy. Ken Tweedie was the first to launch this law into the statistical world. I have had a brief correspondence with Ken while he was in a nursing home in Liverpool (I learned this from C.A.B.Smith) and was hoping to visit him on my return from Australia. Alas that was not to be since I received word from his daughter that Ken had passed away peacefully in April 1996.

Montreal, June 1998   *V. Seshadri*

ACKNOWLEDGMENTS

I am grateful to numerous publishers for their kind permission to reproduce several figures and tables from their journals. I list below the respective publishers, the journals, and the appropriate references.

AMERICAN STATISTICAL ASSOCIATION
JASA(1976) vol.71(356)-table 1,p 826,table3,pg 827 (Tables 7.7, 7.8)
JASA(1983) vol.78(383)-table 1,p 661 (Table 2.1)
JASA(1983) vol.78(384)-tables 2,3 p 825 (Tables B.6, B.7)
Technometrics(1963) vol.5(3)-table 1,p376 (Table 6.1)
AMERICAN SOCIETY FOR QUALITY
Journal of Quality Technology(1985) vol.17(3)-fig.1,p151 (Figure 2.1)
BLACKWELL PUBLISHERS,OXFORD
JRSSA(1972) vol.135(2)-tables 1,2,3,pp 265,267,269 (Tables H.3-H.6)
JRSSA(1979) vol.142(iv)-table 1,p 471 (Tables H.1,H.2)
JRSSC(1982) vol.31-table 5,p 200 (Table 7.5)
JRSSC(1986) vol.35(1)-table 1,p 12 (Table N.1)
JRSSD(1997) vol.46(3)-tables 5,6,p 334 (Tables T.1,T.2)
SCAND.J.STATIST(1978) vol.5-table 1,p 202 (Table 3.1)
SCAND.J.STATIST(1986) vol.13-table 1b,p 214 (Table 7.9)
BLACKWELL SCIENCE,OXFORD
Journal of Animal Ecology(1970) vol.39-table 1,p 33 (Table R.1)
CALIFORNIA FISH AND
GAME(1980) vol.66-table 1,pp 42-3 (Table Q.1)
CANADIAN JOURNAL OF
STATISTICS(1992) vol.20-table 1,p390 (Table 6.1)
ELSEVIER SCIENCE LTD,OXFORD
Arch. of Oral Biol.(1990) vol.35(1)-fig.4,p41 (Figure D.1)
Arch. of Oral Biol.(1990) vol.35(1)-tables 1,2,pp38,41 (Tables D.1,D.2)
Microelectronics and Reliability(1994) vol.34(1)-table 1,p21 (Table 7.14)
Microelectronics and Reliability(1994) vol.34(1)-fig.1,p189 (Fig.5.4)
Microelectronics and
Reliability(1994) vol.34(2)-figs.1,2,6,pp253,258 (Figs.5.1-5.3)
ELSEVIER SCIENCE,AMSTERDAM,THE NETHERLANDS
Ecological modelling(1984) vol.24-table 1,p199 (Table B.3)

Ecological modelling(1984) vol.24-figs.2-7,pp199-201 (Figs.B.1-B.6)
ENTOMOLOGICAL SOCIETY OF AMERICA
Environmental Ecology(1970) vol.21(6)-table 1,p.1234 (Table R.2)
GORDON BREACH PUBLISHERS,LAUSANNE,SWITZERLAND
Jour. Statist.Comp.and Sim.(1994) vol.51-figs.1,2,pp24-5 (Figs 3.1-3.2)
Jour.Math.Soc.(1977) vol.5,p275 (Table J.1)
IEEE PUBLISHING SERVICES
IEEE Trans.on Elec.InsulationEI-6(1971)-fig.1,p166 (Table G.1)
IEEE Trans.on Rel.R(1986) vol.35(4),pp407-8 (Tables 3.3-3.5)
INSTITUTE OF MATHEMATICAL STATISTICS
Lecture Notes-Mon.series(1982) vol.2-table 1,p116 (Table G.2)
MARCELL DEKKER,NEW YORK
Comm.in Statist-Th.and Meth.(1988) vol.17(1)-fig.3,p566 (Fig.7.15)
Comm.in Statist-Th.
and Meth.(1990) vol.19(4)-tables 1,2 pp1463-4 (Tables 4.2-4.3)
OXFORD UNIVERSITY PRESS
Biometrika(1930) vol.30-table D p 387 (Table 7.5)
Biometrika(1965) vol.52-table 1,p 389 (Table 7.3)
SCANDINAVIAN UNIVERSITY PRESS
Scandinavian Actuarial
Journal(1987) vol.18(4)-table 1, p123 Table 7.4
SOUTH AFRICAN STATISTICAL
JOURNAL(1978) vol.12-table 1,p115 (Table 4.1)
WORLD SCIENTIFIC PUBLISHING COMPANY
Int.Jour.of Rel.Quality and
Safety Eng.(1994) vol.1(3)-figs.1,2,pp382-3 (Figs 5.5,5.6)

I am also grateful to Q.Liao for his kind permission to include tables and figures from his Ph.D thesis whch are to be found in Section K, and Arthur Fries and Gouri Bhattacharyya for permitting me to include Table 2 (page 98) appearing in their article from Reliability and Quality Control,1986 and which appears as Table O.2.

Contents

PART I STATISTICAL THEORY

1. **Distribution theory**
   - 1.0 Introduction 1
   - 1.1 Limit laws 5
   - 1.2 Sampling distributions 7
   - 1.3 Conditional distributions 15
   - 1.4 Bayesian sampling distributions 19

2. **Estimation**
   - 2.0 Introduction 23
   - 2.1 Estimation 24
   - 2.2 A shifted model 26
   - 2.3 Maximum likelihood estimation 28
   - 2.3 Estimation under truncation 37

3. **Significance tests**
   - 3.0 Introduction 38
   - 3.1 Likelihood-ratio tests. One sample case 38
   - 3.2 Brownian motion process 43
   - 3.3 Power considerations 44
   - 3.4 Two sample tests 45
   - 3.5 Interval estimation 52
   - 3.6 Examples 58
   - 3.7 Tolerance limits 61
   - 3.8 Tests of separate families 65
   - 3.9 Bahadur efficient tests 70

4. **Sequential methods**
   - 4.0 Introduction 73
   - 4.1 Sequential probability ratio test 73
   - 4.2 Sequential test for the mean and asymptotics 78
   - 4.3 Tests with known coefficient of variation 84
   - 4.4 Asymptotically risk-efficient sequential estimation 87
   - 4.5 Control charts 90

5. **Reliability & Survival analysis**
   - 5.0 Introduction 92

    5.1 Estimation of reliability 92
    5.2 Confidence bounds and tolerance limits 96
    5.3 Hazard rate 102
    5.4 Estimation of critical time 104
    5.5 Confidence intervals for hazard rate. 106

6. **Goodness-of-fit**
    6.0 Introduction 114
    6.1 Modified Kolmogorov-Smirnov test 114
    6.2 Anderson-Darling Statistic 116

7. **Compound laws & mixtures**
    7.0 Introduction 121
    7.1 Poisson-inverse Gaussian 121
    7.2 Inference 125
    7.3 Examples 133
    7.4 A compound inverse Gaussian model 136
    7.5 Normal-gamma mixture 139
    7.6 Normal inverse Gaussian mixture 141
    7.7 A mixture inverse Gaussian 143
    7.8 Exponential-inverse Gaussian mixtures 150
    7.9 Birnbaum-Saunders distribution 154
    7.10 Linear models and the P-IG law 159
    7.11 P-IG regression model 163

PART II   APPLICATIONS

A. **Actuarial science** 167
    Claim cost analysis 167

B. **Analysis of reciprocals** 172
    One-way classification 172
    An application in environmental sciences 175
    Analysis of two factor experiments 181
    Tests for model adequacy 185
    The analysis of reciprocals 187

C. **Demography** 191

D. **Histomorphometry** 194

E. **Electrical networks** 198

F. **Hydrology** 203
   Emptiness of a dam 204

G. **Life tests** 206
   Shelf life failures 206
   Accelerated life tests 207
   Least squares 210
   Variable stress accelerated tests 214

H. **Management science** 220
   Labour turnover 220
   Duration of strikes 223

I. **Meteorology** 230

J. **Mental health** 232

K. **Physiology** 235
   Tracer dilution curves 235
   Pharmacokinetics 237
   Interspike train interval analysis 239

L. **Remote sensing** 252
   Photogrammetry 252
   Cookie cutter detection 257

M. **Traffic noise intensity** 259
   Model assumptions 259

N. **Market research** 262

O. **Regression** 265
   Asymptotics 270
   Analysis of Reciprocals (revisited) 273
   Regression diagnostics 275
   Strong consistency and bookstrap 280

P. **Slug lengths in pipelines** 284

Q. **Ecology** 286
   Time till extinction 286
   Endangered species 289

R. **Entomology** 298
   A stochastic model 298
   Estimation and model adequacy 300

S. **Small area estimation** 305

T. **CUSUM** 309
   Cusum charts 309

U. **Plutonium Estimation** 314
   Model development 314

   References 317
   Author Index 340
   Subject Index 345
   Glossary 347

PART I STATISTICAL THEORY

CHAPTER 1

DISTRIBUTION THEORY

1.0 Introduction

The inverse Gaussian distribution has a history dating back to 1915 when Schrödinger and Smoluchowski presented independent derivations of the density of the first passage time distribution of Brownian motion with positive drift. The drift free case had already been published by Bachelier in 1900 in his doctoral thesis on the theory of speculation. Among the early advocates of this distribution one should single out Hadwiger (1940a, 1940b, 1941, 1942) and Halphen. The first of Hadwiger's expositions dealt with the inverse Gaussian law as a solution to a functional equation in renewal theory. The other papers dealt with applications of the distribution to the reproduction functions arising in the study of population growth. Halphen is credited with the first formulation of what is now known as the generalized inverse Gaussian distribution. His discovery arose from the need to model hydrologic data whose behaviour was subject to decay for both large and small values. A general discussion of the early history can be found in Seshadri (1993) and Chhikara and Folks (1988). The modern day statistical community became aware of this law through the pioneering work of Tweedie (1941, 1945, 1946, 1947, 1956, 1957a, 1957b). The very name "inverse Gaussian" is Tweedie's creation and is based on his observation that the cumulant function of this law is the inverse of the cumulant function of the normal law. From the point of view of probability and mathematical statistics the distribution can be regarded as a natural exponential family generated by the one-sided stable law with index $\frac{1}{2}$. Thus if $\lambda > 0$ and

$$\mu(dx) = \sqrt{\frac{\lambda}{2\pi x^3}} \exp\left(-\frac{\lambda}{2x}\right) 1_{R^+}(x) dx,$$

then the Laplace transform of $\mu(dx)$ is

$$\int_0^\infty e^{\theta x} \mu(dx) = -\sqrt{-2\lambda\theta} \quad \text{for } \theta \in (-\infty, 0].$$

Thus the natural exponential family generated by $\mu(dx)$ is

$$\sqrt{\frac{\lambda}{2\pi x^3}} \exp\left(-\frac{\lambda}{2x} + \theta x + \sqrt{-2\lambda\theta}\right) 1_{R^+}(x) dx. \qquad (1.1).$$

Writing $\theta = -\frac{\lambda}{2\mu^2}$, $\mu > 0$ we obtain the familiar form in which Tweedie presented this law (abbreviated $IG(\mu, \lambda)$), namely,

$$f(x; \mu; \lambda) = \sqrt{\frac{\lambda}{2\pi x^3}} \exp\left(-\frac{\lambda}{2\mu^2} \frac{(x-\mu)^2}{x}\right) 1_{R^+}(x). \qquad (1.2).$$

It is clear from (1.1) and (1.2) that the Laplace transform is given by

$$\exp\left[\frac{\lambda}{\mu}\left\{1 - \left(1 + \frac{2\mu^2}{\lambda}\theta\right)^{\frac{1}{2}}\right\}\right].$$

Tweedie also provided three other versions corresponding to the parametrizations where (μ, λ) is replaced by (α, λ), (μ, ϕ) and (ϕ, λ) with $\lambda = \mu\phi, \mu = (2\alpha)^{-\frac{1}{2}}, \alpha > 0$. The respective densities are

$$f_1(x; \alpha, \lambda) = \sqrt{\frac{\lambda}{2\pi x^3}} \exp\left(-\frac{\lambda}{2x} - \alpha\lambda x + \lambda\sqrt{2\alpha}\right) 1_{R^+}(x),$$

$$f_2(x; \mu, \phi) = \sqrt{\frac{\mu\phi}{2\pi x^3}} \exp\left(-\frac{\mu\phi}{2x} - \frac{\phi x}{2\mu} + \phi\right) 1_{R^+}(x), \quad \text{and}$$

$$f_3(x; \phi, \lambda) = \sqrt{\frac{\lambda}{2\pi x^3}} \exp\left(-\frac{\lambda}{2x} - \frac{\phi^2 x}{2\lambda} + \phi\right) 1_{R^+}(x).$$

Each of these forms is suitable for data analysis in different disciplines. One can readily verify the following relationships.

$$f(x; \mu, \lambda) = \mu^{-1} f_2\left(\frac{x}{\mu}; 1, \phi\right) = \lambda^{-1} f_3\left(\frac{x}{\lambda}; \phi, 1\right)$$

Thus the parameters μ and λ are of the same physical dimension as the random variable X. When $\mu = 1$, the density is referred to as the standardized inverse Gaussian law or more simply as Wald's distribution (Zigangirov,1962), since the same family emerges as the limiting law of the sample size in a special case of Wald's sequential analysis (1944). The distribution function $F(x)$ is

$$F(x) = \Phi(\alpha(x)) + \exp\left(\frac{2\lambda}{\mu}\right) \Phi(\bar{\alpha}(x))$$

where

$$\alpha(x) = \frac{x-\mu}{\mu}\sqrt{\lambda/x}, \quad \bar{\alpha}(x) = -\frac{x+\mu}{\mu}\sqrt{\lambda/x}.$$

Introduction

The parameter ϕ determines the shape of the distribution and the density is highly skewed for moderate values of ϕ. As ϕ increases the inverse Gaussian tends towards the normal law. Some densities are presented in Figures 1.1 and 1.2 for various values of $\mu(\lambda = 1)$ and $\lambda(\mu = 1)$. The density is unimodal with the mode located at

$$X_{\text{mode}} = \mu\{(1 + \frac{9}{4\phi^2})^{\frac{1}{2}} - \frac{3}{2\phi}\}.$$

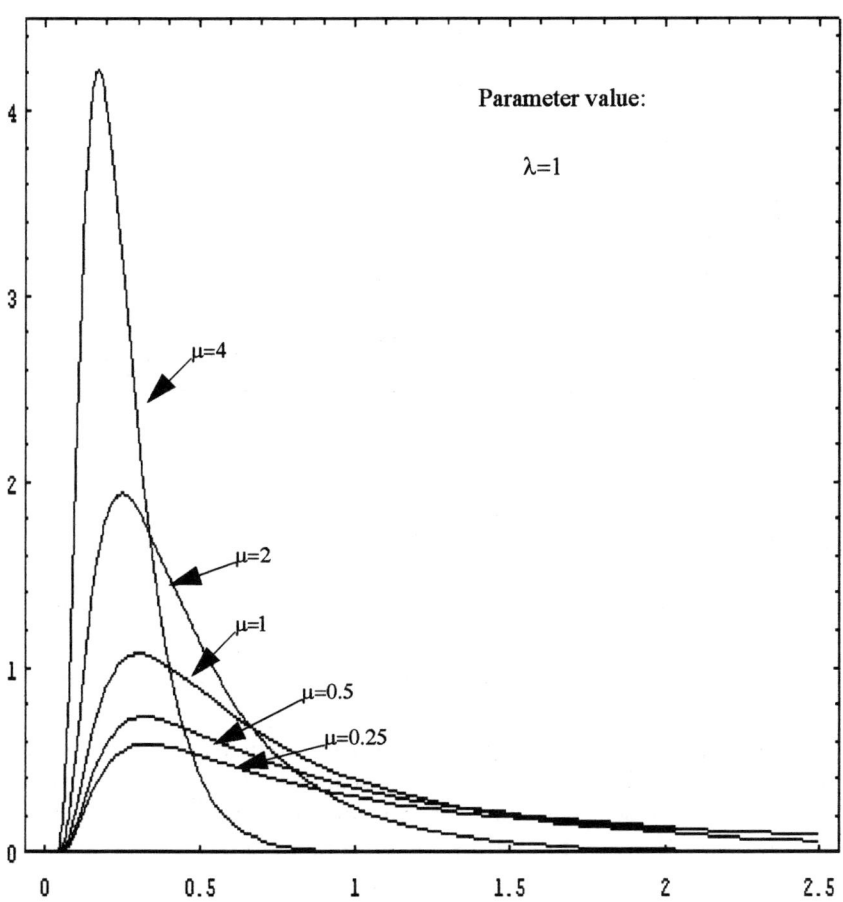

Figure 1.1 *Density functions of $IG(\mu, \lambda)$ for fixed λ and increasing μ*

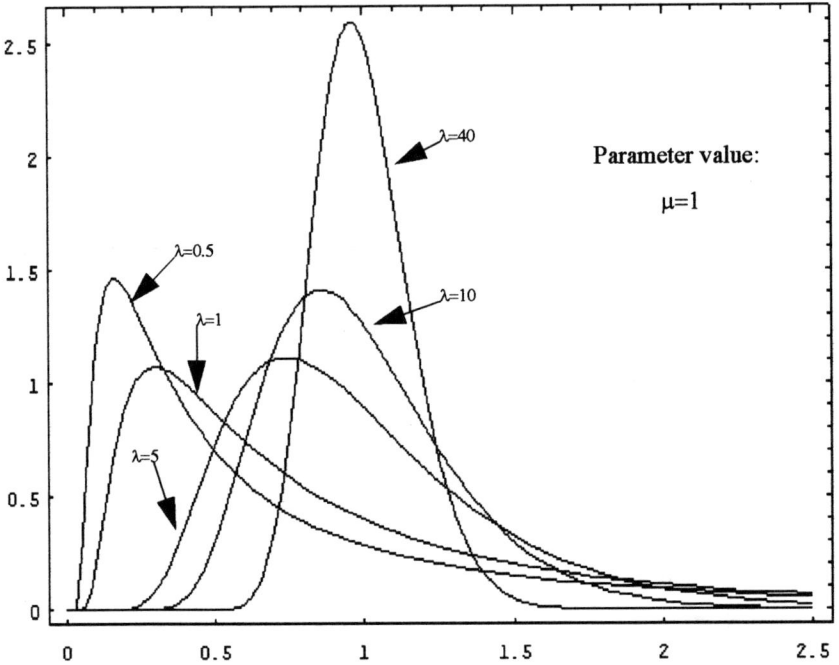

Figure 1.2 *Density functions of $IG(\mu, \lambda)$ for fixed μ and increasing λ*

Moreover the Laplace transform of $\frac{1}{X}$ has the form

$$(1+\frac{2\theta}{\lambda})^{-\frac{1}{2}} \exp\left[\frac{\lambda}{\mu}\left\{1 - \left(1 + \frac{2\theta}{\lambda}\right)^{\frac{1}{2}}\right\}\right]$$

which is the product of two Laplace transforms, namely, that of a gamma law, $\Gamma(\frac{1}{2}, \frac{\lambda}{2})$ and that of an inverse Gaussian law, $IG(\frac{1}{\mu}, \frac{\lambda}{\mu^2})$. Thus the IG law has both positive and negative moments and it can be shown that for real k they are related by the formula

$$\mathbb{E}\left(\frac{\mu}{X}\right)^k = \mathbb{E}\left(\frac{X}{\mu}\right)^{k+1}.$$

We will now briefly examine some limit laws of X and X^{-1} when $\mathcal{L}(X) = IG(\mu, \lambda)$.

1.1 Limit Laws

From the Laplace transform of X, namely

$$\mathbb{E}\{\exp(-\theta X)\} = \exp\left[\frac{\lambda}{\mu}\left\{1 - \left(1 + \frac{2\mu^2}{\lambda}\theta\right)^{\frac{1}{2}}\right\}\right]$$

we have upon writing $\phi = \frac{\lambda}{\mu}$

$$\mathbb{E}\{\exp(-\theta X)\} = \exp\left[\phi\left\{1 - \left(1 + \frac{2\theta}{\phi}\right)^{\frac{1}{2}}\right\}\right]$$

Suppose $\phi \to \infty$. Then

$$1 - \left(1 + \frac{2\theta\mu}{\phi}\right)^{\frac{1}{2}} = -\frac{\theta\mu}{\phi} + \frac{\theta^2\mu^2}{\phi^2}[1 + O(1)].$$

Hence

$$\mathbb{E}\left\{\exp\left(-\frac{\theta X}{\mu}\right)\right\} \to \exp(-\theta).$$

This shows that as $\phi \to \infty$ $\frac{x}{\mu} \to 1$ in probability. Note that this is independent of how λ and μ behave individually, as long as $\phi \to \infty$.

When μ is fixed and $\lambda \to \infty$, we get $X \to \mu$ in probability as $X^{-1} \to \mu^{-1}$ in probability.

On the other hand we could have $\lambda, \mu \to \infty$ or $\lambda \to \infty$, $\mu \to 0$ or $\lambda, \mu \to 0$.

(a) limit law of $\sqrt{\phi}(\frac{X}{\mu} - 1) = Y_1$.

$$\mathbb{E}\left[\exp\left\{-\theta\sqrt{\phi}\left(\frac{X}{\mu} - 1\right)\right\}\right] = \exp(\theta\sqrt{\phi})\mathbb{E}\left[\exp\left\{-\frac{\theta\sqrt{\phi}X}{\mu}\right\}\right]$$

$$= \exp(\theta\sqrt{\phi})\exp\left\{\phi - \phi\left(1 + \frac{2\theta\sqrt{\phi}\mu^2}{\mu\lambda}\right)^{\frac{1}{2}}\right\}$$

$$= \exp(\theta\sqrt{\phi})\exp\left\{\phi - \phi\left(1 + \frac{2\theta}{\sqrt{\phi}}\right)^{\frac{1}{2}}\right\}$$

$$= \exp\left\{\theta\sqrt{\phi} - \theta\sqrt{\phi} + \frac{\theta^2}{2}(1 + O(1))\right\}$$

$$\Rightarrow \exp\frac{\theta^2}{2} \quad \text{as } \phi \to \infty.$$

Pakes' continuity theorem (1978) for moment generating functions now yields $\sqrt{\phi}\left(\frac{X}{\mu} - 1\right) \to N(0, 1)$.

(b) limit law of $Y_2 = \sqrt{\phi}\left(\frac{\mu}{X} - 1\right)$. We can write $Y_1 = \sqrt{\phi}\left(\frac{X}{\mu} - 1\right)$ as $X\sqrt{\phi}\left(\frac{1}{\mu} - \frac{1}{X}\right)$. If μ remains fixed then from (a) and Slutsky's lemma, Y_2 behaves like $\mu\sqrt{\phi}\left(\frac{1}{\mu} - \frac{1}{X}\right)$ so that $\mu\sqrt{\phi}\left(\frac{1}{X} - \frac{1}{\mu}\right) \to N(0,1)$. This suggests $\sqrt{\phi}\left(\frac{\mu}{X} - 1\right) \to N(0,1)$, if $\phi \to \infty$, regardless of how μ behaves. Direct verification involves computation of

$$\mathbb{E}\left[\exp\left(-\frac{\theta\mu}{X}\right)\right] = \left(1 + \frac{2\theta}{\phi}\right)^{-\frac{1}{2}} = \exp\left[\phi\left\{1 - \left(1 + \frac{2\theta}{\phi}\right)^{\frac{1}{2}}\right\}\right].$$

Whitmore and Yalovsky (1978) observed that the convergence to normality of $\sqrt{\phi}\left(\frac{X}{\mu} - 1\right)$ was not fast enough. From the expression for the distribution function of X, namely $F(x;\mu,\lambda)$ we have for $x = 1$, a power series expansion as $\phi \to \infty$ yielding

$$F(1;\phi) = \frac{1}{2} + \exp(2\phi)\Phi(-2\sqrt{\phi})$$
$$\approx \frac{1}{2} + \frac{1}{\sqrt{8\pi\phi}}.$$

Thus the transformation $X \mapsto \sqrt{\phi}(\frac{X}{\mu} - 1)$ results in an error of at least $\frac{1}{\sqrt{8\pi\phi}}$. This error exceeds 0.006 unless $\phi \gg 1000$. Hence Whitmore and Yalovsky propose the transformation

$$Y = \frac{1}{2\sqrt{\phi}} + \sqrt{\phi}\log\left(\frac{X}{\mu}\right)$$

to obtain a much closer approximation to normality. Denoting by X_α and Z_α the $100\alpha^{\text{th}}$ percentile of the IG and the standard normal law respectively the above transformation provides the following approximate relations between X_α and Z_α.

$$X_\alpha \approx \mu \exp\left(\frac{Z_\alpha}{\sqrt{\phi}} - \frac{1}{2\phi}\right).$$

1.2 Sampling distributions

We now develop some basic results concerning the sampling distributions associated with the inverse Gaussian distribution. These distributions are used in subsequent chapters that deal with statistical inference. Throughout this book we use the notation $\mathcal{L}(X) = IG(\mu, \lambda)$ to indicate that a random variable X follows the inverse Gaussian law with mean parameter μ and λ a secondary parameter. Since the IG law is a special member of the Halphen family, known more commonly as the generalized inverse Gaussian law, we will, as the occasion demands, use the notation $GIG(\alpha, \chi, \psi)$ to denote the Halphen family. We remind the readers that when $\alpha = -\frac{1}{2}, \chi = \lambda$ and $\psi = \frac{\lambda}{\mu^2}$ we obtain the $IG(\mu, \lambda)$ law, and when $\alpha = \frac{1}{2}, \chi = \lambda$ and $\psi = \frac{\lambda}{\mu^2}$ we obtain the reciprocal inverse Gaussian law which is abbreviated as $RIG(\mu, \lambda)$. For a random sample of fixed size n from $IG(\mu, \lambda)$, denoted by $X_1, \ldots, X_n$, the log likelihood $l(\mu, \lambda)$ is proportional to

$$l(\mu, \lambda) \propto \frac{n}{2} \log \lambda + \frac{n\lambda}{\mu} - \frac{n\lambda \bar{x}}{2\mu^2} - \frac{n\lambda \bar{x}_-}{2} \qquad (1.3)$$

where $\bar{X} = (X_1 + .. + X_n)/n$ and $\bar{X}_- = (\frac{1}{X_1} + \cdots + \frac{1}{X_n})/n$. Differentiating (1.3) with respect to μ and λ we obtain the maximum likelihood estimates $\hat{\mu} = \bar{X}$ and $\hat{\lambda}^{-1} = \frac{1}{n}\Sigma(\frac{1}{X_i} - \frac{1}{\bar{X}})$. These results readily follow from exponential family theory. Moreover $T = (\bar{X}, \bar{X}_-)$ forms a minimal sufficient statistic for (μ, λ). Not only is T complete for the family but by an application of the Lehmann-Scheffe theorem $(\hat{\mu}, \hat{\lambda}^{-1})$ are minimum variance unbiased estimators of (μ, λ^{-1}). To generalize this slightly we consider a random sample $(X_1, \ldots, X_n)$ such that $\mathcal{L}(X_i) = IG(\mu, \lambda_i)$ where $\lambda_i = \lambda_0 \omega_i$, with $\omega_i > 0$ and known and $\lambda_0 > 0$. Then it follows that

$$\hat{\mu} = \sum_{i=1}^{n} \omega_i X_i / \sum_{i=1}^{n} \omega_i \qquad (1.4)$$

and

$$\hat{\lambda}_0^{-1} = \frac{1}{n} \sum_{i=1}^{n} \omega_i \left(\frac{1}{X_i} - \frac{1}{\hat{\mu}} \right). \qquad (1.5)$$

To obtain the distribution of $\hat{\mu}$ we consider the Laplace transform

$L_{\hat{\mu}}(\theta)$.

$$L_{\hat{\mu}}(\theta) = \prod_{i=1}^{n} L_{x_i}\left(\frac{\omega_i \theta}{\sum \omega_i}\right)$$

$$= \prod_{i=1}^{n} \exp\left(\frac{\lambda_0 \omega_i}{\mu}\left(1 - \left(1 + \frac{2\mu^2 \theta}{\lambda_0 \Sigma \omega_i}\right)^{\frac{1}{2}}\right)\right)$$

$$= \exp\left(\frac{\lambda_0 \Sigma \omega_i}{\mu}\left(1 - \left(1 + \frac{2\mu^2 \theta}{\lambda_0 \Sigma \omega_i}\right)^{\frac{1}{2}}\right)\right).$$

This formula says that

$$\mathcal{L}(\hat{\mu}) = IG(\mu, \lambda_0 \Sigma \omega_i)$$

yielding us the following proposition.

Proposition 1.1 *The sampling distribution of the mean of n independent observations from $IG(\mu, \lambda_0 \omega_i)$, $\mu, \lambda_0, \omega_i > 0$ is $IG(\mu, \lambda_0 \Sigma \omega_i)$.*

Next we examine the sampling distribution of $Q = \frac{1}{n} \sum \omega_i (\frac{1}{X_i} - \frac{1}{\hat{\mu}})$. The joint distribution of $(X_1, \ldots, X_n)$ is

$$\left(\frac{\lambda_0}{2\pi}\right)^{\frac{n}{2}} \prod_{i=1}^{n} \left(\frac{\omega_i}{x^3}\right)^{\frac{1}{2}} \exp\left(\frac{\lambda_0 \sum \omega_i}{\mu}\right) \exp\left\{-\frac{\lambda_0 \sum^n \omega_i x_i}{2\mu^2} - \frac{\lambda_0}{2} \sum_{i=1}^{n} \frac{\omega_i}{x_i}\right\}$$

over $(\mathbb{R}^+)^n$ while that of $\hat{\mu}$ is

$$\left(\frac{\lambda_0 \Sigma \omega_i}{2\pi \hat{\mu}^3}\right)^{\frac{1}{2}} \exp\left(\frac{\lambda_0 \Sigma \omega_i}{\mu}\right) \exp\left\{-\frac{\lambda_0 \sum_{i=1}^n \omega_i x_i}{2\mu^2} - \frac{\lambda_0 (\sum_{i=1}^n \omega_i)^2}{2 \sum_{i=1}^n \omega_i x_i}\right\}$$

so that conditionally given $\hat{\mu}$ the joint law of $(X_1, \ldots, X_n)$ is

$$\left(\frac{\lambda_0}{2\pi}\right)^{\frac{n-1}{2}} \frac{\prod_{i=1}^n \omega_i^{\frac{1}{2}} \hat{\mu}^{\frac{3}{2}}}{(\sum \omega_i)^{\frac{1}{2}} \prod x_i^{\frac{3}{2}}} \exp\left(-\frac{\lambda_0}{2} \sum \omega_i \left(\frac{1}{x_i} - \frac{1}{\hat{\mu}}\right)\right)$$

The Laplace transform of $\lambda_0 \sum \omega_i(\frac{1}{X_i} - \frac{1}{\hat{\mu}})$ in this conditional distribution does not contain $\hat{\mu}$ and is $(1 + 2\theta)^{-\frac{(n-1)}{2}}$ so that $\lambda_0 \sum \omega_i (\frac{1}{X_i} - \frac{1}{\hat{\mu}}) \perp\!\!\!\perp \hat{\mu}$ (is independent of) and follows a χ^2_{n-1} law.

This gives us the next proposition.

Proposition 1.2 *Suppose that $X = (X_1, \ldots, X_n)$ where the X_i is a sample from $IG(\mu, \lambda_0 \omega_i)$, where $\mu, \lambda_0, \omega_i > 0$. Then $\hat{\mu} = \frac{\sum_{i=1}^n \omega_i X_i}{\sum_{i=1}^n \omega_i}$ is independent of*

$$Q = \lambda_0 \sum_{i=1}^{n} \left(\frac{\omega_i}{X_i} - \frac{(\sum \omega_i)^2}{\sum \omega_i X_i}\right) \quad \text{and} \quad \mathcal{L}(Q) = \chi^2_{n-1}$$

Sampling distributions 9

When $\mu = \alpha t_i$ and $\lambda_i = \alpha t_i^2$ the distributional result was given in Seshadri (1993). Under the hypothesis of Proposition 1.2 it is a simple exercise to verify the next proposition.

Proposition 1.3 *Let* $Q_0 = \frac{\lambda_0}{\mu^2} \frac{(\sum_{i=1}^{n} w_i X_i - \mu) \sum_{i=1}^{n} w_i)}{\sum_{i=1}^{n} w_i X_i}$ *and*

$$\bar{Q} = \lambda_0 \sum_{i=1}^{n} \frac{w_i (X_i - \mu)^2}{\mu^2 X_i}; \text{ then } \mathcal{L}(\bar{Q}) = \chi_n^2 \text{ and } \mathcal{L}(Q_0) = \chi_1^2$$

It is easily verified that $\bar{Q} = Q + Q_0$ and the decomposition parallels that for the normal law. In fact Q admits of a further decomposition as stated in the following theorem. Its proof is similar to the case when $\mathcal{L}(X_i) = IG(\alpha t_i, \alpha t_i^2)$ for $\alpha, t_i > 0$ and can be found in Seshadri (1993).

Theorem 1.1 *Let* $\mathcal{L}(X_1, \ldots, X_n) = IG(\mu, \lambda_0 w_1) \otimes \cdots \otimes IG(\mu, \lambda_0 w_1)$ *and define*

$$Q_k = \lambda_0 \left(\frac{\left(\sum_{i=1}^{k} w_i\right)^2}{\sum_{i=1}^{k} w_i X_i} + \frac{w_{k+1}}{X_{k+1}} - \frac{\left(\sum_{i=1}^{k+1} w_i X_i\right)^2}{\sum_{i=1}^{k+1} w_i X_i} \right),$$

$k = 1, 2, \ldots, n - 1$. *Then* $(Q_1, \ldots, Q_{n-1}, \sum_{i=1}^{k+1} w_i X_i)$ *is a sequence of independent random variables and* $\mathcal{L}(Q_k) = \chi_1^2$.

Suppose that $\mathcal{L}(X) = IG(\mu, \lambda)$. What is the distribution of $Y = \sqrt{\lambda}(X - \mu)/\mu\sqrt{X}$? This is given in the next theorem, due to Chhikara and Folks (1974).

Theorem 1.2 *Let* $\mathcal{L}(X) = IG(\mu, \lambda)$ *and define* $Y = \sqrt{\lambda}(X - \mu)/\mu\sqrt{X}$. *Then the density of* Y *is*

$$g(y) = \left(1 - \frac{y}{\sqrt{y^2 + 4\lambda/\mu}}\right) \frac{1}{\sqrt{2\pi}} \exp\left(-\frac{y^2}{2}\right).$$

Proof The map $X \to Y$ represents a bijection from $\mathbb{R}^+$ to R. Now $dy = \frac{\sqrt{\lambda}(x+\mu)}{2\mu x^{3/2}}$. So if we let $t = \sqrt{x}$ we obtain the equation

$$\sqrt{\lambda} t^2 - \mu y t - \sqrt{\lambda}\mu = 0$$

whose admissible solution is

$$t = \frac{\mu y + \sqrt{\mu^2 y^2 + 4\lambda\mu}}{2\sqrt{\lambda}}.$$

Since $x+\mu = t^2 + \mu = \frac{(\mu^2+4\lambda\mu)+\mu y\sqrt{\mu^2 y^2+4\lambda\mu}}{2\lambda}$, writing $\Delta = \mu^2 y^2 + 4\lambda\mu$ we have

$$\frac{2\mu}{x+\mu} = \frac{4\lambda\mu}{t^2+\mu} = \frac{4\lambda\mu(\Delta-\mu y\sqrt{\Delta})}{(\Delta-\mu^2 y^2)\Delta} = \left(1-\frac{\mu y}{\sqrt{\Delta}}\right).$$

Hence

$$g(y) = \sqrt{\frac{\lambda}{2\pi x^3}} \exp\left(-\frac{y^2}{2}\right) \sqrt{\frac{x^3}{\lambda}} \left(1 - \frac{y}{\sqrt{y^2+\frac{4\lambda}{\mu}}}\right)$$

$$= \frac{1}{\sqrt{2\pi}} \left(1 - \frac{y}{\sqrt{y^2+\frac{4\lambda}{\mu}}}\right) \exp\left(-\frac{y^2}{2}\right). \quad \clubsuit \qquad (1.6)$$

This shows that Y has a weighted standard normal law where the weight $w(y)$ satisfies $w(y) + w(-y) =$ constant.

Theorem 1.2 leads to the following corollary.

Corollary 1.1 *For a random sample $(X_1,\ldots,X_n)$ from $IG(\mu,\lambda)$, the density of $\sqrt{n\lambda}(\overline{X}-\mu)/\mu\sqrt{\overline{X}}$ is the density given in (1.6) with $n\lambda$ replacing λ.*

Chhikara and Folks also studied the sampling distributions of a few other statistics which we now develop.

Mimicking the results for the normal law it is possible to study the density of the ratio

$$T = \frac{\sqrt{n\lambda}(\overline{X}-\mu)}{\mu\sqrt{\overline{X}}} \bigg/ \sqrt{\frac{\lambda V}{n-1}} \qquad (1.7)$$

where

$$V = \sum_{i=1}^{n} \left(\frac{1}{X_i} - \frac{1}{\overline{X}}\right).$$

Theorem 1.3 *Let T be defined as in (1.7) based on a random sample from $IG(\mu,\lambda)$. The density of T is*

$$\frac{(1-h(t))(1+\frac{t^2}{n-1})^{-\frac{n}{2}}}{\sqrt{n-1}Be(\frac{1}{2},\frac{n-1}{2})}$$

where

$$h(t) = \frac{t}{\Gamma(\frac{n}{2})} \int_0^\infty \frac{x^{\frac{n-1}{2}}\exp(-x)}{2(n-1+t^2)n\phi+t^2 x} dx$$

Sampling distributions

with $\phi = \lambda/\mu$. ($Be(\alpha, \beta)$ refers to the Beta function).

Proof The joint law of $Z = \frac{\sqrt{n}\lambda(\overline{X}-\mu)}{\mu\sqrt{\overline{X}}}$ and $U = \lambda V$ is, using (1.6),

$$p_1(z,u) = \frac{1}{\sqrt{2\pi}2^{\frac{n-1}{2}}\Gamma(\frac{n-1}{2})}\left(1 - \frac{z}{\sqrt{z^2+4n\phi}}\right) u^{\frac{n-3}{2}} \exp\left(-\frac{v}{2}-\frac{z^2}{2}\right).$$

With the change of variables $v = u$, $t = \frac{a}{\sqrt{u/(n-1)}}$ the Jacobian being $\sqrt{\frac{v}{n-1}}$, the density is

$$p(t,v) = \frac{\left(\Gamma\left(\frac{n}{2}\right)\right)^{-1}}{2^{n/2} Be\left(\frac{1}{2}, \frac{n-1}{2}\right)\sqrt{n-1}} \left(1 - \frac{t\sqrt{v}}{\sqrt{t^2 v + 4(n-1)n\phi}}\right)$$

$$\times \exp\left\{-\frac{v}{2}\left(1 + \frac{t^2}{n-1}\right)\right\}$$

with support ($t \in \mathbb{R}, v \in \mathbb{R}^+$). Hence integrating with respect to v, we obtain

$$f_T(t) = \frac{1}{\sqrt{n-1}Be(\frac{1}{2}, \frac{n-1}{2})}$$

$$\times \int_0^\infty \left(1 - \frac{t\sqrt{v}}{\sqrt{t^2 v + 4(n-1)n\phi}}\right) \frac{v^{n/2-1}}{\Gamma(\frac{n}{2})2^{n/2}}$$

$$\times \exp\left(-\frac{v}{2}\left(1 + \frac{t^2}{n-1}\right)\right) dv.$$

A final change of variable $y = \frac{v}{2}\left(1 + \frac{t^2}{n-1}\right)$ gives

$$f_T(t) = \frac{\left(1 + \frac{t^2}{n-1}\right)^{-n/2}}{\sqrt{n-1}Be(\frac{1}{2}, \frac{n-1}{2})} (1 - w(t)) \qquad (1.8)$$

where

$$w(t) = 1 - \frac{t}{\Gamma(\frac{n}{2})}\int_0^\infty \frac{y^{\frac{n-1}{2}}e^{-y}dy}{\sqrt{t^2 y + 2n\phi(t^2+n-1)}}. \quad \clubsuit$$

Equation (1.8) shows that T has a weighted Student's law with $(n-1)$ degrees of freedom, the weight satisfying $w(t) + w(-t) = $ constant. Observe that the density is a function of n and ϕ. As $\phi \to \infty$ $|T|$ has a folded 't' distribution while T^2 has the F distribution with 1 and $(n-1)$ degrees of freedom.

The next sampling distribution we consider is that of $Z = \overline{X}V/(n-1)$. The statistic Z was shown (Seshadri 1988) to be a U-statistic. Indeed we have

$$z = \frac{\overline{x}}{n-1}\left(\sum_{i=1}^n \frac{1}{x_i} - \frac{n}{\overline{x}}\right)$$

$$= \frac{1}{n(n-1)}\left(\sum_i^n \sum_{j\neq i}^n \frac{x_j}{x_i} - n(n-1)\right)$$

$$= \frac{1}{n(n-1)}\left(\sum_{i<j}^n \frac{x_i}{x_j} + \frac{x_j}{x_i} - 2\right)$$

$$= \frac{1}{\binom{n}{2}}\sum_{i<j}^n \varphi(x_i, x_j)$$

where $\varphi(a,b) = \frac{(a-b)^2}{2ab}$ is the symmetric kernel of degree 2. From the Lehmann-Scheffé theorem it follows that Z is the uniformly minimum variance unbiased estimator of $\frac{\mu}{\lambda}$. We now derive the distribution of Z.

The joint law of $(\overline{X}, \lambda V)$ is, over $(\mathbb{R}^+)^2$

$$\left(\frac{n\lambda}{2\pi\overline{x}^3}\right)^{1/2} \frac{v^{\frac{n-3}{2}}\exp\left(-\frac{\lambda v}{2}\right)}{2^{\frac{n-1}{2}}\Gamma(\frac{n-1}{2})} \exp\left(-\frac{n\lambda}{2\mu^2}\frac{(\overline{x}-\mu)^2}{\overline{x}}\right).$$

Letting $Z_1 = \overline{X}V$ the density $p(z_1)$ is

$$\left(\frac{n\lambda^n}{\pi 2^n}\right) \frac{\exp\left(\frac{n\lambda}{\mu}\right) z_1^{\frac{n-3}{2}}}{\Gamma\left(\frac{n-1}{2}\right)} \int_0^\infty x^{-\frac{n}{2}-1} \exp\left(-\frac{n\lambda x}{2\mu^2} - \frac{\lambda(n+z_1)}{2x}\right) dx.$$

Using the definition of the modified Bessel function we obtain the density of Z_1 as

$$2\left(\frac{n}{\pi}\frac{\lambda^n}{2^n\mu^n}\right)^{1/2} \frac{\exp\left(\frac{n\lambda}{\mu}\right)}{\Gamma\left(\frac{n-1}{2}\right)} z_1^{\frac{n-3}{2}} \left(1 + \frac{z_1}{n}\right)^{\frac{-n}{4}} K_{\frac{n}{2}}\left(\frac{n\lambda}{\mu}\sqrt{1 + \frac{z_1}{n}}\right).$$

Finally the density of Z with support $\mathbb{R}^+$ is

$$2(n-1)\left(\frac{n}{\pi}\frac{\lambda^n}{2^n\mu^n}\right)^{1/2} \frac{\exp\left(\frac{n\lambda}{\mu}\right)}{\Gamma\left(\frac{n-1}{2}\right)}((n-1)Z)^{\frac{n-3}{2}}$$

$$\left(1 + \frac{(n-1)Z}{n}\right)^{\frac{-n}{4}} K_{\frac{n}{2}}\left(\frac{n\lambda}{\mu}\sqrt{1 + \frac{z_1(n-1)}{n}}\right). \qquad (1.9)$$

Using the formula 6.596:3 (page 705) from Gradshteyn and Ryzhik(1963), the moments of Z are easily computed. In particular

$$\mathbb{E}(Z) = \frac{\mu}{\lambda}, \quad Var(Z) = \frac{\mu^2}{(n-1)\lambda^2}\left(2 + \frac{(n+1)\mu}{n\lambda}\right).$$

We have thus proved

Theorem 1.4 *Let $\overline{X}$ be the sample mean of n independent identically distributed random variables from $IG(\mu, \lambda)$ and define $V = \sum_{i=1}^{n}\left(\frac{1}{X_i} - \frac{1}{\overline{X}}\right)$. Then $Z = \overline{X}V/(n-1)$ is a U-statistic whose density is given in (1.9).*

The maximum likelihood estimate of $\phi = \lambda/\mu$ can be shown to be $W = (n/\overline{X}V)$ whose distribution depends on μ and λ only through the ratio $\phi = \lambda/\mu$. Its density can be directly obtained from (1.9) by writing $W = n/(n-1)Z$. The next theorem due to Folks (1972) gives the density of W.

Theorem 1.5 *The sampling distribution of the maximum likelihood estimator of $\phi = \lambda/\mu$ based on a random sample from $IG(\mu, \lambda)$ is*

$$\frac{(n\phi)^\nu \exp(n\phi) w^{\frac{(n+2)}{4}}}{\sqrt{\pi}\,\Gamma\left(\frac{n-1}{2}\right) 2^{\frac{n-1}{2}} (1+w)^{\nu/2}(n-1)} K_\nu(n\phi\sqrt{1+w})\mathbf{1}_{\mathbf{R}^+}(w)$$

where $\nu = n/2$.

We leave the verification as an exercise. However, we give an expression for the distribution function $H_\phi(w)$ based on a result due to Hsieh (1990).

$$P[W \le w] = H_\phi(w) = \frac{\exp((n\phi)(1 - \sqrt{1+w^{-1}}))}{\sqrt{1+w^{-1}}} S(w)$$

where

$$S(w) = \sum_{i=0}^{k-1}\left(\frac{(n\phi)^i}{i![2w\sqrt{1+w^{-1}}]^i} \sum_{s=0}^{i} \frac{(i+s)!}{s!(i-s)![2n\phi\sqrt{1+w^{-1}}]^s}\right).$$

Let $P[W < w_\alpha \mid \phi = \phi_0] = \alpha$, $(0 < \alpha < 1)$. Then for any fixed $\phi < \phi_0$ $H_\phi(w_\alpha) \to 1$ as $n \to \infty$. Moreover $H_\phi(w)$ is decreasing in ϕ. These properties are discussed by Hsieh who shows that tests of hypotheses on ϕ based on W are consistent (see Chapter 4 Section 4.4). Our next result concerns the ratio $R = \frac{\overline{X}}{X_-}$.

Theorem 1.6 *The distribution function of $R = \frac{\overline{X}}{\overline{X}_-}$ based on a random sample of size n from $IG(\mu, \lambda)$ is*

$$F_R(r) = 1 - \exp\left(-\frac{n\phi}{\sqrt{r-1}}\right) \sum_{i=0}^{n-3} \frac{(r-1)^i}{i!} \sum_{j=0}^{i} \frac{(i+j)!(n\phi)^{i-j}}{j!(i-j)!2^{i+j}r^{\frac{i+j+1}{2}}}$$

if n is odd and ≥ 3,

$$= \int_0^\infty G_Y\left(\frac{n(r-1)\phi}{z}\right) \left(\frac{n\phi}{2\pi z^3}\right)^{\frac{1}{2}} \exp\left(-n\phi\left(\frac{1}{2z} + \frac{z}{2} - 1\right)\right) dz$$

if n is even.

Here $G_Y(\cdot)$ is the distribution function of a chi-squared variable with $(n-1)$ degrees of freedom. This theorem is due to Patil and Kovner (1976). (There are some minor misprints in their paper).

Proof Rewrite $Y = \lambda V$ as $Y = n\lambda\left(\frac{1}{\overline{X}_-} - \frac{1}{\overline{X}}\right)$ or $Y = n\mu\left(\frac{R-1}{\overline{X}}\right)\phi$. Then, we have, from the independence of Y and $\overline{X}$,

$$P(Y\overline{X} \leq t) = \int_0^\infty \int_0^{\frac{t}{x}} f_{\overline{X}}(x) f_Y(y) dy dx$$

$$= \int_0^\infty G_Y\left(\frac{t}{x}\right) f_{\overline{X}}(x) dx.$$

If $n \geq 3$ and is odd, $G_Y(\cdot)$ has the following closed form expansion

$$G_Y\left(\frac{t}{x}\right) = 1 - \sum_{i=0}^{(n-3)/2} \frac{(t/2x)^i}{i!} \exp\left(-\frac{t}{2x}\right).$$

This enables the calculation of $F_{Y\overline{X}}(t)$ as

$$1 - \left(\frac{n\lambda}{2\pi}\right)^{\frac{1}{2}} e^{n\phi} \sum_{i=0}^{(n-3)/2} \frac{(t/2)^i}{i!} \int_0^\infty \frac{\exp\left(-\frac{(t+n\lambda)}{2x} - \frac{n\phi x}{2\mu}\right)}{x^{i+3/2}} dx$$

$$= 1 - \left(\frac{n\lambda}{2\pi}\right)^{\frac{1}{2}} \sum_{i=0}^{(n-3)/2} \frac{e^{n\phi}(t/2)^i}{i!\left(\mu + \frac{t}{n\phi}\right)^i} \sum_{j=0}^{i} \frac{(i+j)!}{(i-j)!j!} \left(2n\phi\sqrt{1+\frac{t}{n\lambda}}\right)^{-j}$$

$$\times \frac{\exp\left(-n\phi\sqrt{1+\frac{t}{n\lambda}}\right)}{\sqrt{n\phi\left(1+\frac{t}{n\lambda}\right)}}$$

$$= 1 - \exp\left(-n\phi\sqrt{1+\frac{t}{n\lambda}}\right) \sum_{i=0}^{(n-3)/2} \frac{(t/2\mu)^i}{i!\left(1+\frac{t}{n\lambda}\right)^{\frac{i+1}{2}}} \mathbb{E}(Z)^{-i}$$

where
$$\mathcal{L}(Z) = IG\left(\mu\sqrt{1+\frac{t}{n\lambda}}, n\lambda\left(1+\frac{t}{n\lambda}\right)\right).$$

Finally we note that
$$F_R(r) = F_{Y\overline{X}}(n\lambda(r-1))$$

since $Y\overline{X} \leq t \Rightarrow R \leq 1 + \frac{t}{n\lambda}$. This proves the result for n odd and ≥ 3. If n is even

$$F_{Y\overline{X}}(t) = \int_0^\infty F_{Y\overline{X}}\left(\frac{t}{x}\right)\left(\frac{n\lambda}{2\pi x^3}\right)^{\frac{1}{2}} \exp\left(-\frac{n\phi x}{2\mu} - \frac{n\lambda}{2x} + n\phi\right) dx.$$

Letting $z = \frac{x}{\mu}$, $t = n\lambda(r-1)$ we obtain the result. ♣

1.3 Conditional distributions

Consider the following transformations $T_1 = n\overline{X}$, $T_2 = n(\overline{X_-} + \overline{X})$ on the sample space and let us write $\theta_1 = \frac{\lambda}{2}\left(1 - \frac{1}{\mu^2}\right)$, $\theta_2 = -\frac{\lambda}{2}$ so that

$$\Theta = \{(\theta_1, \theta_2) \mid \theta_1 \in \mathbb{R}, \theta_2 \in \mathbb{R}^-\}.$$

We state and prove a theorem concerning the conditional law of T_1 given T_2.

Theorem 1.7 *The conditional law of T_1 given T_2 where $T_1 = n\overline{X}$ and $T_2 = n(\overline{X} + \overline{X}_-)$ based on a sample from $IG(\mu = 1, \lambda)$ is*

$$\frac{n}{Be\left(\frac{1}{2}, \frac{n-1}{2}\right)\sqrt{t_1^3(t_2 - 2n)}}\left(1 - \frac{(t_1 - n)^2}{t_1(t_2 - 2n)}\right)^{\frac{n-3}{2}} 1_S(t_1)$$

where
$$S = \left\{t_1 \middle| 0 < \frac{(t_1 - n)^2}{t_1(t_2 - 2n)} < 1\right\}. \quad (1.10)$$

Proof Since $\overline{X}$ is independent of λV, the joint law of $(\overline{X}, \lambda V)$ is, over $(\mathbb{R}^+)^2$

$$\sqrt{\frac{n\lambda}{2\pi \overline{x}^3}} e^{n\phi} \exp\left(-\frac{n\lambda}{2\mu^2}\overline{x} - \frac{n\lambda}{2\overline{x}}\right) \frac{v^{\frac{n-3}{2}}}{2^{\frac{n-1}{2}}\Gamma\left(\frac{n-1}{2}\right)} \exp\left(-\frac{\lambda v}{2}\right).$$

We write $t_1 = n\overline{x}$, $t_2 = n(\overline{x}_- + \overline{x}) = t_1 + v + \frac{n^2}{t_1}$ and transform from $(\overline{x}, \lambda v)$ to (t_1, t_2) obtaining for $0 < \frac{t_1}{n} + \frac{n}{t_1} < \frac{t_2}{n} < \infty$, the joint law of (t_1, t_2) as

$$\frac{n^{\frac{n-1}{2}}(\lambda/2)^{\frac{n}{2}}t_1^{-\frac{3}{2}}}{\Gamma(\frac{1}{2})\Gamma(\frac{n-1}{2})}\left(\frac{t_2}{n}-\frac{t_1}{n}-\frac{n}{t_1}\right)^{\frac{n-3}{2}}$$

$$\times \exp\left[\lambda\left\{\frac{n}{\mu}-\frac{t_2}{2}+\frac{t_1}{2}\left(1-\frac{1}{\mu^2}\right)\right\}\right]. \quad (1.11)$$

We first consider the case $\mu = 1$. What is the marginal law of T_2? Setting $\mu = 1$ in (1.11) we need to evaluate the integral (apart from the constants)

$$\int_{0<\frac{t_1}{n}+\frac{n}{t_1}<\frac{t_2}{n}} t^{-\frac{3}{2}}\left(\frac{t_2}{n}-\frac{t_1}{n}-\frac{n}{t_1}\right)^{\frac{n-3}{2}} dt_1.$$

Solving $t_2^2 + n^2 - t_1 t_2 = 0$ we obtain $t_1 = \frac{t_2 \pm \sqrt{(t_2-2n)(t_2+2n)}}{2}$. Clearly $t_2 > 2n$ and if we let $z = t_2 - 2n$ we have for the region of integration

$$D_{t_1} = \left\{t_1 \mid \sqrt{z+2n} - \sqrt{z^2+4nz} < t_1 < \sqrt{z+2n} + \sqrt{z^2+4nz}\right\}.$$

Thus we need to evaluate

$$I = \left(\frac{z}{n}\right)^{\frac{n-3}{2}} \int_{D_{t_1}} t^{-\frac{3}{2}}\left(1 - \frac{(t_1-n)^2}{zt_1}\right)^{\frac{n-3}{2}} dt_1.$$

Let $v = t_1^{-1/2}$. Then $D_v = \{v \mid \sqrt{z+4n}-\sqrt{z} < v < \sqrt{z+4n}+\sqrt{z}\}$ and

$$I = 2\left(\frac{z}{n}\right)^{\frac{n-3}{2}} \int_{D_v} \left(1 - \frac{(nv-v^{-1})^2}{z}\right)^{\frac{n-3}{2}} dv.$$

Finally we let $r = \frac{nv-v^{-1}}{\sqrt{z}}$; then $D_r = \{r \mid |r| \leq 1\}$, and $dv = \frac{\sqrt{z}}{n+v^{-2}}dr$. Solving $nv^2 - r\sqrt{z}v - 1 = 0$ we obtain

$$v = \frac{r\sqrt{z}+\sqrt{r^2z+4n}}{2n} \quad \text{or} \quad v^2 = \frac{(r^2z+2n)+r\sqrt{z}\sqrt{r^2z+4n}}{2n^2}.$$

Therefore

$$v^{-2} = \frac{r^2z+2n-r\sqrt{z}\sqrt{r^2z+4n}}{2} \quad \text{and} \quad \frac{n+v^{-2}}{\sqrt{z}} = \frac{P-r\sqrt{z}\sqrt{P}}{2\sqrt{z}}$$

where $P = r^2z + 4n$. This gives us

$$\frac{\sqrt{z}}{n+v^{-2}} = \frac{2\sqrt{z}(\sqrt{P}+r\sqrt{z})}{\sqrt{P}(P-r^2z)} = \frac{2\sqrt{z}}{4n}\left(1-\frac{r\sqrt{z}}{\sqrt{r^2z+4n}}\right).$$

Hence
$$I = 2\left(\frac{z}{n}\right)^{\frac{n-3}{2}} \int_{-1}^{1} \frac{\sqrt{z}}{2n}(1-r^2)^{\frac{n-3}{2}} dr.$$

Since $\frac{r\sqrt{z}}{\sqrt{r^2z+4n}}$ is an odd function of r we can write $y = 1 - r^2$ and easily show that the integral reduces to a Beta integral. We then have

$$I = \left(\frac{z}{n}\right)^{\frac{n-3}{2}} \frac{\sqrt{z}}{n} Be\left(\frac{1}{2}, \frac{n-1}{2}\right).$$

Thus when $\mu = 1$, the marginal law of T_2 is

$$f_0(t_2) = \left(\frac{\lambda}{2}\right)^{\frac{n}{2}} \frac{(t_2 - 2n)^{\frac{n}{2}-1}}{\Gamma\left(\frac{n}{2}\right)} \exp\left(-\frac{\lambda(t_2 - 2n)}{2}\right) 1_{(0,2n)}(t_2). \quad (1.12)$$

From this one obtains the conditional law of T_1 given $T_2 = t_2$ for $\mu = 1$, as given by (1.10). When $\mu \neq 1$, the marginal law of T_2 is complicated and involves the parameter μ. We give an integral expression for this density $f_{\theta_1}(t_2)$ as

$$\frac{2n\left(\frac{\lambda}{2}\right)^{\frac{n}{2}} (t_2 - 2n)^{\frac{n-3}{2}} \exp\left(-\frac{\lambda}{2}(t_2 - 2n)\right)}{\Gamma\left(\frac{1}{2}\right)\Gamma\left(\frac{n-1}{2}\right)} 1(\theta_1) \quad (1.13)$$

where

$$I(\theta_1) = \int_{D_v} \left(1 - z^{-1}(nv - v^{-1})\right)^{\frac{n-3}{2}} \exp\left(\frac{\theta_1}{v^2}\right) dv. \quad \clubsuit$$

The next three theorems deal with the conditional laws of X_1 given various statistics. They are due to Chhikara and Folks.

Theorem 1.8 *Let $X = (X_1, \ldots, X_n)$ be a random sample from $IG(\mu, \lambda)$. Then the conditional law of X_1 given $\overline{X}$ is for $0 < x_1 < n\overline{x}$*

$$\sqrt{\frac{n\lambda}{2\pi}}(n-1)\left(\frac{\overline{x}}{x_1(n\overline{x} - x_1)}\right)^{\frac{3}{2}} \exp\left(-\frac{n\lambda(x_1 - \overline{x})^2}{2x_1\overline{x}(n\overline{x} - x_1)}\right) \quad (1.14).$$

Proof We start with the joint law of X_1 and $\overline{X}_1 = \sum_{i=2}^{n} X_i/(n-1) = \frac{n\overline{X} - X_1}{n-1}$, over $(\mathbb{R}^+)^2$ and transform to $(X_1, \overline{X})$ obtaining

$$p(x_1, \overline{x}) = \frac{n(n-1)\lambda}{2\pi(x_1(n\overline{x} - x_1))^{3/2}}$$

$$\times \exp\left[-\frac{\lambda}{2\mu^2}\left\{\frac{(x_1 - \mu)^2}{x_1} + \frac{((n\overline{x} - x_1) - \mu(n-1))^2}{n\overline{x} - x_1}\right\}\right].$$

After division by the density of $\overline{X}$ we obtain after substantial simplification for $0 < x_1 < n\overline{x}$ the required density. ♣

The distribution of X_1 given $Z = \sum_{i=1}^{n} (X_i - \mu)^2 X_i$ is provided by

Theorem 1.9 *Let $X_1, \ldots, X_n$ be a random sample from $IG(\mu, \lambda)$. Then the conditional law of X_1 given $Z = \sum_{i=1}^{n} \frac{(X_i - \mu)^2}{X_i}$ is*

$$f(x_1 \mid z) = \frac{\mu}{(x_1 z^3)^{1/2} Be\left(\frac{1}{2}, \frac{n-1}{2}\right)} \left(1 - \frac{(x_1 - \mu)^2}{z x_1}\right)^{\frac{n-3}{2}} 1_D(x_1)$$

where

$$D = \{x_1 \mid 0 < \frac{(x_1 - \mu)^2}{z x_1} < 1\}. \tag{1.15}$$

Proof Let $T_1 = \sum_{i=2}^{n} \frac{(X_i - \mu)^2}{X_i}$. Then the joint law of X_1 and T_1 is (since $\mathcal{L}\left(\frac{\lambda}{\mu^2} T_1\right) = \chi^2_{n-1}$)

$$\left(\frac{\lambda}{2}\right)^{1/2} \frac{x_1^{-\frac{3}{2}}}{\Gamma\left(\frac{1}{2}\right)} \exp\left(-\frac{\lambda}{2\mu^2} \frac{(x_1 - \mu)^2}{x_1}\right)$$

$$\times \left(\frac{\lambda}{2}\right)^{\frac{n-1}{2}} \frac{t_1^{\frac{n-3}{2}}}{\mu^{n-1} \Gamma\left(\frac{n-1}{2}\right)} \exp\left(-\frac{\lambda t_1}{2\mu^2}\right) 1_{(\mathbb{R}^+)^2}(x_1, t_1),$$

while that of (X_1, Z) is

$$\left(\frac{\lambda}{2}\right)^{\frac{n}{2}} \frac{x_1^{-\frac{3}{2}} \left(z - \frac{(x_1-\mu)^2}{x_1}\right)^{\frac{n-3}{2}}}{\Gamma\left(\frac{1}{2}\right)\Gamma\left(\frac{n-1}{2}\right)\mu^{n-1}} \exp\left(-\frac{\lambda z}{2\mu^2}\right) 1_D(x_1, z)$$

where $D = \{0 < \frac{(x_1-\mu)^2}{x_1} < z\}$. Since $\mathcal{L}\left(\frac{\lambda}{\mu^2} Z\right) = \chi^2_n$, we at once have

$$f(x_1 \mid z) = \frac{\mu x_1^{-3/2} \left(z - \frac{(x_1-\mu)^2}{x_1}\right)^{\frac{n-3}{2}} z^{-\frac{(n-2)}{2}}}{Be\left(\frac{1}{2}, \frac{n-1}{2}\right)} 1_D(x_1)$$

$$= \frac{\mu \left(1 - \frac{(x_1-\mu)^2}{z x_1}\right)^{\frac{n-3}{2}}}{(z x_1^3)^{1/2} Be\left(\frac{1}{2}, \frac{n-1}{2}\right)} 1_D(x_1). \quad ♣$$

The next theorem concerns the conditional law of X_1 given $(\overline{X}, V)$.

Theorem 1.10 Let $X_1, \ldots, X_n$ be a random sample from $IG(\mu, \lambda)$. Then the conditional law of X_1 given $(\overline{X}, V)$ is for $x_1 \in D$

$$\frac{\sqrt{n}(n-1)}{Be\left(\frac{1}{2}, \frac{n-1}{2}\right)} \left(\frac{\overline{x}^3}{v x_1^3 (n\overline{x} - x_1)^3}\right)^{\frac{1}{2}} \left(1 - \frac{n(x_1 - \overline{x})^2}{v x_1 \overline{x}(n\overline{x} - x_1)}\right)^{\frac{n-4}{2}} 1_D(x_1) \tag{1.16}$$

where $D = \{x_1 \mid L < x_1 < U\}$ and L, U are the 2 roots of the equation $n(x_1 - \overline{x})^2 - v x_1 \overline{x}(n\overline{x} - x_1) = 0$.

We omit the proof.

1.4 Bayesian sampling distributions

Consider the $IG(\mu, \lambda)$ law with the parametrization $\delta = 1/\mu$. The likelihood function based on a random sample of size n, $X = (X_1, \ldots, X_n)$ is

$$\mathcal{L}(\delta, \lambda) = \left(\frac{\lambda}{2\pi}\right)^{\frac{n}{2}} \prod_{i=1}^{n} x_i^{-\frac{3}{2}} \exp\left(-\frac{\lambda}{2} \sum_{i=1}^{n} \frac{(\delta x_i - 1)^2}{x_i}\right) \tag{1.17}.$$

Banerjee and Bhattacharyya (1979) obtain some Bayesian results with the use of a prior, known as a vague prior

$$\pi(\delta, \lambda) = \frac{c}{\lambda} 1_{\mathbf{R}^+}(\lambda)$$

for some arbitrary constant c. With this prior, the posterior distribution of X is

$$\frac{\lambda^{\frac{n}{2}-1} \exp\left[-\frac{\lambda v}{2}\left(1 + \frac{n(\delta \overline{x} - 1)^2}{v\overline{x}}\right)\right]}{\int_0^\infty \int_0^\infty \lambda^{\frac{n}{2}-1} \exp\left[-\frac{\lambda v}{2}\left(1 + \frac{n(\delta \overline{x} - 1)^2}{v\overline{x}}\right)\right] d\lambda d\delta}$$

where, as usual $v = \sum_{i=1}^{n} \left(\frac{1}{X_i} - \frac{1}{\overline{X}}\right)$. The denominator when first integrated with respect to λ gives

$$\Gamma\left(\frac{n}{2}\right) \left(\frac{2}{v}\right)^{\frac{n}{2}} \int_0^\infty \frac{d\delta}{\left(1 + \frac{n(\delta \overline{x} - 1)^2}{v\overline{x}}\right)^{\frac{n}{2}}}$$

which, with the transformation $\frac{t^2}{n-1} = \frac{n(\delta \overline{x} - 1)^2}{v\overline{x}}$ yields

$$\frac{\Gamma\left(\frac{n-1}{2}\right) \Gamma\left(\frac{1}{2}\right) S_{n-1}\left(\sqrt{\frac{n(n-1)}{v\overline{x}}}\right)}{\left(\frac{v}{2}\right)^{\frac{n}{2}} \left(\frac{n\overline{x}}{v}\right)^{\frac{1}{2}}}$$

where $S_{n-1}(\cdot)$ is the distribution function of Student's 't' with $(n-1)$ degrees of freedom. We thus obtain the following theorem.

Theorem 1.11 *Suppose that $X = (X_1, \ldots, X_n)$ is a random sample from $IG(\frac{1}{\delta}, \lambda)$ and we use a vague prior $\pi(\delta, \lambda) \propto \lambda^{-1}$; then the posterior distribution of (δ, λ) given the data X is*

$$\frac{v^{\frac{n}{2}-1}\sqrt{n\bar{x}}\lambda^{\frac{n}{2}-1}\exp\left[-\frac{\lambda v}{2}\left(1+\frac{n(\delta\bar{x}-1)^2}{v\bar{x}}\right)\right]}{2^{\frac{n}{2}}\Gamma\left(\frac{1}{2}\right)\Gamma\left(\frac{n-1}{2}\right)S_{n-1}\left(\sqrt{\frac{n(n-1)}{v\bar{x}}}\right)}1_D(\delta,\lambda) \qquad (1.18)$$

where $D = \{(\delta, \lambda) \mid \lambda \in \mathbb{R}, \ \delta > 0\}$.

Corollary 1.2 *With the same prior as in Theorem 1.11, the posterior distribution of δ is*

$$f_\delta(\delta \mid X) = \frac{\sqrt{\frac{n\bar{x}}{v}}\left(1+\frac{n(\delta\bar{x}-1)^2}{v\bar{x}}\right)^{-\frac{n}{2}}}{\Gamma\left(\frac{1}{2}\right)\Gamma\left(\frac{n-1}{2}\right)S_{n-1}\left(\sqrt{\frac{n(n-1)}{v\bar{x}}}\right)}. \qquad (1.19)$$

It is clear from the density that we have a truncated Student's t (truncated from the left at zero) with location parameter $\frac{1}{\bar{x}}$ and scale parameter $\sqrt{\frac{v}{n(n-1)\bar{x}}}$.

Corollary 1.3 *With the same prior as in Theorem 1.11, the posterior distribution of λ is*

$$f_\lambda(\lambda \mid X) = \left(\frac{v}{2}\right)^{\frac{n-1}{2}}\lambda^{\frac{n-3}{2}}\frac{\Phi\left(\sqrt{\frac{n\lambda}{\bar{x}}}\right)}{\Gamma\left(\frac{n}{2}\right)S_{n-1}\left(\sqrt{\frac{n(n-1)}{v\bar{x}}}\right)}\exp\left(-\frac{\lambda v}{2}\right)1_{\mathbb{R}^+}(\lambda).$$

$$(1.20)$$

Banerjee and Bhattacharyya call this law a modified gamma and denote it by $\Gamma\left(\frac{v}{2}, \frac{n-1}{2}, \frac{n}{\bar{x}}\right)$. They show that this law is unimodal. To obtain a family which is a natural conjugate family for (δ, λ) they consider a family of priors which is proportional to the likelihood (1.17) thus obtaining

$$\pi_C(\delta, \lambda) = K\lambda^{\frac{v}{2}-1}\exp\left(-\frac{\lambda\mu}{2}\left(1+\frac{v(\delta z-1)^2}{uz}\right)\right)1_{(\mathbb{R}^+)^2}(\delta,\lambda) \qquad (1.21)$$

Estimation

where $\mu > 0$, $\nu > 1$, $z > 0$ and K is a suitable normalizing constant. By comparing with (1.18) we have

$$K = \frac{\sqrt{\nu z}\, u^{\frac{\nu}{2}-1}}{2^{\frac{\nu}{2}} \Gamma\left(\frac{1}{2}\right) \Gamma\left(\frac{n-1}{2}\right) S_{\nu-1}\left(\sqrt{\frac{\nu(\nu-1)}{uz}}\right)}.$$

We are then lead to

Theorem 1.12 *With a natural conjugate family for (δ, λ) as defined in (1.21), the posterior density of (δ, λ) given the data X is*

$$K_c \lambda^{\frac{m}{2}-1} \exp\left[-\frac{\lambda w}{2}\left(1 + \frac{(\delta y - 1)^2}{wy}\right)\right] 1_{(\mathbf{R}^+)^2}(\delta, \lambda) \tag{1.22}$$

where $m = n + (\nu - 1)$, $y = \frac{n\bar{x}+\nu z}{n+(\nu-1)}$ and $w = v + u + \frac{n}{\bar{x}} + \frac{\nu}{z} - \frac{n+\nu}{(n\bar{x}+z)^2}$; K_c is the suitable normalizing constant.

It is interesting to note the similarities between (1.18) and (1.22). This should account for the marginal posterior of the latter exhibiting a behaviour analogous to the former. One can further show that the conditional posterior of δ given λ is a truncated normal (truncated to the left at the origin), the mean being $\frac{1}{y}$ and the variance $\frac{my}{\lambda}$. Banerjee and Bhattacharyya also point out that these results are similar to the Bayesian results for the normal model.

Betró and Rotondi (1991) have derived Bayesian sampling laws using a prior not belonging to any conjugate family. This method provides parameter estimates whose computation relies heavily on numerical integration. Their parametrization of $IG(\mu, \lambda)$ gives

$$f(x \mid \mu, \phi) = \sqrt{\frac{\mu\phi}{2\pi x^3}} \exp\left[-\frac{\phi(x-\mu)^2}{2\mu x}\right] 1_{\mathbf{R}^+}(x)$$

which can be rewritten as

$$f(x \mid \mu, \phi) = \sqrt{\frac{\mu\phi}{2\pi x^3}} e^{\phi} \exp\left[-\frac{\phi}{2}\left(\frac{x}{\mu} + \frac{\mu}{x}\right)\right] 1_{\mathbf{R}^+}(x).$$

The prior chosen by these authors is

$$\pi(\mu, \phi) = \pi_1(\mu \mid \phi)\pi_2(\phi)$$

where

$$\pi_1(\mu \mid \phi) = \sqrt{\frac{\eta \phi w}{2\pi \mu^3}} \exp(\phi w) \exp\left[-\frac{\phi w}{2}\left(\frac{\eta}{\mu} + \frac{\mu}{\eta}\right)\right] 1_{\mathbf{R}^+}(\mu)$$

and

$$\pi_2(\phi) = \frac{a^\gamma}{\Gamma(\gamma)} \phi^{\gamma-1} \exp(-a\phi) 1_{\mathbf{R}^+}(\phi)$$

with $a, \eta, w, \gamma > 0$.

CHAPTER 2

ESTIMATION

2.0 Introduction

We have seen briefly that the principal parameters μ and λ are estimated by their maximum likelihood estimators. The moment estimate for the variance m^3/λ is $s^2 = \sum_{i=1}^{n}(x_i - \bar{x})^2/(n-1)$. As a result it is possible to show that as a consistent estimator of λ^{-1}, $s^2/\bar{x}^3$ has an asymptotic efficiency of $\phi/(\phi+3)$. For small ϕ, the estimator is not reliable. The reciprocal of X can be used to estimate $1/\mu$, but has a bias equal to $1/\lambda$, and a mean squared error equal to $(\phi+3)/\lambda^2$. We reproduce in Table 2.1 uniformly minimum variance unbiased estimators (UMVUE) of several parametric functions of μ and λ as given by Iwase and Setô (1983) as well as Korwar (1980). Uniformly minimum variance unbiased estimators for cumulants and the density itself were given by Park et al. (1988). We state two propositions in the next section that provide these estimators. Also presented is Table 2.2 indicating estimates of the reliability $R(t) = 1 - F(t)$.

Quite recently Balakrishnan and Chen (1997) have prepared extensive tables for the determination of means, variances and covariances of order statistics from the inverse Gaussian law. These quantities are very useful in the derivation of best linear unbiased estimators of the location and scale parameters. They present formulas for their derivation and furthermore discuss the best linear unbiased estimation methods for complete as well as Type-II censored samples. Extensive tables are devoted to the variances and covariances of these best linear unbiased estimators for sample sizes upto 25 for various values of the shape parameter at various levels of censoring. The use of these tables is demonstrated with numerical examples. Finally they present a useful application of the means of order statistics in constructing Q-Q plots as well as developing formal correlation-type goodness-of-fit tests.

2.1 Estimation

Table 2.1 *UMVU estimators*

Parameter	Uniformly minimum variance unbiased estimator
λ	$^*(n-3)/V$
λ^{-1}	$V/(n-1)$
$\sqrt{(m^3/\lambda)}$	$\dfrac{\Gamma((n-1)/2)}{\sqrt{2}\,\Gamma(n/2)}\sqrt{\overline{X}_n^3 V} \times F\left(\tfrac{1}{4}, \tfrac{3}{4}; \tfrac{n}{2}; -\overline{X}_n V n\right)$
$\phi = \lambda/m$	$^*(n-3)/\overline{X}_n V - 1/n$
m	$\overline{X}_n$
m^3/λ	$n\overline{X}_n^2\left\{1 - F(\tfrac{1}{2}, 1; \tfrac{n-1}{2}; -\overline{X}_n V n)\right\}$
m^{-1}	$(1/\overline{X}_n) - V/n(n-1)$
λ/m^3	$^*(n-3)/\overline{X}_n^3 V - 6/n\overline{X}_n^2 + 3V/n^2(n-1)\overline{X}_n$
$\dfrac{3}{\sqrt{\phi}}$	$\dfrac{3\Gamma((n-1)/2)}{\sqrt{2}\,\Gamma(n/2)}(\overline{X}_n V)^{\tfrac{1}{2}} \times F\left(-\tfrac{1}{4}, \tfrac{1}{4}; \tfrac{n}{2}; -\dfrac{\overline{X}_n V}{n}\right)$
$\dfrac{15}{\phi}$	$15\overline{X}_n V/(n-1)$
$\sqrt{\dfrac{m}{\lambda}}$	$\dfrac{\Gamma((n-1/2)}{\sqrt{2}\,\Gamma(n/2)}(\overline{X}_n V)^{\tfrac{1}{2}} F\left(-\tfrac{1}{4}, \tfrac{1}{4}; \tfrac{n}{2}; -\dfrac{\overline{X}_n V}{n}\right)$
κ_r	$\dfrac{\Gamma\left(\tfrac{r-1}{2}\right)}{\sqrt{\pi}}\dfrac{\Gamma\left(\tfrac{n-1}{2}\right)}{\Gamma\left(\tfrac{n-1}{2}+r-1\right)}\overline{X}_n^{2r-1} V^{r-1}$ $\times F\left(r-1, \tfrac{2r-1}{2}; \tfrac{n+2r-3}{2}; -\dfrac{\overline{X}_n V}{n}\right)$
κ_{-r}	$\dfrac{\Gamma((n-1)/2)}{\Gamma((n-1)/2+r)}\overline{X}_n^{-1} V^{r-1}$ $\times \left\{\dfrac{\Gamma(r-1/2)}{\sqrt{\pi}}\left(\tfrac{n-1}{2}+r\right) + \left(\dfrac{(r-1)!}{2} - \dfrac{(r-1/2)}{2n\sqrt{\pi}}\right)\overline{X}_n V\right\}$

$n \geq 4$ for * and $n \geq 2$ elsewhere.

Estimation 25

Table 2.2 *Umvu estimators of $R(t)$*

Parameter status	umvue

λ known
$$\hat{R}(x;m) = \begin{cases} 0 & x > n\overline{X}_n \\ 1 & x < 0 \\ \Phi(-w_1) - \frac{n-2}{n} e^{\frac{2(n-1)\lambda}{n\overline{x}_n}} \Phi(-w_2) & \text{otherwise} \end{cases}$$

where

$$w_1 = \frac{\sqrt{n\lambda}(x - \overline{X}_n)}{\sqrt{x(n\overline{X}_n - x)\overline{X}_n}}, \quad w_2 = \frac{\sqrt{\lambda}\{n\overline{X}_n + (n-2)x\}}{\sqrt{n\overline{X}_n(n\overline{X}_n - x)x}}.$$

m known
$$\hat{R}(x;\lambda) = \begin{cases} 0, & x > \frac{1}{2}\{(2m+t) + \sqrt{4mt + t^2}\} \\ 1, & x < \frac{1}{2}\{(2m+t) - \sqrt{4mt + t^2}\} \\ G_{n-1}(w_1) - \left\{\frac{t+4m}{t}\right\}^{\frac{n-2}{2}} G_{n-1}(w_2) & \text{otherwise} \end{cases}$$

where

$$w_1 = \frac{\sqrt{n-1}(x-m)}{\sqrt{tx - (x-m)^2}}, \quad w_2 = \frac{\sqrt{n-1}(x+m)}{\sqrt{tx - (x-m)^2}}, \quad t = \sum \frac{(x_i - m)^2}{x_i}$$

and G denotes the right tail of Student's t with $(n-1)$ degrees of freedom.

m, λ unknown
$$\hat{R}(x;m,\lambda) = \begin{cases} 0, & x > U \\ 1, & x < L \\ G_{n-2}(w_1) - \frac{n-2}{n}\left[1 + \frac{4(n-1)}{nV\overline{X}_n}\right]^{\frac{n-3}{2}} \\ \times G_{n-2}(w_2) \end{cases}$$

where
$$w_1 = \frac{\sqrt{n(n-2)}(x - \overline{X}_n)}{\sqrt{V x \overline{X}_n(n\overline{X}_n - x) - n(x - \overline{X}_n)^2}},$$

$$w_2 = \frac{\sqrt{n(n-2)}\left[\overline{X}_n + \frac{n-2}{n}x\right]}{\sqrt{V x \overline{X}_n(n\overline{X}_n - x) - n(x - \overline{X}_n)^2}},$$

$$L = \frac{\overline{X}_n}{2(n + V\overline{X}_n)}\left\{n(2 + V\overline{X}_n) - \sqrt{4n(n-1)V\overline{X}_n + n^2V^2\overline{X}_n^2}\right\}$$

$$U = \frac{\overline{X}_n}{2(n + V\overline{X}_n)}\left\{n(2 + V\overline{X}_n) + \sqrt{4n(n-1)V\overline{X}_n + n^2V^2\overline{X}_n^2}\right\},$$

and G denotes the right tail of Student's t with $(n-2)$ degrees of freedom.

Proposition 2.1 *Uniformly minimum variance unbiased estimators for the rth cumulant of $IG(\mu, \lambda_0)$, where λ_0 is known is given by*

$$K_n(\overline{x}) = \left(\frac{2}{\lambda_0}\right)^{2r-2} \frac{1}{\sqrt{\pi}} \overline{x}^{3r-2} n^{3-4r} \sum_{k=0}^{\infty} \binom{r-2+k}{k} \Gamma(r+k-2)$$

for $r = 1, 2, \ldots$

Proposition 2.2 *A uniformly minimum variance unbiased estimator of the density $f(x; \mu, \lambda_0)$ of the inverse Gaussian law is*

$$\left(\frac{(n-1)^5 \lambda_o}{2\pi n^5 \overline{x}^3}\right)^{\frac{1}{2}} \left(\frac{\overline{x}}{\overline{x}-x}\right)^{\frac{3}{2}} \exp\left(-\frac{\lambda_0(\overline{x}-nx)}{2x(\overline{x}-x)\overline{x}}\right) 1_{(0,\overline{x})}(x).$$

The Fisher information based on a single observation from $IG(\mu, \lambda)$ is

$$i(\theta_1, \theta_2) = \frac{1}{2} \begin{pmatrix} \frac{(-\theta_2)^{1/2}}{(-\theta_1)^{3/2}} + \frac{1}{\theta_1^2} & \frac{-1}{(-\theta_1)^{1/2}(-\theta_2)^{1/2}} \\ \frac{-1}{(-\theta_1)^{1/2}(-\theta_2)^{1/2}} & \frac{(-\theta_2)^{1/2}}{(-\theta_1)^{3/2}} \end{pmatrix}$$

where $\theta_1 = -\frac{\lambda}{2}$ and $\theta_2 = -\frac{\lambda}{2\mu^2}$. Expressed in terms of μ, λ we have

$$\begin{pmatrix} \frac{1}{\mu\lambda} + \frac{2}{\lambda^2} & -\frac{\mu}{\lambda} \\ -\frac{\mu}{\lambda} & \frac{\mu^3}{\lambda} \end{pmatrix}.$$

Table 2.1 can be used to estimate quantities like the coefficient of variation $\sqrt{\phi^{-1}}$, the skewness $3\sqrt{\phi^{-1}}$ and kurtosis $15\sqrt{\phi^{-1}}$.

If $\hat{\kappa}_2$ denotes the uniformly minimum variance unbiased estimator of the variance of the inverse Gaussian law $IG(\mu, \lambda)$ then it is known that the variance of the estimator, for large values of n, is approximately equal to

$$(m^6/n\lambda^2)(2 + 9m/\lambda) + O(n^{-2}).$$

2.2 A shifted model

Padgett and Wei (1979) extended the $IG(\mu, \lambda)$ law by introducing a threshold parameter α. Thus if $\mathcal{L}(X) = IG(\mu, \alpha, \lambda)$ then

$$f(x) = \left(\frac{\lambda}{2\pi(x-\alpha)^3}\right)^{\frac{1}{2}} \exp\left(-\frac{\lambda(x-\alpha-\mu)^2}{2\mu^2(x-\alpha)}\right) 1_{(\alpha,\infty)}(x). \quad (2.1)$$

The extra parameter α varies in $(-\infty, \infty)$. For this law λ is a scale parameter and $\mu^* = \mu + \alpha$ is the mean. This law has been proposed as an alternative to the three parameter lognormal, gamma and Weibull laws investigated by several authors. One peculiarity with respect to these models is that maximum likelihood estimators can yield estimators that may be inconsistent and do not have asymptotic normality since the usual regularity assumptions are not satisfied. The likelihood may become infinite as the threshold parameter α tends to the minimum order statistic. Cheng and Amin (1981) show that for the $IG(\mu, \alpha, \lambda)$ model maximum likelihood estimation is numerically straightforward and cannot produce inconsistent estimators. First we consider moment estimators of α, μ and λ. The cumulant function of $IG(\mu, \alpha, \lambda)$ is $\phi\left(1 - \left(1 + \frac{2\mu t}{\phi}\right)^{\frac{1}{2}}\right) - \alpha t$ and therefore the first three cumulants are $K_1 = \mu + \alpha$, $K_2 = \frac{\mu^3}{\lambda}$ and $K_3 = \frac{3\mu^5}{\lambda^2}$.

Hence

$$\tilde{\mu} = \overline{X} - \tilde{\alpha}$$

$$\tilde{\lambda} = \frac{n\overline{X}^3}{\sum_{i=1}^{n}(X_i - \overline{X})^2} = \frac{\overline{X}^3}{S^2} \qquad (2.2)$$

$$\frac{3(S^2)^2}{\frac{1}{n}\sum_{i=1}^{n}(X_i - \overline{X})^3} = \overline{X} - \tilde{\alpha} = \tilde{\mu}$$

The following theorem due to Padgett and Wei gives the asymptotic normality of the vector $(\tilde{\alpha}, \tilde{\mu}, \tilde{\lambda})$ of moment estimators.

Theorem 2.1 Let $(\tilde{\alpha}, \tilde{\mu}, \tilde{\mu})$ be as defined in (2.2). Then $(\tilde{\alpha} - \alpha, \tilde{\mu} - \mu, \tilde{\lambda} - \lambda)^t$ converges in distribution to $N_3(\underline{0}, \Sigma)$ where $\Sigma = G\Lambda G^t$,

$$\Lambda = \begin{pmatrix} \frac{\mu^3}{\lambda} & \mu_3 - \mu_1\mu_2 & \mu_4 - \mu_1\mu_3 \\ \mu_3 - \mu_1\mu_2 & \mu_4 - \mu_2^2 & \mu_5 - \mu_2\mu_3 \\ \mu_4 - \mu_1\mu_3 & \mu_5 - \mu_2\mu_3 & \mu_6 - \mu_3^2 \end{pmatrix}$$

$$G = \left(\frac{\partial g_i(\mu_1, \mu_2, \mu_3)}{\partial \mu_j}\right)_{i,j=1,2,3}$$

and

$$g_1(\mu_1, \mu_2, \mu_3) = \alpha = \mu_1 - \frac{3(\mu_2 - \mu_1^2)^2}{\mu_3 - 3\mu_1\mu_2 + 2\mu_2^2}$$

$$g_2(\mu_1, \mu_2, \mu_3) = \mu = \mu_1 - \alpha$$

$$g_3(\mu_1, \mu_2, \mu_3) = \lambda = \frac{\mu_1^3}{(\mu_2 - \mu_1)^2}.$$

Proof By the Central Limit Theorem, $\sqrt{n}(\frac{\sum^n X_i}{n} - \mu_1, \frac{\sum^n X_i^2}{n} - \mu_2, \frac{\sum^n X_i^3}{n} - \mu_3)^t$ converges in law to $N_3(\underline{0}, \Lambda)$. The functions g_i are differentiable and an application of Rao's theorem (1973) shows that

$$\left[\sqrt{n}\left(g_1\left(\frac{\sum X_i}{n}, \frac{\sum X_i^2}{n}, \frac{\sum X_i^3}{n}\right) - g_1(\mu_1, \mu_2, \mu_3)\right), \ldots, \right.$$
$$\left. -g_3\left(\frac{\sum X_i}{n}, \frac{\sum X_i^2}{n}, \frac{\sum X_i^3}{n}\right) - g_3(\mu_1, \mu_2, \mu_3)\right)\right]$$

converges in law to $N_3(\underline{0}, G\Lambda G^t)$. ♣

One can simplify Σ_{ME} (the covariance matrix of the moment estimator) to obtain

$$\mu^2 \begin{bmatrix} 2(\phi+6)^2\phi^{-1} & -2(\phi+6)^2\phi^{-1} & -2(\phi+5)(\phi+8) \\ -2(\phi+6)^2\phi^{-1} & \phi^{-1}+2(\phi+6)^2\phi^{-1} & 2(\phi+5)(\varphi+8) \\ -2(\phi+5)(\phi+8) & 2(\phi+5)(\phi+8) & 2\phi(\phi+3(\phi+5)(\phi+9)) \end{bmatrix}.$$

Using consistent estimates of μ and ϕ an estimate of Σ_{ME} can be obtained.

2.2.1 Maximum likelihood Estimation

The log-likelihood corresponding to $IG(\mu, \alpha, \lambda)$ is

$$l(\mu, \alpha, \lambda) = \frac{n}{2}\log\left(\frac{\lambda}{2\pi}\right) - \frac{3}{2}\sum_{i=1}^n \log(x_i - \alpha) - \frac{\lambda}{2\mu^2}\sum \frac{(x_i - \mu - \alpha)^2}{x_i - \alpha}. \tag{2.3}$$

For every fixed α, the maximum of $l(\mu, \alpha, \lambda)$ can be obtained as

$$\hat{\mu} = \overline{x} - \alpha$$
$$\hat{\lambda}^{-1}(\alpha) = n^{-1}\left(\sum_{i=1}^n (x_i - \alpha)^{-1} - \hat{\mu}^{-1}(\alpha)\right). \tag{2.4}$$

Now let us write
$$\mathcal{L}^*(\alpha) = \mathcal{L}_X(\alpha; \hat{\mu}(\alpha), \hat{\lambda}(\alpha))$$
where
$$\mathcal{L}_X(\alpha; \hat{\mu}(\alpha), \hat{\lambda}(\alpha)) = \left(n^{-1}\sum\left((x_i - \alpha)^{-1} - (\overline{x} - \alpha)^{-1}\right)\right)^{-\frac{3}{2}}$$
$$\times \prod_{i=1}^n (x_i - \alpha)^{-\frac{3}{2}} 1_{[\alpha < X_{(1)}]}(x_1, \ldots, x_n) \tag{2.5}$$

Maximum likelihood estimation

and where $X_{(1)}$ is the smallest order statistic. Observe that (2.5) represents the likelihood function using the estimates $\hat{\mu}$ and $\hat{\lambda}$. The maximum likelihood estimate of α is obtained by maximizing $\mathcal{L}^*(\alpha)$. Once the maximizing value of $\hat{\alpha}$ of α is determined, $\hat{\mu}(\hat{\alpha})$ and $\hat{\lambda}(\hat{\alpha})$ are obtained by substitution.

Returning to (2.5) we can now write

$$\mathcal{L}^*(\alpha) = \frac{(x_{(1)} - \alpha)^{\frac{n-3}{2}}}{\left(\frac{1}{n}\sum_{i=1}^{n}\frac{x_{(1)}-\alpha}{x_i-\alpha} - \frac{x_{(1)}-\alpha}{\bar{x}-\alpha}\right)^{\frac{n}{2}} \prod_{i=2}^{n}(x_{(i)} - \alpha)^{\frac{3}{2}}}.$$

When $\alpha \to x_{(1)}$ the denominator can be shown to converge to a constant. Therefore if $n > 3$ $\log \mathcal{L}^*(\alpha) \to -\infty$ as $\alpha \to x_{(1)}$. When $n > 3$, for a fixed set of observations it turns out that $\mathcal{L}^*(\alpha)$ converges to $\left(\frac{1}{n}\sum(x_i - \bar{x})^2\right)^{-\frac{n}{2}}$ as $\alpha \to -\infty$. To show this let us consider a series expansion of the log-likelihood function (2.3). Observe first that

$$\frac{1}{n}\sum_{i=1}^{n}\frac{1}{x_i - \alpha} \approx -\frac{1}{\alpha} + \frac{\bar{x}}{\alpha^2} - \frac{\sum_{i=1}^{n} x_i^2}{n\alpha^3}$$

while

$$\frac{1}{\bar{x} - \alpha} \approx -\frac{1}{\alpha} + \frac{\bar{x}}{\alpha^2} - \frac{\bar{x}}{\alpha^3}.$$

Hence

$$\frac{1}{n}\sum_{i=1}^{n}\frac{1}{x_i - \alpha} - \frac{1}{\bar{x} - \alpha} \approx -\frac{1}{n\alpha^3}\sum(x_i - \bar{x})^2.$$

Next

$$-\frac{3}{2}\sum_{i=1}^{n}\log(x_i - \alpha) \approx -\frac{3n}{2}\log(-\alpha) + \frac{3n}{2\alpha}\bar{x},$$

so that the first two terms of (2.3) give approximately

$$-\frac{n}{2}\log(2\pi s^2) + \frac{3n}{2\alpha}\bar{x} + O\left(\frac{1}{\alpha^2}\right).$$

The third term is roughly equal to

$$-\frac{1}{2s^2}\left(1 + \frac{2\bar{x}}{\alpha}\right)\left(\sum_{i=1}^{n}(x_i - \bar{x})^2\left(1 + \frac{x_i}{\alpha}\right)\right) + O\left(\frac{1}{\alpha^2}\right)$$

which reduces to

$$-\frac{s^{-2}}{2}\left(ns^2 + \frac{n\bar{x}^3}{\alpha} + \frac{\sum_{i=1}^{n} x_i^3}{\alpha}\right) + O\left(\frac{1}{\alpha^2}\right).$$

Thus

$$l(\mu, \alpha, \lambda) = -\frac{n}{2}(1 + \log 2\pi s^2) - \frac{n\bar{s}^2}{2\alpha}\left(\bar{x}^3 + \frac{1}{n}\sum_{i=1}^n x_i^3 - 3\bar{x}s^2\right) + O\left(\frac{1}{\alpha^2}\right)$$

$$= -\frac{n}{2}(1 + \log 2\pi s^2) - \frac{n\bar{s}^2}{2\alpha}\left(\frac{\sum_{i=1}^n x_i^3}{n} + 3\frac{(\sum_{i=1}^n x_i^2)\bar{x}}{n} + 2\bar{x}^3\right)$$
$$+ O\left(\frac{1}{\alpha^2}\right)$$

$$= -\frac{n}{2}(1 + \log 2\pi s^2) - \frac{n}{2\alpha}sg_1 + O\left(\frac{1}{\alpha^2}\right)$$

where g_1 is the sample skewness. Finally we can write

$$l(\mu, \alpha, \lambda) = l_0 + \frac{1}{\alpha}l_1 + O\left(\frac{1}{\alpha^2}\right).$$

As $\alpha \to -\infty$ we note that $l(\mu, \alpha, \lambda) \to$ a constant l_0. So for $-\infty < \alpha < x_{(1)}$, $l(\mu, \alpha, \lambda)$ is bounded above. Furthermore if the sample skewness $g_1 > 0$, $l(\mu, \alpha, \lambda)$ decreases as $\alpha \to -\infty$. Thus $l(\mu, \alpha, \lambda)$ considered as a function of α must attain a global maximum at a stationary point where $\frac{\partial l}{\partial \alpha} = 0$. When $g_1 < 0$, $l(\mu, \alpha, \lambda)$ can achieve its overall maximum at $\alpha = -\infty$. Padgett and Wei(1979) report that empirical evidence points to a monotone decrease of the likelihood function. Cheng and Iles (1990) show that in this case (when $\alpha \to -\infty$) l_0 is just the log-likelihood function corresponding to a normal model, called an embedded two-parameter model. A similar situation obtains for the three-parameter gamma and log normal models. The embedded two-parameter case corresponds to infinite parameter values in the original model. We will now describe following Cheng and Iles the relation between the parameters of the 3-parameter model and the embedded two parameter model and the recipe for deciding which model is appropriate.

Let $f(x - \alpha, \theta)$ be a probability defined for $x > \alpha > -\infty$, α being a threshold parameter and θ, a vector of parameters excluding α. Denote by $\mathcal{L}(\alpha, \theta)$ the log-likelihood based on $X = (X_1, \ldots, X_n)$ a random sample. To examine the behaviour of $\mathcal{L}(\alpha, \theta)$ as $\alpha \to -\infty$, one should first look for a transformation $\psi = \binom{\mu}{\sigma}$ where $\mu = \mu(\alpha, \theta)$ and $\sigma = \sigma(\alpha, \theta)$ are respectively the mean and standard deviation of $f(x - \alpha, \theta)$. Let $\mathcal{L}(\alpha, \psi)$ be the log-likelihood after the parametrization and assume that we can expand $\mathcal{L}$ in an asymptotic series in α in $-\infty < \alpha < X_{(1)}$ (including the limit as $\alpha \to -\infty$), $X_{(1)}$ being the minimum order statistic. Thus let

$$\mathcal{L}(\alpha, \psi) = \mathcal{L}_0(\psi) + \alpha^{-1}\mathcal{L}_1(\psi) + \mathcal{R}(\psi, \alpha). \tag{2.6}$$

Maximum likelihood estimation

In the above we will further assume that $\mathcal{L}_0$, $\mathcal{L}_1$ are twice continuously differentiable in ψ and $\mathcal{R}(\psi, \alpha)$ is continuously differentiable in ψ and $\mathcal{R}(\psi, \alpha)$ and its derivative are of the order $O(\alpha^{-2})$ as $\alpha \to -\infty$ locally uniformly in ψ. Cheng and Iles establish the following theorem which we state without proof.

Theorem 2.2 *Let the log-likelihood of the three-parameter model have the structure (2.6). Then*
(a) $\mathcal{L}_0(\psi)$ is the log-likelihood of a two-parameter embedded distribution,
(b) if $\mathcal{L}^*(\alpha) = \max_\psi \mathcal{L}(\alpha, \psi)$ and $\hat{\psi}$ is a stationary maximum of $\mathcal{L}_0(\psi)$ and the matrix of second partial derivatives $\frac{\partial^2 \mathcal{L}}{\partial \psi^2}$ is non-singular in the neighbourhood of $\hat{\psi}$, there exists $\hat{\psi}(\alpha)$ maximizing $\mathcal{L}(\alpha, \psi) = \mathcal{L}(\alpha, \hat{\psi}(\alpha))$ such that $\hat{\psi}(\alpha) \to \hat{\psi}$ as $\alpha \to -\infty$ and moreover $\mathcal{L}^*(\alpha) = \mathcal{L}_0(\hat{\psi}) + \alpha^{-1}\mathcal{L}_1(\hat{\psi}) + O(\alpha^{-2})$.

In the case of the three-parameter $IG(\mu, \alpha, \lambda)$ model, we first let $\lambda = \beta\gamma$ and then parameterize

$$\psi = \begin{pmatrix} \mu(\alpha, \theta) \\ \sigma(\alpha, \theta) \end{pmatrix} = \begin{pmatrix} \alpha + \gamma \\ |\gamma|\sqrt{\beta} \end{pmatrix}$$

where $\beta > 0$ and $x > \alpha$ if $\gamma > 0$ while $x < \alpha$ if $\gamma < 0$. The density (2.1) takes the form

$$\left(\frac{\beta\gamma}{2\pi(x-\alpha)^3}\right)^{\frac{1}{2}} \exp\left(-\frac{\beta(x-\alpha-\gamma)^2}{2\gamma(x-\alpha)}\right).$$

On expanding the log-likelihood it turns out that

$$\mathcal{L}_0(\psi) = -\frac{n}{2}\log(2\pi) - n\log\sigma - \frac{\sum_{i=1}^n (x_i - \mu)^2}{2\sigma^2}$$

and

$$\mathcal{L}_1(\psi) = \frac{3}{2}\sum_{i=1}^n (x_i - \mu) - \frac{\sum_{i=1}^n (x_i - \mu)^3}{2\sigma^2}.$$

Furthermore $\mathcal{L}_1(\hat{\psi}) = -\frac{n}{2}sg_1$ where s is the sample standard deviation and g_1 the sample skewness. The question as to when to fit the IG model and when to fit the normal model is settled by examining g_1. If $g_1 > k\sqrt{\frac{6}{n}}$ for $k > 0$ (to be determined) the IG model is appropriate. Otherwise the normal model is fitted. If the underlying law is the IG model it turns out that $g_1 \sim N\left(0, \frac{6}{n}\right)$. The constant k is chosen so that the probability of fitting the IG model when the true model is Gaussian is small. For $k = 1.64$ the probability is about 5%. The estimate for α is obtained by solving the implicit equation $\frac{\partial \mathcal{L}^*(\alpha)}{\partial \alpha} = 0$. A recommended method due to Cheng and Amin (1981) is to use an iterative procedure starting with

$$\alpha_0 = \alpha_{(1)} - (2s^2 \log n)^{-1}(\bar{x} - x_{(1)})^3$$

$$\alpha_{k+1} = \alpha_k + \left(\frac{3}{n}\sum(x_i - \alpha_k)^{-1} + \lambda_k \left(m_k^{-2} - \frac{\sum(x_i - \alpha_k)^{-2}}{n}\right)\right)$$
$$\times \left(3m_k^{-1}\lambda_k^{-1} + 12\lambda_k^{-2}\right), \quad k = 0, 1, \ldots$$

where m_k, λ_k are the estimates evaluated at $\alpha = \alpha_k$. For further details the reader should consult Cheng and Amin. They also show that as $n \to \infty$, there exists in law a stationary point $(\hat{\alpha}, \hat{\mu}, \hat{\lambda})$ of $\mathcal{L}$ such that $\sqrt{n}(\hat{\alpha}-\alpha, \hat{\mu}-\mu, \hat{\lambda}-\lambda)$ is asymptotically $N_3(\underline{0}, \sum_{\text{MLE}})$ where the elements of $\sum_{\text{MLE}}$ are given by $d^{-1}\sigma_{ij}$, when

$$d = \frac{3}{4}\lambda^{-2}\mu^{-4} + 3\lambda^{-3}\mu^{-3},$$

$$\sigma_{11} = \frac{1}{2}\lambda^{-1}\mu^{-3},$$

$$\sigma_{22} = \frac{1}{2}\lambda^{-1}\mu^{-2} + \frac{3}{4}\lambda^{-3}\mu^{-1} + 3\lambda^{-4},$$

$$\sigma_{33} = \frac{9}{2}\lambda\mu^{-5} + \frac{21}{2}\mu^{-4} + \frac{21}{2}\lambda^{-1}\mu^{-3},$$

$$\sigma_{12} = -\frac{1}{2}\lambda^{-1}\mu^{-3},$$

$$\sigma_{13} = -\frac{3}{2}(\mu^{-4} + \lambda^{-1}\mu^{-3}), \text{ and}$$

$$\sigma_{23} = -\sigma_{13}.$$

More simply

$$\sum_{\text{MLE}} = A\begin{pmatrix} 1 & -1 & -B \\ -1 & C & B \\ -B & B & B \end{pmatrix}$$

where $A = \frac{2\phi^2\mu^2}{3(\phi+4)}$, $B = 3(\phi+1)$, $C = \frac{2\phi^3+3\phi+12}{2\phi^3}$, and $D = 3(2\phi^2 + 7\phi + 7)$.

The above expressions are due to Jones and Cheng (1984) who also found the joint asymptotic efficiency of the moment estimators as defined by

$$\frac{\det \sum_{\text{MLE}}}{\det \sum_{\text{ME}}} = \phi^4 \left((\phi+4)\left(\phi(\phi+6)^2 + 12(\phi+5)\right)^{-1}\right).$$

The asymptotic relative efficiencies (ARE) of the moment estimators of α, μ, λ with respect to the maximum likelihood estimators are given as follows.

$$\text{ARE}(\hat{\alpha}/\tilde{\alpha}) = \phi^3 \left(3(\phi+4)(\phi+6)^2\right)^{-1}$$
$$\text{ARE}(\hat{\mu}/\tilde{\mu}) = (2\phi^3 + 3\phi + 12) \left(3(\phi+4)\left(1+2(\phi+6)^2\right)\right)^{-1}$$
$$\text{ARE}(\hat{\lambda}/\tilde{\lambda}) = \phi^3(3\phi^2+7\phi+7)\left((\phi+4)(\phi+3(\phi+5)(\phi+4))\right)^{-1}.$$

When $\gamma_1 = 3/\sqrt{\phi}$, the population skewness lies in $(0.4, 4)$, the relative efficiencies of the moment estimators appear quite low; when $\gamma_1 \to 0$ the relative efficiency reaches 1. This happens because the $IG(\mu, \lambda)$ tends to the normal law as $\phi \to \infty$ (see section 1.1). In this instance both estimators perform equally well.

An alternative version of $IG(\mu, \alpha, \sigma)$

Using the parametrization $\lambda = \frac{\mu^2}{\sigma^2}$, $\sigma > 0$, Chan (1982), Chan, Cohen and Whitten (1983, 1984) and Cohen, Whitten (1985) discuss estimation procedures and bias reducing techniques using the coefficient of skewness γ_2, the first order statistic $x_{(1)}$ and the sample size n. We now consider some of the issues discussed by these authors.

The distribution function of the standardized variable $Z = \frac{X-\alpha-\mu}{\sigma}$ is

$$G(z;\beta) = \Phi\left(\frac{z}{\sqrt{1+\beta z}}\right) + \exp\left(\frac{2}{\beta^2}\right) \Phi\left(\frac{-(z+\frac{2}{\beta})}{\sqrt{1+\beta z}}\right)$$

where

$$\beta = \frac{\gamma_2}{3}.$$

Since $\mathbb{E}\left[F(X_{(1)})\right] = \frac{1}{n+1}$, one can obtain values of γ_2 as a function of the standardized first order statistic $Z_{(1)}$ and the sample size n from tables of $G(z;\beta)$. Armed with an estimate of γ_2 (and hence of β) one now solves the moment equations $\alpha + \mu = \overline{X}$, $\sigma^2 = s^2 = \sum_{i=1}^n (X_i - \overline{X})^2/(n-1)$, and $\frac{\sigma}{\mu} = \frac{\gamma_2}{3} = \beta$, to obtain the modified moment estimators. These values are then used in an iterative way to obtain modified maximum likelihood estimators. The maximum likelihood estimators are the solutions to

$$n^2 + 3a^2b^2 - 3nab - na^2c = 0$$
$$\hat{\mu} = a \qquad (2.7)$$
$$\hat{\sigma} = a[(ab/n) - 1]^{\frac{1}{2}}$$

where $a = \overline{X} - \hat{\alpha}$, $b = \sum_{i=1}^n (X_i - \hat{\alpha})^{-1}$, and $c = \sum_{i+1}^n (X_i - \hat{\alpha})^{-2}$.

The equations (2.7) are solved iteratively for $\hat{\alpha}$ and the estimates $\hat{\mu}$ and $\hat{\sigma}$ obtained from the last two equations in (2.7).

The asymptotic variances and covariances as given by Chan et al. are

$$\frac{\sigma^2}{n}\begin{pmatrix} \phi_{11} & \phi_{12} & \phi_{13} \\ \phi_{21} & \phi_{22} & \phi_{23} \\ \phi_{31} & \phi_{32} & \phi_{33} \end{pmatrix}$$

where

$$\phi_{11} = \frac{2}{D}, \quad \phi_{12} = -\phi_n, \quad \phi_{13} = 3\beta^3$$
$$\phi_{22} = \phi_{11} + 1, \quad \phi_{23} = -3\beta(C - AE)/D$$
$$\phi_{33} = (BC - E^2)/D$$

and

$$A = \beta^2 + 1, \quad B = \frac{9\beta^2}{2} + 1, \quad C = B + \frac{21\beta^4 A}{2}$$
$$E = \frac{9\beta^2 A}{2} + 1, \quad D = 2(BC - E^2) + 9\beta^2(2AE - A^2 B - C).$$

Cohen and Whitten caution that these quantities are strictly applicable only for the maximum likelihood estimates. However, simulations seem to indicate that they approximate the corresponding quantities for the moment estimates. A final warning is given that the above variances and covariances are not to be used unless $\gamma_2 > 1 (3\beta > 1)$.

In the version studied by Balakrishnan and Chen (1997) they use the symbol k for the shape parameter. Thus their k corresponds to 3β, or the coefficient of skewness.

We conclude the discussion with two examples provided by Cohen and Whitten, which is illustrative of the computational aspects of these procedures. We also present the best linear unbiased estimators of the mean and the standard deviation as well as the standard errors of the estimators for some selected values of the coefficient of skewness.

Example 2.1 The maximum flood levels in millions of cubic feet per second for the Susquehanna River at Harrisburg, Pennsylvania over 20 four-year periods from 1890-1969 are

0.654	0.613	0.315	0.449	0.297
0.402	0.379	0.423	0.379	0.3235
0.269	0.740	0.418	0.412	0.494
0.416	0.338	0.392	0.484	0.265.

From the above data we have

$$X_{(1)} = 0.265, \quad \overline{X} = 0.423125, \quad n = 20$$
$$\hat{\sigma} = s = 0.1253, \quad Z_{(1)} = -1.262.$$

Maximum likelihood estimation

From Fig 2.1 for $n = 20$ and $Z_{(1)} = -1.262$ we find $\gamma_2 = 1.25$ that

$$\hat{\alpha} = 0.423125 - 0.301 = 0.122 \quad \hat{\mu} = 33/\gamma_2 = 0.301.$$

Computer calculations using a FORTRAN program developed by Chan et al.,(1983) confirm the closeness of the approximations of the estimates.

The best linear unbiased estimates of the mean and standard deviations using the order statistic approach together with their standard errors are given below for a few chosen values of γ_2, the coefficient of skewness.

Table 2.1 *Best Linear Unbiased Estimates*

γ_2	μ^*	σ^*	s.e(μ^*)	s.e(σ^*)
1.0	0.42202	0.11892	0.02655	0.02082
1.5	0.42321	0.13231	0.02938	0.02511
2.0	0.42733	0.15147	0.03327	0.03169

Example 2.2 Data from fatigue life in hours of 10 bearings of a certain type reported by McCool (1974) are

152.7	172.0	172.5	173.3	193.0	204.7
216.5	234.9	262.6	422.6		

For this data $X_{(1)} = 1.52.7$, $\overline{X} = 220.48$ $s = 78.406$ and $Z_{(1)} = -0.864$. From Figure 2.1 the estimated value of $\gamma_2 = 2.45$. This gives a value of $\hat{\mu} = 96.0$ and hence $\hat{\alpha} = 124.5$. These values are quite close to the exact values as shown by Cohen and Whitten (1985).

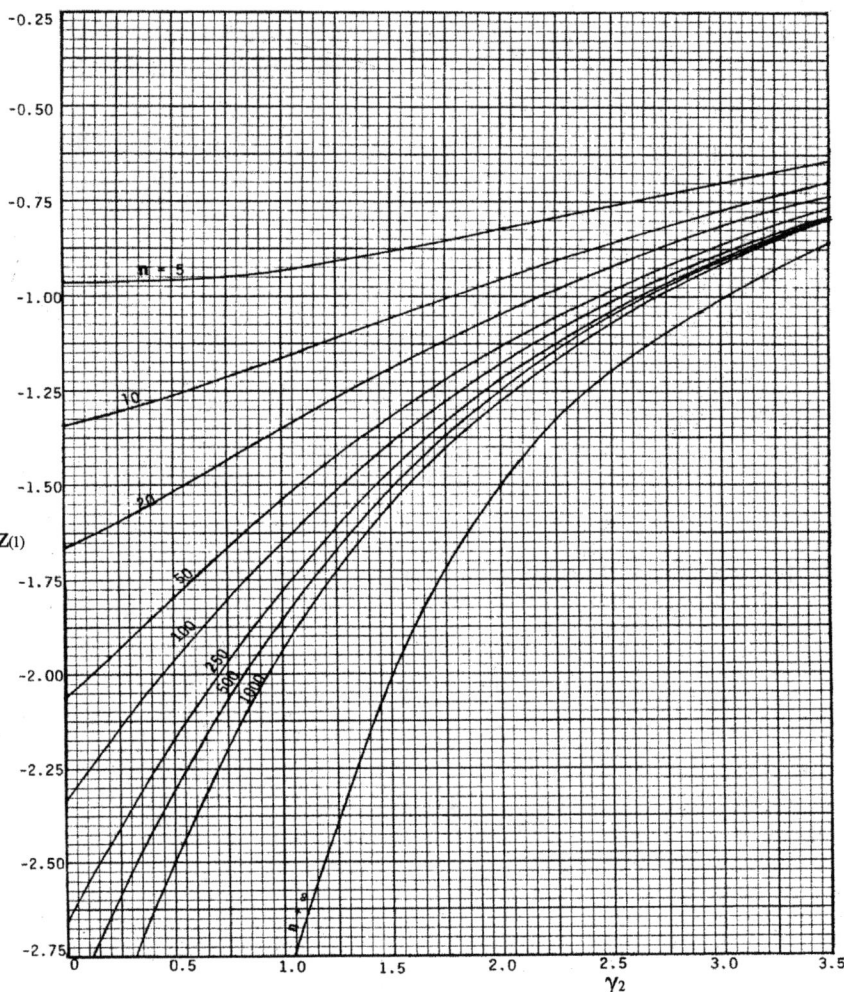

Figure 2.1 *Graphs of γ_2 as a function of $z_{(1)}$ and n*

2.3 Estimation under truncation

Assuming small and large observations from $IG(\mu, \lambda)$ are truncated, Patel (1965) has examined the problem of parameter estimation. Let x_1 and x_2 be two numbers such that $0 < x_1 < x_2 < \infty$. If $IG(\mu, \lambda)$ is truncated at the points x_1 and x_2, the density has the form

$$K x^{-\frac{3}{2}} \exp\left(-\frac{\lambda x}{2\mu^2} - \frac{\lambda}{2x}\right) 1_{(x_1 \leq x \leq x_2)}(x)$$

where K is the normalizing constant. If $f(x)$ is the untruncated density, then differentiating $f(x)$ with respect to x we obtain

$$\frac{\partial f}{\partial x} = \left(-\frac{3}{2x} + \frac{\lambda}{2x^2} - \frac{\lambda}{2\mu^2}\right) f(x).$$

Now multiply both sides by x^s and integrate from x_1 to x_2 to obtain

$$\frac{\lambda}{\mu^2}\mu'_s - \lambda\mu'_{s-2} - 2(x_1^s f(x_1) - x_2^s f(x_2)) = (2s-3)\mu'_{s-1} \qquad (2.8)$$

where

$$\mu'_s = \int_{x_1}^{x_2} x^s f(x) dx, \quad s = 0, \pm 1, \ldots$$

When the moments μ'_s are estimated by the sample counterparts, denoting by M, h and c

$$M = \begin{bmatrix} 1 & -m'_{-2} & -2 & 2 \\ m'_1 & -m'_{-1} & -2x_1 & 2x_2 \\ m'_2 & -1 & -2x_1^2 & 2x_2^2 \\ m'_3 & -m'_1 & -2x_1^3 & 2x_2^3 \end{bmatrix}, \quad h = \begin{bmatrix} \frac{\lambda}{\mu^2} \\ \lambda \\ f(x_1) \\ f(x_2) \end{bmatrix}, \quad c = \begin{bmatrix} -3m'_{-1} \\ -1 \\ m'_1 \\ 3m'_2 \end{bmatrix}$$

we obtain the equation

$$Mh = c$$

whose solutions provide the estimates of μ and λ.

CHAPTER 3

SIGNIFICANCE TESTS

3.0 Introduction

This chapter is devoted to tests of hypotheses for parameters of $IG(\mu, \lambda)$. We derive the likelihood ratio tests for the mean parameter as well as the lambda parameter in the one and two sample cases; we also consider tests for the Brownian motion process. The power of these tests is examined briefly. We study interval estimation from both the frquentist as well as the Bayesian points of view. Prediction intervals and tolerance limits are examined in detail and numerous illustrative examples are provided. A section is devoted to tests of separate families first considered by Cox (1961) and we illustrate this with applications to simulated and physiological data to test inverse Gaussian against the lognormal and vice versa. Finally we discuss Bahadur efficient tests.

3.1 Likelihood ratio tests — one sample case

(i) Tests for the mean μ (λ known).

Let $X = (X_1, \ldots, X_n)$ be a random sample from $IG(\mu, \lambda)$. The null and alternative hypotheses are

$$H_0\colon \mu = \mu_0, \; (\lambda \text{ known}) \; vs \; H_A\colon \mu \neq \mu_0, \; (\lambda \text{ known}).$$

Writing

$$\Omega = \{(\mu, \lambda) \mid 0 < \mu < \infty, \; 0 < \lambda < \infty\}, \text{ and}$$
$$\Omega_0 = \{(\mu, \lambda) \mid \mu = \mu_0, \; 0 < \lambda < \infty, \}$$

the standard procedure calls for rejection of H_0 for small values of

$$\Lambda = \frac{L(\hat{\Omega}_0)}{L(\hat{\Omega})}$$

where $L(\hat{\Omega}_0)$, $L(\hat{\Omega})$ are the maxima of the likelihoods under H_0 and $\{H_0 \cup H_A\}$ respectively. Routine calculation gives us

$$\Lambda = \exp\left(\lambda \left(\sum_{i=1}^{n} \frac{(x_i - \bar{x})^2}{2x_i \bar{x}^2} - \sum_{i=1}^{n} \frac{(x_i - \mu_0)^2}{2x_i \mu_0^2}\right)\right)$$

Likelihood ratio tests

so that H_0 is rejected at level α when the difference between the two sums is small. Upon simplification we have the result that H_0 is rejected if $-\frac{n(\bar{x}-\mu_0)^2}{\bar{x}\mu_0^2}$ is small. The test is therefore equivalent to rejecting H_0 for large values of $\left|\frac{\sqrt{n}(\bar{x}-\mu_0)}{\mu_0\sqrt{\bar{x}}}\right|$. Since λ is known, this says that we reject H_0 when

$$|Y| = \left|\frac{\sqrt{n}(\bar{x}-\mu_0)}{\mu_0\sqrt{\bar{x}}}\right| > \text{constant}$$

where the constant is determined subject to $P(|Y| > \text{constant} \mid H_0) \leq \alpha$. From Corollary 1.1 we see that this rejection rule provides a critical region given by $\{Y < k_1\} \cup \{Y > k_2\}$ where

$$\int_{k_1}^{k_2} \left(1 - \frac{y}{\sqrt{y^2 + \frac{4n\lambda}{\mu_0}}}\right) \frac{\sqrt{n\lambda}}{\sqrt{2\pi}} \exp\left(-\frac{y^2}{2}\right) dy = 1 - \alpha. \quad (3.1)$$

Chhikara and Folks (1976) use Lehmann's criterion for uniformly most powerful unbiasedness of the test and show that k_1 and k_2 should also satisfy

$$\int_{k_1}^{k_2} \bar{x} f(\bar{x}) d\bar{x} = \mu_0(1-\alpha). \quad (3.2)$$

Evaluating (3.2) (see Theorem 1.2) we have upon writing $\frac{\sqrt{n\lambda}(\bar{x}-\mu_0)}{\mu_0\sqrt{\bar{x}}} = y$, since $\frac{\sqrt{n\lambda}}{\sqrt{\bar{x}}}\frac{d\bar{x}}{dy} = \frac{2\mu_0\bar{x}}{\bar{x}+\mu_0}$ and $\frac{2\bar{x}\mu_0}{\bar{x}+\mu_0} = \mu_0\left(1 + \frac{\mu_0 y}{\sqrt{\mu_0^2 y^2 + 4n\lambda\mu_0}}\right)$. Hence

$$\int_{k_1}^{k_2} \sqrt{\frac{n\lambda}{2\pi\bar{x}}} \exp\left(-\frac{n\lambda(\bar{x}-\mu_0)^2}{2\mu_0^2 \bar{x}}\right) d\bar{x} = \mu_0(1-\alpha)$$

$$\Rightarrow \int_{k_1}^{k_2} \frac{\sqrt{n\lambda}}{\sqrt{2\pi}} \left(1 + \frac{\mu_0 y}{\sqrt{\mu_0^2 y^2 + 4\lambda n\mu_0}}\right) \exp\left(-\frac{n\lambda y^2}{2}\right) dy = (1-\alpha).$$

Thus

$$\int_{k_1}^{k_2} \left(1 + \frac{y}{\sqrt{y^2 + 4\frac{n\lambda}{\mu_0}}}\right) \sqrt{\frac{n\lambda}{2\pi}} \exp\left(-\frac{n\lambda y^2}{2}\right) dy = 1 - \alpha. \quad (3.3)$$

From (3.1) and (3.3) we readily see that $k_1 = -k_2 = -Z_{1-\frac{\alpha}{2}}$, the $100(1-\frac{\alpha}{2})\%$ point of the standard normal distribution. Therefore the likelihood ratio test of H_0 against H_A recommends rejection of H_0 at level α if $|Y| > Z_{1-\frac{\alpha}{2}}$. One sided hypotheses $\mu \leq \mu_0$ ($\mu \geq \mu_0$) against $\mu > \mu_0$ ($\mu < \mu_0$) can be handled using the same statistic Y. These tests are uniformly most powerful at level α.

(ii) Tests for the mean (λ unknown).

As in (i) let us consider the two-sided alternative $H_A: \mu \neq \mu_0$. Routine computations give us the maximum likelihood estimate of λ under H_0 ($\mu = \mu_0$) as $\dfrac{n}{\sum_{i=1}^n \frac{(x_i - \mu_0)^2}{\mu_0^2 x_i}}$, while under $\Omega = \{H_0 \cup H_A\}$, $\hat{\mu} = \bar{x}$ and $\hat{\lambda} = \dfrac{n}{\sum_{i=1}^n \left(\frac{1}{x_i} - \frac{1}{\bar{x}}\right)}$. Hence the likelihood ratio is

$$\Lambda = \frac{\sum_{i=1}^n \left(\frac{1}{x_i} - \frac{1}{\bar{x}}\right)}{\sum_{i=1}^n \frac{(x_i - \mu_0)^2}{\mu_0^2 x_i}} = \frac{1}{\left(1 + \dfrac{n(\bar{x} - \mu_0)^2}{\mu_0^2 \bar{x} \sum_{i=1}^n \left(\frac{1}{x_i} - \frac{1}{\bar{x}}\right)}\right)}.$$

The likelihood ratio test therefore recommends rejection of H_0 at level α for large values of $|T^*|$ (see 1.7) where

$$|T^*| = \left| \frac{\sqrt{n}(\bar{x} - \mu_0)}{\mu_0 \sqrt{\bar{x} \sum \left(\frac{1}{x_i} - \frac{1}{\bar{x}}\right)}} \right|.$$

This amounts to rejecting H_0 when

$$|T| = \left| \frac{\sqrt{n\lambda}\left(\frac{\bar{x} - \mu_0}{\mu_0 \sqrt{\bar{x}}}\right)}{\left(\sqrt{\frac{\lambda V}{n-1}}\right)} \right|$$

is large, where $V = \sum_{i=1}^n \left(\frac{1}{X_i} - \frac{1}{\bar{X}}\right) = n/\hat{\lambda}$. The distribution of T is given in Theorem 1.3 and we are led to an equal-tails test as in the previous case, the only difference being that we reject H_0 if $|T| > t_{1-\frac{\alpha}{2}, n-1}$. Alternatively one could reject by using the α-fractile points of a Beta distribution with parameters $\frac{n-1}{2}$ and $\frac{1}{2}$ or using an F test statistic with 1 and $(n-1)$ degrees of freedom, an observation due to Miura (1978). For one-sided tests Chhikara and Folks reduce the tests to a test of

$$H_0': \theta \leq 0 \; (\theta \geq 0) \text{ against } H_A': \theta > 0 \; (\theta < 0)$$

where $\theta = \frac{\lambda}{2}\left(1 - \frac{1}{\mu^2}\right)$ and μ_0 is taken as 1. Uniformly most powerful unbiased tests are obtainable using the conditional law of T_1 given T_2 where $T_1 = n\bar{X}$ and $T_2 = n(\bar{X} + \bar{X}_-)$, (see Theorem 1.7). Indeed we have

$$f(t_1 \mid t_2) = \frac{n(t_1^3(t_2 - 2n))^{-\frac{1}{2}}}{Be\left(\frac{1}{2}, \frac{n-1}{2}\right)} \left(1 - \frac{(t_1 - n)^2}{t_1(t_2 - 2n)}\right)^{\frac{n-3}{2}} 1_D(t_1)$$

where $D = \{t_1 \mid 0 < \frac{(t_1-n)^2}{t_1(t_2-2n)} < 1\}$. With the substitution

$$w = \frac{\sqrt{n-1}(t_1 - n)}{\left(t_1(t_2 - 2n)\left(1 - \frac{(t_1-n)^2}{t_1(t_2-2n)}\right)\right)^{\frac{1}{2}}}$$

which is none other than $w = \frac{\sqrt{n(n-1)}(\overline{x}-1)}{\sqrt{\overline{x}v}}$, one is again led to a weighted Student's law

$$\frac{1}{\sqrt{n-1}Be\left(\frac{1}{2}, \frac{n-1}{2}\right)}\left(1 - \frac{(n-1)w\sqrt{t_2 - 2n}}{\sqrt{4n(n-1) + (t_2 + 2n)w^2}}\right)\left(1 + \frac{w^2}{n-1}\right)^{-\frac{n}{2}}.$$

The critical region is equivalent to $W >$ constant ($W <$ constant) and the constant c can be shown to satisfy the equation

$$S_{n-1}(-c) + \left(\frac{t_2 + 2n}{t_2 - 2n}\right)^{\frac{n-2}{2}} S_{n-1}\left(-\sqrt{4n + c^2(t_2 + 2n)}\right) = \alpha.$$

$S_{n-1}(\cdot)$ being the distribution function of Student's t with $(n-1)$ degrees of freedom. When testing for μ_0 other than 1, one uses the statistic

$$\frac{\sqrt{n(n-1)}(\overline{X} - \mu_0)}{\mu_0\sqrt{\overline{X}V}}$$

and now the critical value c is determined by solving

$$S_{n-1}(-c) + \left(\frac{\sum_{i=1}^{n}(x_i + \mu_0)^2/x_i}{\sum_{i=1}^{n}(x_i - \mu_0)^2/x_i}\right)^{\frac{n-2}{2}}$$

$$S_{n-1}\left(-\sqrt{4m + c^2\mu_0 \sum_{i=1}^{n}\frac{(x_i + \mu_0)^2}{x_i}}\right) = \alpha.$$

When n is large (or $\frac{\lambda}{\mu}$ is large) both the one-sided tests (known λ as well as unknown λ) have critical values well approximated by the normal and Student's table values.

(iii) Tests for λ (μ known).

This is by far the simplest test. From basic theory in Chapter 2 we know that

$$\mathcal{L}\left(\lambda \sum_{i=1}^{n} \frac{(x_i - \mu)^2}{\mu^2 X_i}\right) = \chi_n^2.$$

Hence for testing $H_0: \lambda = \lambda_0$ or equivalently $\frac{1}{\lambda} = \frac{1}{\lambda_0}$, μ known against $H_A: \lambda \neq \lambda_0$ or $\frac{1}{\lambda} \neq \frac{1}{\lambda_0}$, μ known we use the statistic $Y = \sum_{i=1}^{n} \frac{(X_i - \mu)^2}{X_i}$, to obtain a uniformly most powerful unbiased test of size α with critical region of the form $\{Y \leq k_1\}$ or $\{Y \geq k_2\}$ where $g_n(y)$, the density of $\frac{\lambda Y}{\mu^2}$ satisfies

$$\int_{k_1}^{k_2} g_n(y) dy = (1 - \alpha) \qquad (3.4)$$

and

$$\int_{k_1}^{k_2} y g_n(y) dy = n(1 - \alpha). \qquad (3.5)$$

Using the technique used by Lehmann one can show that $y g_n(y) = n g_{n+2}(y)$. Letting $G_n(\cdot)$ denote the distribution function of χ_n^2 law, (3.4) and (3.5) yield

$$G_n(k_2) - G_n(k_1) = G_{n+2}(k_2) - G_{n+2}(k_1) = 1 - \alpha$$

and k_1 and k_2 are then found from the tables. For large n or $\lambda \gg \mu$ an equal-tails test can be used.

(iv) Tests for λ (μ unknown).

We are concerned with the test of

$$H_0: \frac{1}{\lambda} = \frac{1}{\lambda_0}, \; \mu \text{ unknown against } H_A: \frac{1}{\lambda} \neq \frac{1}{\lambda_0}, \; \mu \text{ unknown.}$$

Here $\Omega = \{(\mu, \lambda) \mid 0 < \mu < \infty, \; 0 < \lambda < \infty\}$ and $\Omega_0 = \{(\mu, \lambda) \mid 0 < \mu < \infty, \; \lambda = \lambda_0\}$. The likelihood ratio test can be shown to be based on the familiar statistic $V = \sum_{i=1}^{n} \left(\frac{1}{X_i} - \frac{1}{X}\right)$ and H_0 is rejected either for large values of V or small values of V. The two-tailed test with critical region $\{V \leq k_1\}$ or $\{V \geq k_2\}$ can be handled as in (iii). The one-sided optimum tests are similar in spirit to testing σ^2 in the normal distribution when the mean is unknown.

Remarks Bar-Lev and Reiser (1982) have shown that for some exponential models there exists an exponential sub-family which admits uniformly most powerful unbiased test based on a single test statistic. The $IG(\mu, \lambda)$ law is one such model. In this case we write (see Seshadri 1993)

$$f(x; \theta) = \frac{x^{-\frac{3}{2}}}{\sqrt{2\pi}} \exp\left(\frac{\theta_1}{x} + \theta_2 x - k(\theta_1, \theta_2)\right) 1_{(0,\infty)}(x)$$

where $(\theta_1, \theta_2) = \left(-\frac{\lambda}{2}, -\frac{\lambda}{2\mu^2}\right)$ and $k(\theta) = -2\sqrt{\theta_1 \theta_2} - \frac{1}{2}\log(-\theta)$. If we write $t_1(x) = \frac{1}{x}$, $t_2(x) = x$, then from Theorem 1.7, Seshadri (1993)

$w' = \bar{t}_1 - H(\bar{t}_2)$ is independent of $\bar{t}_2$ where $H(a) = \frac{1}{a}$, $\bar{t}_1 = \frac{1}{n}\sum_{i=1}^{n}\frac{1}{X_i}$ and $\bar{t}_2 = \sum_{i=1}^{n} X_i$. Moreover the distribution of w' depends only on θ_1 and therefore the $IG(\mu, \lambda)$ family admits of a uniformly most powerful unbiased test for testing hypotheses of the form
(a) $\theta_1 \leq \theta_1^0$ versus $\theta_1 > \theta_1^0$
(b) $\theta_1 \leq \theta_1^1$ or $\theta_1 \geq \theta_1^2$ versus $\theta_1^1 < \theta_1 < \theta_1^2$
(c) $\theta_1^1 \leq \theta_1 \leq \theta_1^2$ versus $\theta_1 < \theta_1^1$ or $\theta_1 > \theta_1^2$
(d) $\theta_1 = \theta_1^0$ versus $\theta_1 \neq \theta_1^0$. Note that w' is $\frac{V}{n} = \frac{1}{n}\sum_{i=1}^{n}\left(\frac{1}{X_i} - \frac{1}{\bar{X}}\right)$.

3.2 Brownian motion process

Suppose that $W(x)$ is a Brownian motion process with drift ν and diffusion constant σ^2, with $W(0) = 0$. The first passage time X to a (barrier) state a follows the $IG(\mu, \lambda)$ law where $\mu = \frac{a}{\nu}$ and $\lambda = \frac{a^2}{\sigma^2}$. When the drift $\nu = 0$, $\mu = \infty$ and conversely (unless we have negative drift). Thus for a fixed a hypotheses tests concerning ν can be translated to tests for μ. The special case of zero drift was first investigated by Nádas (1973) for known λ and by Seshadri and Shuster (1974) for unknown λ.

(a) Tests on ν; (λ known). Consider testing

$$H_0: \nu = 0, \lambda \text{ known against } H_A: \nu > 0, \lambda \text{ known}.$$

Clearly in this case the test statistic is $Y = \frac{\sqrt{n\lambda}(\bar{X}-\mu)}{\mu\sqrt{\bar{X}}}$, which reduces to ($\nu = 0 \Rightarrow \mu = \infty$) $Y_0 = -\sqrt{\frac{n\lambda}{\bar{X}}}$. Now $P(|Y| < y) = G_{|Y|}(y)$ can be differentiated to yield

$$g_{|Y|}(y) = \sqrt{\frac{2}{\pi}} \exp\left(-\frac{y^2}{2}\right) 1_{R^+}(y)$$

the half normal law. Hence Y_0 has this density with support on $\mathbb{R}^-$. Using standard techniques, a uniformly most-powerful level α test of $\mu = \infty$ against $\mu < \infty$ can be shown to give a critical region of the form $\left\{-\sqrt{\frac{n\lambda}{\bar{X}}} < Z_{\frac{\alpha}{2}}\right\}$ or equivalently $\left\{\bar{X} > \frac{n\lambda^2}{Z_{1-\frac{\alpha}{2}}}\right\}$.

(b) Tests on ν; (λ unknown).

When λ is unknown the test-statistic is based on T, and for $\mu = \infty$ we have as statistic $T_0 = \sqrt{\frac{n(n-1)}{\bar{X}V}}$. Hence the critical region of a uniformly most powerful level α test of

$$H_0: \nu = 0, \lambda \text{ unknown against } H_A: \nu > 0, \lambda \text{ unknown}$$

is given by $\left\{\bar{X}V \leq \frac{n(n-1)}{F_{1,n-1,\alpha}}\right\}$, $F_{1,n-1,\alpha}$ being the $100\alpha\%$ point of the F distribution with one and $(n-1)$ degrees of freedom.

3.3 Power considerations

We consider briefly the power of the optimum tests for the mean μ. Patil and Kovner (1979) have studied the power of tests of
(a) $\mu \leq 1$ against $\mu > 1$, as well as
(b) $\mu = 1$ against $\mu \neq 1$.

Case (a) is best handled by using the conditional law T_1 given T_2 (see Theorem 1.7 as well as (ii) of section 3.1) $f_{\theta_1}(t_1 \mid t_2)$ which is

$$\left(\frac{n}{t_2 - 2n}\right)^{\frac{n-3}{2}} \frac{t_i^{-\frac{3}{2}} \left(\frac{t_2}{n} - \frac{t_1}{n} - \frac{n}{t_1}\right)^{\frac{n-3}{2}} \exp(\theta_1 t_1)}{2 \int_D \left(1 - \frac{(nv - v^{-1})^2}{t_2 - 2n}\right)^{\frac{n-3}{2}} \exp\left(\frac{\theta_1}{v^2}\right) dv} 1_D(t_1)$$

where $D = \left\{t_1 \mid (t_2 - 2n) - \sqrt{t_2^2 - 4n^2} < t_1 < (t_2 - 2n) + \sqrt{t_2^2 - 4n^2}\right\}$.

Writing $F_{\theta_1}(\cdot \mid t_2)$ for the distribution function of $f_{\theta_1}(\cdot \mid t_2)$ we can express $P(T_1 \leq c \mid t_2)$ as $P(V > \frac{1}{\sqrt{c}})$ under the transformation $T_1^{-\frac{1}{2}} = V$. Then letting

$$G(c) = \int_{\frac{1}{\sqrt{c}}}^{\frac{\delta}{2\pi}} \left(1 - \frac{(nv - v^{-1})^2}{t_2 - 2n}\right)^{\frac{n-3}{2}} \exp\left(\frac{\theta_1}{v^2}\right) dv$$

where $\delta = \sqrt{t_2 + 2n} + \sqrt{t_2 - 2n}$ we see that

$$F_{\theta_1}(c \mid t_2) = \frac{G(c)}{G\left(\frac{\delta^2}{4}\right)}$$

when n is odd and > 3 the exponential term should be expanded as a power series. Then the second term in the integral expression for $G(c)$ is developed using the binomial expansion; upon integrating with respect to v, $G(c)$ has the form

$$\sum_{j=0}^{\infty} \sum_{k=0}^{\frac{n-3}{2}} \sum_{i=0}^{2k} (-1)^{k+1} \binom{\frac{n-3}{2}}{k} \binom{2k}{l} \left(\frac{\theta_1}{2}\right)^j \frac{n^{2k-l}}{j!\beta(t_2 - 2n)^k} \left(\left(\frac{\delta}{2n}\right)^\beta - c^{-\frac{\beta}{2}}\right)$$

for c such that $c \in D$ and $\beta = 2(k - j - l) + 1$. When n is even the second sum (on k) becomes an infinite sum. Now the power of the test of $\mu \leq 1$ against $\mu > 1$ is given by

$$\pi(\theta_1) = 1 - F_{\theta_1}(k^* \mid t_2)$$

k^* being the critical value given by

$$\int_{k^*}^{\infty} f_{\theta_1 = 0}(t_1 \mid t_2) dt_1 = \alpha.$$

For the two-sided test the power is given by

$$\pi(\theta_1) = F_{\theta_1}(k_2^* \mid t_2) - F_{\theta_1}(k_1^* \mid t_2)$$

where $\int_{k_1^*}^{k_2^*} f_0(t_1 \mid t_2) = 1 - \alpha$. The power of the likelihood ratio test of $\mu = \mu_0$ against $\mu \neq \mu_0$ when λ is unknown has been investigated by Miura (1978) using simulation studies for $n = 5, 10, 15, 20$ and 30, and $(\mu_0 = 1, \lambda = 1)$ at level $\alpha = 0.05$. Miura's results are given in the following table.

Table 3.1 *Power of the LR test-a simulated study*

n	$\mu = 0.50$	1.67	2.50	5.00	6.67	10.00
5	0.21	0.14	0.28	0.59	0.68	0.67
10	0.59	0.27	0.51	0.87	0.86	0.94
15	0.76	0.43	0.79	0.96	0.97	0.99
20	0.92	0.54	0.92	0.99	0.99	0.99
30	0.99	0.69	0.98	1.00	1.00	1.00

3.4 Two sample tests

We now consider two independent random samples $X = (X_1, \ldots, X_m)$ and $Y = (Y_1, \ldots, Y_n)$ where $\mathcal{L}(X_i) = IG(\mu_1, \lambda)$ and $\mathcal{L}(Y_j) = IG(\mu_2, \lambda)$, $i = 1, \ldots, m$ and $j = 1, \ldots, n$.

(i) Tests on μ_i (λ known).

The likelihood ratio test of

$$H_0 : \mu_1 = \mu_2, \; (\lambda \text{ known}) \text{ against } H_A : \mu_1 \neq \mu_2, \; (\lambda \text{ known})$$

is based on Λ where

$$\Lambda = \exp\left(-\frac{\lambda}{2}\left(\frac{(m+n)^2}{m\overline{X} + n\overline{Y}} - \frac{m}{\overline{X}} - \frac{n}{\overline{Y}}\right)\right).$$

To see this observe that under H_0 the maximum likelihood estimate of $m_1 = m_2$ is $\frac{m\overline{X} + n\overline{Y}}{m+n}$, while under $(H_0 \cup H_A)$ it is given by $\hat{\mu}_1 = \overline{X}$, $\hat{\mu}_2 = \overline{Y}$. Therefore rejecting H_0 for small Λ is equivalent to rejecting when

$$\left|\frac{\sqrt{mn}(\overline{X} - \overline{Y})}{\sqrt{\overline{X}\overline{Y}(m\overline{X} + n\overline{Y})}}\right| > \text{constant} = c \; (\text{say}).$$

Chhikara (1975) obtained this result when he considered uniformly most powerful unbiased tests of size α for testing one-sided tests on μ_i, namely

$$H_0' : \mu_1 \leq \mu_2 \; (\mu_1 \geq \mu_2) \text{ against } H_0' : \mu_1 > \mu_2 \; (\mu_1 < \mu_2).$$

The constant c for these one-sided tests is obtained as the solution of the equation

$$\Phi(c) + \frac{n-m}{n+m}\exp\left(\frac{2mn\lambda}{m\overline{X}+n\overline{Y}}\right)(1-\Phi(c')) = 1-\alpha$$

where $c' = \left(c^2 + \frac{4mn\lambda}{m\overline{X}+n\overline{Y}}\right)^{\frac{1}{2}}$. In the case of the two-sided test the constant c reduces to $Z_{1-\frac{\alpha}{2}}$, the $100(1-\frac{\alpha}{2})\%$ point of a standard normal law. When $m = n$, note that the constant is $Z_{1-\alpha}$ for a two-tailed test and $t_{1-\alpha}$ for one-sided tests.

(ii) Tests on μ_i (λ unknown).

We consider the following hypotheses.

$H_0: \mu_1 = \mu_2$, (λ unknown) against $H_A: \mu_1 \neq \mu_2$, (λ unknown).

Under H_0, it is easily shown that

$$\hat{\mu}_1 = \hat{\mu}_2 = \hat{\mu} = \frac{m\overline{X}+n\overline{Y}}{m+n}, \quad \hat{\lambda} = \frac{m+n}{\sum_{i=1}^{m}\frac{(X_i-\hat{\mu})^2}{X_i\hat{\mu}} + \sum_{i=1}^{n}\frac{(Y_i-\hat{\mu})^2}{Y_i\hat{\mu}}}.$$

Under $(H_0 \cup H_A)$, $\hat{\mu}_1 = \overline{X}$, $\hat{\mu}_2 = \overline{Y}$ while $\hat{\lambda} = \frac{m+n}{V_1+V_2}$ where $V_1 = \sum_{i=1}^{m}\left(\frac{1}{X_i} - \frac{1}{\overline{X}}\right)$ and $V_2 = \sum_{i=1}^{n}\left(\frac{1}{Y_i} - \frac{1}{\overline{Y}}\right)$. Routine computations now give

$$\Lambda = \frac{V_1 + V_2}{\sum_{i=1}^{m}\frac{(X_i-\hat{\mu})^2}{X_i\hat{\mu}^2} + \sum_{i=1}^{n}\frac{(Y_i-\hat{\mu})^2}{Y_i\hat{\mu}^2}} < \text{constant}$$

as the critical region. The denominator of Λ can be expressed as

$$V_1 + V_2 + \left(\frac{m}{\overline{X}} + \frac{n}{\overline{Y}} - \frac{m+n}{\hat{\mu}}\right)$$

which simplifies as

$$V_1 + V_2 + \frac{mn(\overline{X}-\overline{Y})^2}{\overline{X}\overline{Y}(m\overline{X}+n\overline{Y})}.$$

Letting

$$Q = \frac{mn(\overline{X}-\overline{Y})^2}{\overline{X}\overline{Y}(m\overline{X}+n\overline{Y})(V_1+V_2)}$$

we obtain $\Lambda = \frac{1}{1+Q}$. Thus the likelihood ratio test calls for rejection of H_0 for large values of $|Q|^{\frac{1}{2}}$.

Using similar regions to construct uniformly most powerful unbiased tests Chhikara shows that one obtains the same statistic. We leave this as an exercise.

In fact $((m+n-2)Q)^{\frac{1}{2}}$ has the well-known t distribution with $(m+n-2)$ degrees of freedom. Thus the recipe for the two-sided testing problem is to reject H_0 if $|(m+n-2)^{\frac{1}{2}}Q^{\frac{1}{2}}| > t_{1-\frac{\alpha}{2}}$, the $100(1-\frac{\alpha}{2})\%$ point of Student's t law with $m+n-2$ degrees of freedom.

(iii) Tests of λ_i (μ_1, μ_2 known).

We suppose that the two independent random samples X and Y are such that $\mathcal{L}(X_i) = IG(\mu_1, \lambda_1)$ $i = 1, \ldots, m$ while $\mathcal{L}(Y_i) = IG(\mu_2, \lambda_2)$ $i = 1, \ldots, n$. We consider the likelihood ratio test of

$$H_0 : \lambda_1 = \lambda_2; \ (\mu_1, \mu_2 \text{ known}) \text{ against } H_A : \lambda_1 \neq \lambda_2; \ (\mu_1, \mu_2 \text{ known}).$$

Under H_0, λ the common value of λ_1 and λ_2 is estimated by $\frac{m+n}{S_1+S_2}$ where $S_1 = \sum_{i=1}^{m} \frac{(X_i-\mu_1)^2}{X_i\mu_1^2}$ and $S_2 = \sum_{i=1}^{n} \frac{(Y_i-\mu_2)^2}{Y_i\mu_2^2}$, while under $(H_0 \cup H_A)$ we have

$$\hat{\lambda}_1 = \frac{m}{S_1}, \quad \hat{\lambda}_2 = \frac{n}{S_2}.$$

Therefore

$$\Lambda = \frac{(m+n)^{\frac{m+n}{2}}}{m^{\frac{m}{2}} n^{\frac{n}{2}}} \frac{S_1^{\frac{m}{2}} S_2^{\frac{n}{2}}}{(S_1+S_2)^{\frac{m+n}{2}}}.$$

Thus apart from a constant factor Λ is a function of $Q = \frac{S_2}{S_1}$, namely $Q^{\frac{n}{2}} \left(\frac{1}{1+Q}\right)^{\frac{m+n}{2}}$. Moreover, it can be shown that Λ is increasing for $Q < \frac{n}{m}$ and decreasing for $Q > \frac{n}{m}$, so that the likelihood ratio test is based on the F ratio where F has n and m degrees of freedom. One-sided uniformly most powerful unbiased tests of size α can also be shown to be based on the statistic Q.

(iv) Tests on λ_i (μ_1, μ_2 unknown).

We assume that the X and Y samples are distributed as in (iii) and consider the likelihood ratio test of

$$H_0 : \lambda_1 = \lambda_2 \text{ against } H_A : \lambda_1 \neq \lambda_2; \ (\mu_1, \mu_2 \text{ unknown}).$$

Under H_0, $\hat{\mu}_1 = \overline{X}$, $\hat{\mu}_2 = \overline{Y}$ while the common value of λ is estimated by $\frac{m+n}{V_1+V_2}$. Under $(H_0 \cup H_A)$, $\hat{\mu}_1 = \overline{X}$, $\hat{\mu}_2 = \overline{Y}$, $\hat{\lambda}_1 = \frac{m}{V_1}$, $\hat{\lambda}_2 = \frac{n}{V_2}$. Therefore we have

$$\Lambda = \frac{(m+n)^{\frac{m+n}{2}}}{m^{\frac{m}{2}} n^{\frac{n}{2}}} \left(\frac{V_1}{V_1+V_2}\right)^{\frac{m}{2}} \left(\frac{V_2}{V_1+V_2}\right)^{\frac{n}{2}}.$$

Once again, apart from a constant factor writing $Q = \frac{V_2}{V_1}$, Λ becomes $= Q^{\frac{n}{2}}(1+Q)^{-\frac{m+n}{2}}$. Davis (1980) shows that the first derivative of $\ln \Lambda$ with respect to Q is > 0 if $Q < \frac{n}{m}$ and < 0 if $Q > \frac{n}{m}$. Thus rejection of H_0 for small Λ is equivalent to rejection of H_0 for either large or small values of Q. Since Q is the ratio of two independent chi-squares we have another F test based on $(n-1)$, $(m-1)$ degrees of freedom. The likelihood ratio tests are also equivalent to the optimum one-sided tests.

(v) Tests on μ_i (λ_1, λ_2 unknown).

We now turn to the situation where we have $\mathcal{L}(X_i) = IG(\mu_1, \lambda_1)$, $i = 1, 2, \ldots, m$, and $\mathcal{L}(Y_i) = IG(\mu_2, \lambda_2)$, $i = 1, 2, \ldots, n$, the X and Y samples being independent, and we wish to test for equality of the means when λ_1 and λ_2 are unknown.

Samanta (1985) has studied this problem in detail. We now describe his procedure. We denote by $\Theta = \{\mu_1, \mu_2, \lambda_1, \lambda_2\}$ and note that for testing

$$H_0 : \mu_1 = \mu_2, \; (\lambda_1 = \lambda_2 \text{ unknown})$$

against

$$H_A : \mu_1 \neq \mu_2, \; (\lambda_1, \lambda_2 \text{ unknown and different}).$$

the estimates of $\mu_1, \mu_2, \lambda_1 = \lambda_2 = \lambda$ under H_0 are given by $\frac{m\overline{X}+n\overline{Y}}{m+n}$, $\frac{m}{V_1}$ and $\frac{n}{V_2}$ respectively while under $(H_0 \cup H_A)$ they are given by $\overline{X}, \overline{Y}$ and $\hat{\lambda} = \frac{m+n}{V_1+V_2+V}$ where $V = \frac{m}{\overline{X}} + \frac{n}{\overline{Y}} - \frac{m+n}{m\overline{X}+n\overline{Y}}$ respectively. Therefore the likelihood ratio statistic

$$\Lambda = \frac{\hat{\lambda}^{\frac{1}{2}}}{\left(\frac{m}{V_1}\right)^{\frac{m}{2}}\left(\frac{n}{V_2}\right)^{\frac{n}{2}}} = \left(\frac{N}{m}\right)^{\frac{m}{2}}\left(\frac{N}{m}\right)^{\frac{n}{2}} \frac{V_1^{\frac{m}{2}} V_2^{\frac{n}{2}}}{(V_1+V_2+V)^{\frac{N}{2}}}$$

where $N = m + n$. Now under H_0, Λ is distributed as $U_1^{\frac{m}{2}}(1-U_1)^{\frac{n}{2}}U_2^{\frac{N}{2}}$ where for independent U_1, U_2, we have $\mathcal{L}(U_1) = Be\left(\frac{m-1}{2}, \frac{n-1}{2}\right)$ and $\mathcal{L}(U_2) = Be\left(\frac{N-2}{2}, \frac{1}{2}\right)$. To see why, note that if $\mu_1 = \mu_2$ and $\lambda_1 = \lambda_2 = \lambda$ (say) $\lambda V_1, \lambda V_2$ and λV have independent χ^2 laws with $m-1$, $n-1$ and one degrees of freedom respectively. Therefore if we let $U_1 = \frac{V_1}{V_1+V_2+V}$, $U_2 = \frac{V_2}{V_1+V_2+V}$, $U_3 = \frac{V}{V_1+V_2+V}$, it follows that (U_1, U_2, U_3) has a Dirichlet law with parameters $(\alpha_1 = \frac{m-1}{2}, \alpha_2 = \frac{n-1}{2}, \alpha_3 = \frac{1}{2})$.

Despite this distributional result the evaluation of $P(\Lambda \leq \text{const} \mid H_0)$ is formidable. Samanta now uses a χ^2 approximation recommended by Box (1949) by considering the statistic $-2\rho \log \Lambda$, where $\rho = 1 - \frac{1}{6}\left(\frac{1}{m}+\frac{1}{n}\right) - \frac{1}{12N}$. It turns out $\mathcal{L}(-2\rho \log \Lambda) = \chi_2^2$ and an approximate level α test rejects H_0 if $-2\rho \log \Lambda > \chi_{2;\alpha}^2$.

An alternative test is based on a method due to Perng and Littel (1976). This involves conducting a preliminary test of the hypothesis

Two sample tests 49

$\lambda_1 = \lambda_2$ against $\lambda_1 \neq \lambda_2$ at level α_1. This test is described in (iv) and is based on the F statistic $\frac{(n-1)V_2}{(m-1)V_1}$. The hypothesis is accepted at level α_1 if and only if $F_{n-1,m-1,1-\frac{\alpha_1}{2}} < F < F_{n-1,m-1,\frac{\alpha_1}{2}}$. Following acceptance one considers the test of

$$H'_0 : \mu_1 = \mu_2 \text{ given } \lambda_1 = \lambda_2 \text{ against } H'_A : \mu_1 \neq \mu_2 \text{ given } \lambda_1 = \lambda_2.$$

In this case Λ reduces to $\frac{V_1+V_2}{V_1+V_2+V}$. One can then use the test statistic $\frac{(n-2)V}{V_1+V_2}$ which under H'_0 has an $F_{1,n-2}$ distribution. Equivalently a 't_{n-2}' can also be used. Then H'_0 is accepted at level α_2 if and only if

$$F_{1,n-2,1-\frac{\alpha_2}{2}} < \frac{(n-2)V}{V_1+V_2} < F_{1,n-2,\frac{\alpha_2}{2}}.$$

These tests have all been shown to be uniformly most powerful unbiased by Chhikara (1975). A test which combines the two procedures described above will now be developed based on Fisher's technique (1950).

To do so we define

$$W = \begin{cases} 2(1 - G(F))) & \text{if } G(F) \geq \frac{1}{2} \\ 2G(F) & \text{otherwise} \end{cases}$$

where $G(\cdot)$ is the distribution function of an F random variable with $n-1$ and $m-1$ degrees of freedom.

$$T = \sqrt{\frac{(n-2)V}{V_1+V_2}}.$$

Then the following two propositions due to Samanta help in using Fisher's method of combining the two tests.

Proposition 3.1 *T and W are independently distributed if $\lambda_1 = \lambda_2$.*

This proposition is proved using Lukacs theorem (1964). Note that $\frac{V_2}{V_1}$ is independent of $V_1 + V_2$ when $\lambda_1 = \lambda_2 = \lambda$. V is a function of $\overline{X}, \overline{Y}$ and $V_1 + V_2$ while W is a function of $\frac{V_2}{V_1}$. Hence T and W are independent.

Proposition 3.2 *$\mathcal{L}(W) = $ uniform $[0, 1]$ if $\lambda_1 = \lambda_2$.*

Conditionally on $G(F) \geq \frac{1}{2}$, $G(F)$ is uniform $[\frac{1}{2}, 1]$ and similarly $G(F)$ is $U[0, \frac{1}{2}]$ conditionally on $G(F) < \frac{1}{2}$. It then follows that $\mathcal{L}(W) = U[0, 1]$.

From the definition of W we have $[W \leq w]$ if and only if $F \geq G^{-1}\left(1 - \frac{w}{2}\right)$ or $F \leq G^{-1}\left(\frac{w}{2}\right)$. Therefore a level α_1 two sided F test of

$\lambda_1 = \lambda_2$ corresponds to rejecting if $W \leq \alpha_1$. A test that combines W and T in testing $\lambda_1 = \lambda_2$ and $\mu_1 = \mu_2$ in succession is based on

$$Q_n = -2\log 2[1 - S_{n-2}(|T|)] - 2\log W,$$

$S_{n-2}(\cdot)$ being the distribution function of Students's t. Under H_0 ($\mu_1 = \mu_2$, $\lambda_1 = \lambda_2$), $\mathcal{L}(Q_n) = \chi_4^2$ and a level α test rejects H_0 if $Q_n > \chi_{4,\alpha}^2$.

(vi) Tests on $\phi = \frac{\lambda}{\mu}$.

Hsieh (1990) considers the likelihood ratio test of ϕ and indicates how similar tests can be extended to tests on the coefficient of variation, skewness and kurtosis. We first consider testing

$$H_0 : \frac{\mu}{\lambda} = k_0 \text{ against } H_A : \frac{\mu}{\lambda} = k_1,$$

where $0 < k_0 < k_1$ and k_0, k_1 are fixed constants. Under H_A, $\mu = k_0\lambda$. Then if $\hat{\lambda}_A$ is the maximum likelihood estimate of λ, we have

$$\hat{\lambda}_A = \frac{\left(1 + \sqrt{1 + \frac{4\overline{X}_-\overline{X}}{k_1^2}}\right)}{2\overline{X}_-} = \frac{1 + I_1}{2\overline{X}_-} \text{ (say)}.$$

Similarly under H_0 the maximum likelihood estimate of λ is

$$\hat{\lambda}_0 = \frac{\left(1 + \sqrt{1 + \frac{4\overline{X}_-\overline{X}}{k_0^2}}\right)}{2\overline{X}_-} = \frac{1 + I_0}{2\overline{X}_-} \text{ (say)}.$$

Now note that $\hat{\lambda}_0, \hat{\lambda}_A$ satisfy the equation

$$\hat{\lambda}_i^{-1} + \overline{X}(k_i\hat{\lambda}_i)^{-2} - \overline{X}_- = 0$$

where $\hat{\lambda}_i$ is $\hat{\lambda}_0$ or $\hat{\lambda}_A$. The log-likelihood ratio $\log \Lambda$, apart from a constant factor, is

$$\frac{n}{2}\log\left(\frac{\hat{\lambda}_0}{\hat{\lambda}_A}\right) + \frac{n}{2}\frac{\overline{X}}{k_1^2\hat{\lambda}_A} + \frac{n\hat{\lambda}_A\overline{X}_-}{2} - \frac{n\overline{X}}{2k_0^2\hat{\lambda}_0} - \frac{n\hat{\lambda}_0\overline{X}_-}{2}$$

which simplifies to

$$\frac{n}{2}\log\left(\frac{\hat{\lambda}_0}{\hat{\lambda}_A}\right) - \frac{n(\hat{\lambda}_0\overline{X}_- - 1)}{2} - \frac{n(\hat{\lambda}\overline{X}_-)}{2} + \frac{n(\hat{\lambda}\overline{X}_- - 1)}{2} + \frac{n\hat{\lambda}_A\overline{X}}{2}.$$

Thus
$$\log \Lambda = \frac{n}{2}\left(\log \frac{1+I_0}{1+I_1} - (I_0 - I_1)\right).$$

The derivative of $\log \Lambda$ with respect to T is $(T = \overline{X}\, \overline{X}_{-})$

$$\frac{n}{2}\left(\frac{dI_0}{dT}\left(-\frac{2}{k_0^2(1+I_0)}\right) + \frac{dI_1}{dT}\left(\frac{2}{k_1^2(1+I_1)}\right)\right)$$
$$= n\left(\frac{1}{k_1^2(1+I_1)} - \frac{1}{k_0^2(1+I_0)}\right) < 0$$

for all $T \geq 0$. This then implies that $\log \Lambda$ is a strictly decreasing function of T. Moreover T is independent of k_1 and $P(T > \text{const})$ is increasing in $\frac{\mu}{\lambda}$ so that the size of the critical region for testing $\frac{\mu}{\lambda} \leq k_0$ against $\frac{\mu}{\lambda} > k_0$ is unaffected. Therefore $W = \left(\frac{n}{\overline{X}V} + 1\right) < \text{constant}$ is the critical region. Hsieh provides tables of critical values of W. In particular, quantiles (1%, 2.5%, 5%, 95%, 97.5%, and 99%) are given for odd n. For even n, one has to use the average of consecutive odd integers $(n-1)$ and $(n+1)$. In summary the likelihood ratio test of size α for testing

$$H_0 : \phi \geq \phi_0 \text{ against } H_a : \phi < \phi_0$$

has critical region of the form $\{W < w_\alpha\}$ where $W = \frac{n}{\overline{X}V} + 1$ and $P_{\phi_0}(W < w_\alpha) = \alpha$. Hsieh also shows that these tests are consistent. By examining the distribution function $H_\phi(w)$ (Theorem 2.5) one can see that if $w > 0$ and $n \geq 3$, $H_\phi(w)$ is decreasing in ϕ. This means that the power of the likelihood ratio test increases as ϕ decreases. The tests can be extended to tests on $\phi^{-\frac{1}{2}}$, $3\phi^{-\frac{1}{2}}$ and $15\phi^{-1}$ since they are monotone fucntions of ϕ.

The various tests that were developed thus far provide the basis for the construction of confidence intervals on the parameters and parametric functions of $IG(\mu, \lambda)$. Table 3.2 summarizes the typical $100(1-\alpha)\%$ confidence intervals for the parameters.

3.5 Interval estimation

Table 3.2 *Interval estimates*

Parameter	Interval
μ (λ known)	$\overline{X}\left(1+\sqrt{\frac{\overline{X}}{n\lambda}}Z_{1-\frac{\alpha}{2}}\right)^{-1}, \overline{X}\left(1-\sqrt{\frac{\overline{X}}{n\lambda}}Z_{1-\frac{\alpha}{2}}\right)^{-1}$ $\left(\sqrt{\frac{\overline{x}}{n\lambda}}Z_{1-\frac{\alpha}{2}}<1\right)$
(n or ϕ large)	$\overline{X}\left(1+\sqrt{\frac{\overline{X}}{n\lambda}}Z_{1-\frac{\alpha}{2}}\right)^{-1}, \infty$ (approximate)
μ (λ unknown)	$\overline{X}\left(1+\sqrt{\frac{\overline{X}V}{n(n-1)}}t_{1-\frac{\alpha}{2}}\right)^{-1}, \overline{X}\left(1-\sqrt{\frac{\overline{X}V}{n(n-1)}}t_{1-\frac{\alpha}{2}}\right)^{-1}$ $\left(\sqrt{\frac{\overline{x}V}{n(n-1)}}t_{1-\frac{\alpha}{2}}<1 \;\; (n-1) \text{ d.f. for } t\right)$
(n or ϕ large)	$\overline{X}\left(1+\sqrt{\frac{\overline{X}V}{n(n-1)}}t_{1-\frac{\alpha}{2}}\right)^{-1}, \infty$ (approximate)
λ (μ known)	$\dfrac{\mu^2 \chi^2_{1-\frac{\alpha}{2}}}{\sum_{i=1}^{n}\frac{(X_i-\mu)^2}{X_i\mu^2}}, \dfrac{\mu^2 \chi^2_{\frac{\alpha}{2}}}{\sum_{i=1}^{n}\frac{(X_i-\mu)^2}{X_i\mu^2}}$ (n d.f. for χ^2)
λ (μ unknown)	$\dfrac{\chi^2_{1-\frac{\alpha}{2}}}{V}, \dfrac{\chi^2_{\frac{\alpha}{2}}}{V}$ (($n-1$) d.f. for χ^2)
$\phi^{-1} = \frac{m}{\lambda}$	$\dfrac{\overline{X}V}{n-1} - \dfrac{\hat{\sigma}}{\sqrt{n-1}}Z_{\frac{\alpha}{2}}, \dfrac{\overline{X}V}{n-1} + \dfrac{\hat{\sigma}}{\sqrt{n-1}}Z_{\frac{\alpha}{2}}$ $(\hat{\sigma} = \hat{\phi}^{-1}\sqrt{2+\hat{\phi}^{-1}}\; n \text{ large})$

Prediction intervals Prediction intervals are quite useful in quality control and reliability studies. A number of results on predictive inference have been derived by Chhikara and Guttman (1982), Padgett (1982) and Padgett and Tsoi (1986). Chhikara and Guttman obtained exact prediction intervals using both the frequentist and Bayesian approaches. Their methods did not always lead to two-sided intervals. On the other hand Padgett's method, through approximate not only provides two-sided intervals but performs better in terms of estimated coverage probabilities and smaller estimated mean widths for sample sizes

Interval estimation

at least as large as 15. We examine their methods in the ensuing discussion and conclude with some comments on Bayes prediction intervals considered by Upadhyay et al,(1994).

Padgett's method Let $X = (X_1, \ldots, X_m)$ be a random sample from $IG(\mu, \lambda)$ and $\overline{Y}$ be the sample mean based on n independent future observations from the same $IG(\mu, \lambda)$. We seek two statistics $\underline{T}(X), \overline{T}(X)$ such that for $0 < \alpha < 1$

$$P(\underline{T}(X) \leq \overline{Y} \leq \overline{T}(X)) = 1 - \alpha.$$

First observe that
  (a) $\mathcal{L}\left(\frac{n\lambda(\overline{Y}-\mu)^2}{\overline{Y}}\right) = \chi_1^2$,
  (b) $\mathcal{L}(\lambda V) = \chi_{m-1}^2$ and
  (c) $V = \sum \left(\frac{1}{X_i} - \frac{1}{\overline{X}}\right) \perp\!\!\!\perp \overline{Y}$. Therefore $\mathcal{L}\left(\frac{(m-1)n(\overline{Y}-\mu)^2}{\mu^2 \overline{Y} V}\right) = F_{1,n-1}$,
and from the tables of the F distribution, F_α can be found such that

$$P\left(\frac{(m-1)n(\overline{Y}-\mu)^2}{\mu^2 \overline{Y} V} \leq F_\alpha\right) = 1 - \alpha.$$

When μ is known the expression inside the brackets can be solved for $\overline{Y}$ and an exact $100(1-\alpha)\%$ prediction interval is obtained. When μ is unknown approximations are to be used. Padgett first replaces μ^2 in the denominator by $\overline{X}^2$. The term μ appearing in $(\overline{Y} - \mu)^2$ is replaced by $\frac{m\overline{X}+n\overline{Y}}{m+n}$ the updated mean incorporating the extra information available and the equation is then solved for $\overline{Y}$. Thus the approximated coverage probability statement is

$$P\left(\overline{Y} + \frac{\overline{X}^2}{\overline{Y}} \leq \frac{(m+n)^2 \overline{X}^2 V F_\alpha}{m^2 n(m-1)} + 2\overline{X}\right) \approx 1 - \alpha.$$

We now solve

$$\overline{Y}^2 - C(X)\overline{Y} + \overline{X}^2 = 0 \text{ where } C(X) = \left(\frac{m+n}{m}\right)^2 \frac{\overline{X}^2 V F_\alpha}{n(m-1)}$$

obtaining two real positive roots $\underline{T}(X)$ and $\overline{T}(X)$.

Chhikara and Guttman's method Chhikara and Guttman obtain an exact prediction interval using a form of the Chi-square decomposition of Theorem 2.1. (There is a subtle flaw in their arguments despite the correct conclusion.) From Theorem 2.1 with $w_i = 1$ let us define P_1 and P_2 by

$$P_1 = \lambda \left(\sum_{i=1}^n \frac{1}{X_i} - \frac{n^2}{\sum_{i=1}^n X_i} \right)$$

$$P_2 = \lambda \left(\frac{1}{Y} + \frac{n^2}{\sum_{i=1}^n X_i} - \frac{(n+1)^2}{Y + \sum_{i=1}^n X_i} \right),$$

where Y is the new observation from $IG(\mu, \lambda)$. Since $P_1 \perp\!\!\!\perp P_2$ with $\mathcal{L}(P_1) = \chi^2_{n-1}$, and $\mathcal{L}(P_2) = \chi^2_1$ it follows that regardless of what μ and λ are

$$P\left(\frac{(n-1)P_2}{P_1} \leq F_\alpha \right) = 1 - \alpha$$

and a rearrangement of terms inside the brackets gives $\underline{T}(X)$ and $\overline{T}(X)$ where

$$\underline{T}(X) = \left(\frac{1}{\overline{X}} + \frac{VF_\alpha}{2(n-1)} + \left(\frac{VF_\alpha}{(n-1)\overline{X}} + \frac{F_\alpha^2 V^2}{4(n-1)} \right)^{\frac{1}{2}} \right)^{-1}, \text{ and}$$

$$\overline{T}(X) = \left(\frac{1}{\overline{X}} + \frac{VF_\alpha}{2(n-1)} - \left(\frac{VF_\alpha}{(n-1)\overline{X}} + \frac{F_\alpha^2 V^2}{4(n-1)} \right)^{\frac{1}{2}} \right)^{-1}.$$

Since there is a positive probability $\overline{T}(X)$ can be negative this method does not always guarantee two-sided intervals. Padgett and Tsoi have done a Monte Carlo simulation study of both the exact and approximate methods and computed the width of the intervals together with the numbers of intervals containing Y. The average interval lengths from 1000 pairs as well as the proportion of intervals containing Y were obtained as estimates of the mean interval lengths and coverage probabilities. This procedure was repeated for many values (n, m, λ) and $\alpha = 0.01$, 0.05 and 0.10. Their results are summarized in Tables 3.3–3.5. They note that for small λ a sizeable proportion of the samples do not yield a two-sided interval using the Chhikara and Guttman (C&G) method. The estimated coverage probabilities were low as well for small n. With increasing $(1 - \alpha)$ the approximate method seems superior in terms of larger estimated coverage probabilities and (or) smaller estimated mean widths.

Interval estimation

Table 3.3 *Simulation Results for $\gamma = 0.90$*

μ	λ	n	Average Width		Coverage Probability	
			C&G	Padgett	C&G	Padgett
1	0.25	5	30.2	46.1	0.705	0.897
3	0.25	5	24.4	739.9	0.738	0.895
1	4	15	2.1	2.0	0.903	0.893
5	4	15	51.4	31.9	0.903	0.887

Table 3.4 *Simulation Results for $\gamma = 0.95$*

μ	λ	n	Average Width		Coverage Probability	
			C&G	Padgett	C&G	Padgett
1	0.25	5	26.0	73.4	0.828	0.942
1	0.25	30	849.1	21.9	0.945	0.962
3	0.25	5	28.1	1441.8	0.759	0.948
3	0.25	30	577.0	226.0	0.932	0.957
1	1	15	12.0	7.2	0.951	0.956
1	1	50	6.4	6.0	0.963	0.960
1	4	5	30.3	4.4	0.946	0.959
1	4	30	2.4	2.4	0.937	0.944
5	1	15	782.4	198.0	0.940	0.973
5	1	30	776.9	138.5	0.948	0.948
5	4	5	201.0	100.6	0.905	0.956
5	4	30	45.3	37.5	0.961	0.942

Table 3.5 *Simulation Results for $\gamma = 0.99$*

μ	λ	n	Average Width		Coverage Probability	
			C&G	Padgett	C&G	Padgett
1	0.25	5	30.2	208.3	0.806	0.983
1	0.25	30	583.2	38.4	0.991	0.989
3	0.25	5	36.5	3430.0	0.933	0.977
3	0.25	30	1895.5	368.1	0.957	0.990
1	1	15	103.8	12.4	0.988	0.993
1	4	5	47.6	9.2	0.986	0.988
1	4	15	4.6	4.1	0.993	0.994
1	4	30	3.7	3.5	0.990	0.991
5	1	15	550.3	332.8	0.947	0.990
5	4	5	403.6	220.0	0.945	0.989
5	4	30	96.1	62.8	0.987	0.992

Example 3.1 Maintenance data on 46 active repair times in hours for an airborne communication transceiver reported by von Alven (1964) have been analyzed by Chhikara and Folks (1977) who conclude that the IG model was a good fit. The data appears below.

0.2	0.3	0.5	0.5	0.5	0.5	0.6	0.6
0.7	0.7	0.7	0.8	0.8	1.0	1.0	1.0
1.0	1.1	1.3	1.5	1.5	1.5	1.5	2.0
2.0	2.2	2.5	2.7	3.0	3.0	3.3	3.3
4.0	4.0	4.5	4.7	5.0	5.4	5.4	7.0
7.5	8.8	9.0	10.3	22.0	24.5		

For the von Alven data set with $m = 46$ and $n = 10$, Padgett reports a 95% prediction interval for the mean of the next 10 repair times to be (1.2009, 10.8311). For $n = 1$, the interval is (0.3185, 40.8393).

Interval estimation

Bayes prediction intervals Bayesian philosophy relies on the use of a prior or even guess values of a parameter to enhance the performance of an estimator in terms of its efficiency. Thompsons's work (1978) pertains to shrinking an estimator towards the guessed value θ_0 of the unknown parameter. A shrinkage estimator $T = \alpha\hat{\theta} + (1-\alpha)\theta_0$ where $0 < \alpha < 1$ is then examined for its efficiency. Upadhyay et al., (1994) consider this approach and introduce a general class of priors which places a constant weight on the guess values of the IG parameters $\delta_0 = \frac{1}{\mu_0}$ and λ_0, while distributing the rest of the mass over a suitable interval. The Bayes predictive procedure is then used to obtain prediction limits for future failure times or even the unused components of a system or process. They compare their procedure with the frequentist approach advanced by Padgett and conclude that their method works well, if not better. We now describe their procedure.

Consider the parametrization $\delta = \frac{1}{\mu}$ and the associated likelihood

$$L(\delta, \lambda) = \left(\frac{\lambda}{2\pi}\right)^{\frac{n}{2}} \prod_{i=1}^{n} x_i^{-\frac{3}{2}} \exp\left(-\frac{\lambda v}{2}\left(1 + \frac{n(\delta\overline{x}-1)^2}{v\overline{x}}\right)\right).$$

Let (δ_0, λ_0) be guess values of (δ, λ) and take $q(\delta, \lambda) = \frac{1}{\lambda}1_{\mathbb{R}^+}(\lambda)$. For a real a such that $0 \leq a \leq 1$ define

$$\varphi(\delta, \lambda) = \begin{cases} 1-a, & \text{if } \delta = \delta_0, \lambda = \lambda_0, \\ aq(\delta, \lambda), & \text{otherwise.} \end{cases}$$

Suppose $X = (X_1, \ldots, X_n)$ represent failure times such that $L(\delta, \lambda)$ corresponds to their likelihood, then the distribution of X is

$$a \int_0^\infty \int_0^\infty L(\delta, \lambda) d\lambda d\delta + (1-a) L(\delta_0, \lambda_0).$$

The integral $a \int_0^\infty \int_0^\infty L(\delta, \lambda) d\lambda d\delta$ gives

$$a \frac{\Gamma\left(\frac{n}{2}\right)}{(2\pi)^{\frac{n}{2}}} \prod_{i=1}^{n} x_i^{-\frac{3}{2}} \frac{1}{v^{\frac{n}{2}}} \int_0^\infty \frac{d\delta}{\left(1 + \frac{n(\delta\overline{x}-1)^2}{v\overline{x}}\right)^{\frac{n}{2}}}.$$

With the substitution $\frac{t^2}{n-1} = \frac{n(\delta\overline{x}-1)^2}{v\overline{x}}$ we obtain

$$a \frac{\Gamma\left(\frac{n}{2}\right)}{(\pi v)^{\frac{n}{2}}} \prod_{i=1}^{n} x_i^{-\frac{3}{2}} \frac{\sqrt{\pi v(n-1)}}{\Gamma\left(\frac{n}{2}\right)} \frac{\Gamma\left(\frac{n-1}{2}\right)}{\sqrt{n(n-1)\overline{x}}} S_{n-1}\left(\sqrt{\frac{n(n-1)}{v\overline{x}}}\right)$$

where $S_{n-1}(\cdot)$ is the distribution function of Students's t with $(n-1)$ degrees of freedom. Upon simplification it reduces to

$$\frac{a\sqrt{\pi}\Gamma\left(\frac{n-1}{2}\right)}{(2\pi)^{\frac{n}{2}}}\frac{v^{-\frac{(n-1)}{2}}}{\sqrt{n\overline{x}}}\prod_{i=1}^{n}x_i^{-\frac{3}{2}}S_{n-1}\left(\sqrt{\frac{n(n-1)}{v\overline{x}}}\right) = \frac{h(x)}{(2\pi)^{\frac{n}{2}}}\prod_{i=1}^{n}x_i^{-\frac{3}{2}}.$$

The posterior of (δ, λ) is therefore

$$\frac{(1-a)\lambda_0^{\frac{n}{2}}\exp\left(-\frac{\lambda_0 v}{2}\left(1+\frac{n(\delta_0\overline{x}-1)^2}{v\overline{x}}\right)\right)}{(1-a)\lambda_0\frac{n}{2}\exp\left(-\frac{\lambda_0 v}{2}\left(1+\frac{n(\delta_0\overline{x}-1)}{v\overline{x}}\right)\right) + h(x)}$$

$$+\frac{a\lambda^{\frac{n}{2}-1}\exp\left(-\frac{\lambda v}{2}\left(1+\frac{n(\delta\overline{x}-1)^2}{v\overline{x}}\right)\right)}{(1-a)\lambda_0^{\frac{n}{2}}\exp\left(-\frac{\lambda_0 v}{2}\left(1+\frac{n(\delta_0\overline{x}-1)}{v\overline{x}}\right)\right) + h(x)}$$

$$= p(\delta, \lambda \mid x) \text{ (say)}.$$

If now Y is a future observation from $IG(\mu, \lambda)$, after parametrization in terms of δ, the predictive density of Y given X is

$$\psi(y \mid x) = \int\int f_Y(y \mid \delta, \lambda) p(\delta, \lambda \mid x) d\delta d\lambda.$$

The $100(1-\alpha)\%$ confidence interval for Y is obtained by solving for constants c_1 and c_2 (for an equal-tail prediction limit) the integrals

$$\int_0^{c_1}\psi(y \mid x)dy = \int_{c_2}^{\infty}\psi(y \mid x)dy = \frac{\alpha}{2}.$$

A similar argument applies for the prediction of the mean of a sample of m future observations. The integrals can be solved by Gauss-Legendre quadrature or other numerical methods.

3.6 Examples

Example 3.1(continued) For the von Alven data set $\overline{X} = 3.61$, $\frac{V}{n-1} = 0.60$, $t_{45,0.95} \approx 2.00$

$$\overline{X}\left(1+\sqrt{\frac{\overline{X}V}{n(n-1)}}t_{1-\frac{\alpha}{2}}\right)^{-1} = 2.518$$

$$\overline{X}\left(1-\sqrt{\frac{\overline{X}V}{n(n-1)}}t_{1-\frac{\alpha}{2}}\right)^{-1} = 6.356$$

and $\qquad \hat{\sigma}^2 = \frac{\overline{X}^3}{\hat{\lambda}} = 28.227.$

Examples

A 95% confidence interval for μ is $(2.518, 6.356)$.
A 95% confidence interval for λ is $(1.052, 2.421)$.

Now consider testing $H_0 : \mu \leq 1$ against $\mu > 1$. $\sum^n X_i = 165.9$, $\sum^n X_i + \sum^n \frac{1}{X_i} = 206.4$. The critical value k^* (see Section 3.3) to 65.49. Patil and Kovner find $\pi(\theta_1 = 0.1) = 0.163$, $\pi(\theta_1 = 0.4) = 0.934$ and $\pi(0.5) = 0.992$.

Upadhyay et al., have used the von Alven data of Example 3.1 to compare their method with that of Padgett, choosing three different values of a $(0.25, 0.50, 0.75)$ and several values of (δ_0, λ_0). Their tabulated values indicate that their methods always yield narrower intervals.

Example 3.2 Nádas (1973) reports that a sample of 10 electronic devices were tested under high stress conditions until all of them failed due to mass depletion at the critical location. The data yielded $\overline{X} = 1.352$, $V = 2.083$. It is desired to test if there was zero drift in the Brownian motion. In terms of the first passage times T of Brownian motion with positive drift ν one wishes to test $\nu = 0$ or equivalently $\mu = \infty$.

Now $\frac{\overline{X}V}{n(n-1)} = 0.0313$ and with $\alpha = 0.05$, $\frac{1}{F_{1,9,0.95}} = 0.195$. Thus the observed value of the test statistic falls in the critical region (see Section 3.2.(b)), and the hypothesis of zero drift at level $\alpha = 0.05$ is rejected.

A 95% confidence interval for μ is $(0.966, 2.249)$.

Chhikara and Folks considered the test of $\mu = 1$ against $\mu \neq 1$. Using the W statistic (see (ii) of section 3.2) namely $W = \sqrt{\frac{n(n-1)}{\overline{X}V}}(\overline{X} - 1)$ we find $W = 1.99$ and at level $\alpha = 0.05$ there is no evidence to reject H_0 ($\mu = 1$). If on the other hand one tests $\mu \leq 1$ against $\mu > 1$, the observed value of $W = 1.99$ should be compared with the critical value c determined by solving

$$S_9(-c) + \left(\frac{43}{3}\right)^4 S_9(-\sqrt{40 + 43c^2}) = 0.05.$$

(Note that $T_2 = 23$.) This value of c is determined to be $c = 1.90$. Hence H_0 is rejected.

The power of the test corresponding to zero drift was considered by Patil and Kovner. The critical value is obtained by numerical integration (using Gaussian approximation) with $\theta_1 = 0$ in the integral (see 3.3)

$$\int_{k^*}^{\infty} f_{\theta_1=0}(t_1 \mid t_2) dt_1 = 0.05.$$

Patil and Kovner find that $k^* = 13.12$. They find by numerical integration of $G(c)$ (section 3.3) that

$$\pi(\theta_1 = 0.5) = 0.121,$$
$$\pi(\theta_1 = 1) = 0.244.$$

Example 3.2 (continued) **Bayesian analysis** Observe that the confidence interval for μ when converted to an interval for $\delta = \frac{1}{\mu}$ becomes (0.444, 1.031). If we are using a Bayesian approach and applying the notion of highest posterior density intervals, then it is best to use the marginal posterior of δ developed in Corollary 1.2. Banerjee and Bhattacharyya have examined this situation and they note that over the interval $\left(0, \frac{2}{\bar{x}}\right)$ the posterior of δ (see equation 1.19) is symmetric and its mode $\delta_{mode} = \frac{1}{\bar{x}}$. Therefore when $P(\delta \leq \frac{2}{\bar{x}} \mid X) \leq 1 - \alpha$, then necessarily $\delta \leq S_{n-1}^{-1}\left(\frac{1}{1+\alpha}\right)$ and conversely. Thus the highest posterior density interval corresponds to

$$\left[0, \frac{1}{\bar{x}} + \sqrt{\frac{v}{n(n-1)\bar{x}}} S_{n-1}^{-1}\left(1 - \alpha S_{n-1}\left(\sqrt{\frac{n(n-1)}{v\bar{x}}}\right)\right)\right]. \quad (3.6a)$$

When $\sqrt{\frac{n}{v\bar{x}}}$ gets large, the highest posterior density interval is centered at the mode and is

$$\left[\frac{1}{\bar{x}} \pm \sqrt{\frac{v}{n(n-1)\bar{x}}} S_{n-1}^{-1}\left(\frac{1 + (1-\alpha)S_{n-1}\left(\sqrt{\frac{n(n-1)}{v\bar{x}}}\right)}{2}\right)\right]. \quad (3.6b)$$

Tables of incomplete Beta integrals are needed to make the calculations. In the frequentist approach the transformed $(\delta = \frac{1}{\mu})$ intervals are

$$\left(0, \frac{1}{\bar{x}} + \sqrt{\frac{v}{n(n-1)\bar{x}}} S_{n-1}^{-1}\left(1 - \frac{\alpha}{2}\right)\right) \text{ if } \sqrt{\frac{n(n-1)}{v\bar{x}}} < S_{n-1}^{-1}\left(1 - \frac{\alpha}{2}\right) \quad (3.7a)$$

and

$$\left(\frac{1}{\bar{x}} \pm \sqrt{\frac{v}{n(n-1)\bar{x}}} S_{n-1}^{-1}\left(1 - \frac{\alpha}{2}\right)\right) \text{ if } \sqrt{\frac{n(n-1)}{v\bar{x}}} \geq S_{n-1}^{-1}\left(1 - \frac{\alpha}{2}\right) \quad (3.7b)$$

These are the uniformly most accurate unbiased confidence intervals as developed by Chhikara and Folks. If n is large $S_{n-1}\left(\sqrt{\frac{n(n-1)}{v\bar{x}}}\right) \to 1$ and we see that when comparing 3.6b and 3.7b both methods give almost identical results. On the other hand comparisons betweens (3.6a) and (3.7a) are not meaningful since for large n, small values of $\sqrt{\frac{n(n-1)}{v\bar{x}}}$ cannot materialize. For the Nádas data, the 95% highest posterior density interval is identical to the interval given by the frequentist approach.

Tolerance limits

Highest posterior intervals on λ are obtained by using (1.20) and solving for λ_L and λ_U where (see Corollary 1.3)

$$\int_{\lambda_U}^{\lambda_L} f_\lambda(\lambda \mid X) d\lambda = 1 - \alpha.$$

Banerjee and Bhattacharyya find the interval to be $(1.419, 9.742)$ as opposed to the frequentist result of $(1.296, 9.132)$.

Example 3.3 Ang and Tang (1975) considered runoff amounts at Jug Bridge, Maryland given by the following data set.

0.17	0.39	0.52	0.66	0.78	1.12
0.19	0.39	0.56	0.70	0.95	1.24
0.23	0.40	0.59	0.76	0.97	1.59
0.33	0.45	0.64	0.77	1.02	1.74
2.92					

This set was analyzed by Folks and Chhikara (1978) who judged it as being well descriptive of an IG model. For this data $\overline{X} = 0.803$, $V = 1.7375$. Using this data Hsieh (1990) considered a test of $H_0 : \phi \geq 2$ against $H_A : \phi < 2$. The likelihood ratio statistic is $W = \frac{n}{\overline{X}V} = 1.792$. From the tables of the quantiles of W the critical value at level $\alpha = 0.01$ being 1.078, H_0 is rejected. Hsieh uses a linear interpolation of the quantile values of W to obtain $(0.875, 2.706)$, $(0.755, 2.943)$ and $(0.627, 3.236)$ as confidence intervals at levels 90%, 95% and 98% respectively. An exact 95% confidence interval for the coefficient of variation $\sqrt{\phi^{-1}}$ is found to be $(0.583, 1.150)$.

3.7 Tolerance limits

The inverse Gaussian law has been proposed as a failure time model in reliability analysis by Chhikara and Folks (1977), fatigue failure models by Bhattacharyya and Fries (1980), general stochastic wear-out by Desmond (1985) and Goh et al. (1989). Its importance in these applications is due to the fact that it represents the first passage time distribution of Brownian motion to a critical level. This makes it an ideal candidate for studying safety limits of equipments. In particular lower tolerance limits can be constructed which will guarantee some kind of safety in the design of equipments. There exists a relationship between the confidence limit of a quantile and a one-sided tolerance limit. Tang and Chang (1994) exploit this link to derive upper (lower) tolerance limits. The following discussion describes their strategy. Denote by $F(x; \mu, \lambda)$ the distribution function of the $IG(\mu, \lambda)$ law. Let x_p denote the $100p$ percentile of the IG law. Thus x_p satisfies

$$F(x_p : \mu, \lambda) = p.$$

Suppose there exists a random variable L such that for $0 \leq \gamma \leq 1$, $P(L \leq x_p) \geq \gamma$, then $P(F(L) \leq p) \geq \gamma$ and conversely. Hence if $\mathcal{L}(X) = IG(\mu, \lambda)$,

$$P[P(X \geq L) \geq 1 - p] \geq \gamma \iff P(L \leq x_p) \geq \gamma.$$

Definition 3.1 L is a one-sided $100\gamma\%$ confidence limit of x_p if and only if it is also a lower $(1-p)$ content one-sided tolerance limit with confidence coefficient $100\gamma\%$.

This implies then that we can restrict attention to the problem of finding a $100\gamma\%$ confidence limit of x_p for constructing tolerance intervals.

(a) Tolerance limits when μ is known and λ is unknown.
When $\mathcal{L}(X) = IG(\mu, \lambda)$ we obtain via the transformation $Y = \frac{X}{\mu}$, the law of Y to be $\mathcal{L}(Y) = IG(1, \phi)$ where $\phi = \frac{\lambda}{\mu}$. The distribution function of Y, written $G(y, \phi)$ expressed in terms of the distribution function of X, namely $F(x; \mu, \lambda)$ is

$$G(y, \phi) = F(y; 1, \phi)$$
$$= 1 - \Phi\left(\sqrt{\frac{\phi}{y}}(1-y)\right) + \exp(2\phi)\Phi\left(-\sqrt{\frac{\phi}{y}}(1+y)\right).$$

We now solve for $y_p = \frac{x_p}{\mu}$ (using this form of $F(y; 1, \phi)$) from the equation

$$F(y_p; 1, \phi) - p = 0.$$

This is done using different values of ϕ and plotting a graph of ϕ versus y_p for values of p ranging from 0.1 to 0.95 (see Figure 3.1). From the figure it is clear that y_p is not a monotonic function of ϕ. The minimum y_p occurs at the end points of any fixed intervals of ϕ. For $p \leq 0.7$ it turns out that y_p is monotone increasing in ϕ.

Let us denote by $L(y_p, \gamma)$ the unique solution given by

$$L(y_p, \gamma) = \min\left[\{y \mid F(y; 1, \phi_L) = p\}, \{y \mid F(y; 1, \phi_U) = p\}\right]$$

where

$$\phi_L = \frac{\chi^2_{n,1-\gamma}}{S}, \quad \phi_U = \frac{\chi^2_{n,\gamma}}{S} \quad \text{and} \quad S = \sum_{i=1}^{n} \frac{(x_i - 1)^2}{x_i}.$$

Then $\mu L(y_p, \gamma)$ is a one-sided $(1-p)$ content tolerance limit with confidence coefficient $100\gamma\%$.

Tolerance limits

For $p \le 0.7$, of course, $L(y_p, \gamma)$ is the unique solution of $\{y \mid F(y; 1, \phi) = p\}$. Note that ϕ_L and ϕ_U are the lower and upper $100\gamma\%$ confidence limits on ϕ.

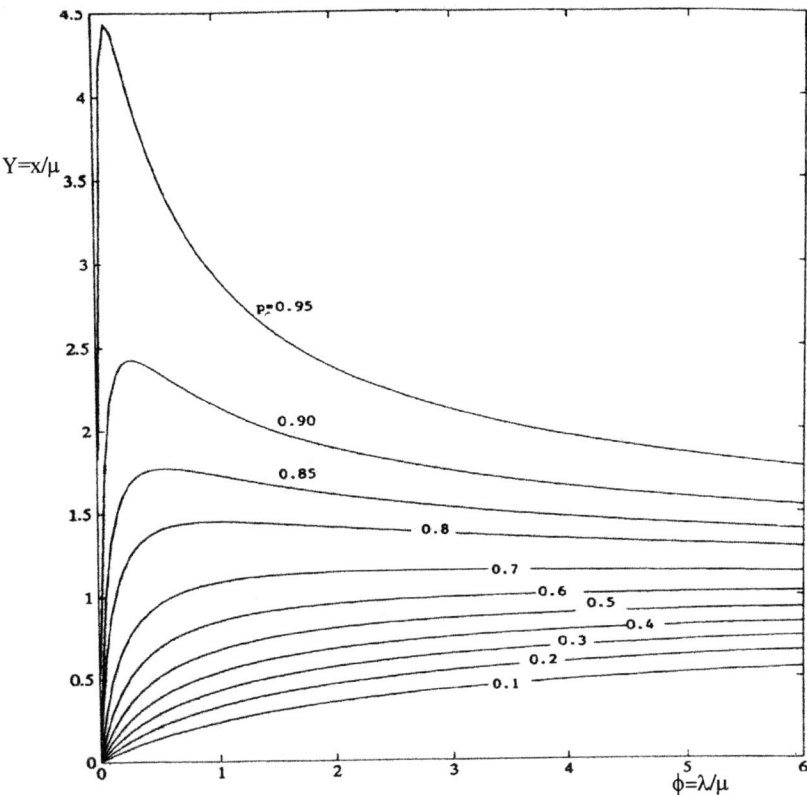

Figure 3.1 The pth percentiles as a fucntion of λ/μ

(b) Tolerance limits when μ is unknown and λ is known.

When $\mathcal{L}(X) = IG(\mu, \lambda)$ the law of $Z = \frac{X}{\lambda}$ is $\mathcal{L}(Z) = IG(\theta, 1)$ where $\theta = \phi^{-1}$ and hence the distribution function $H(z, \theta)$ of Z expressed in terms of that of X is

$$H(z, \phi) = F(z; \theta, 1)$$
$$= 1 - \Phi\left(\sqrt{\frac{1}{z}}\left(1 - \frac{z}{\theta}\right)\right) + \exp\left(\frac{2}{\theta}\right) \Phi\left(-\sqrt{\frac{1}{z}}\left(1 + \frac{z}{\theta}\right)\right).$$

Proceeding as in case (a) we solve for $z_p = \frac{x_p}{\lambda}$ from the equation $F(z; \theta_L, 1) - p = 0$ using various values of θ and p. The resulting graph (see Figure 3.2) shows that z_p is monotone in θ. Then denoting by

$L(z_p, \gamma)$ the unique solution of $F(z; \theta_L, 1) = p$, we see that $\lambda L(z_p, \gamma)$ is a one-sided $(1-p)$ content tolerance limit with confidence coefficient $100\gamma\%$. Here θ_L is a lower $100\gamma\%$ confidence limit on θ, given by

$$\theta_L = \left[\frac{1}{\bar{z}} + \sqrt{\frac{\chi^2_{1,\gamma}}{n\bar{z}}}\right]^{-1} , \quad \bar{z} = \frac{1}{n}\sum_{i=1}^{n} z_i.$$

Note that $\theta \in (\theta_L, \infty) \iff z_p \in (L(z_p, \gamma), \infty)$ since z_p is monotone increasing in θ, so that

$$P[z_p \in (L(z_p, \gamma), \infty)] = P[\lambda z_p \in (\lambda L(z_p, \gamma), \infty)] = P[\theta \in (\theta_L, \infty)] = 1-\gamma.$$

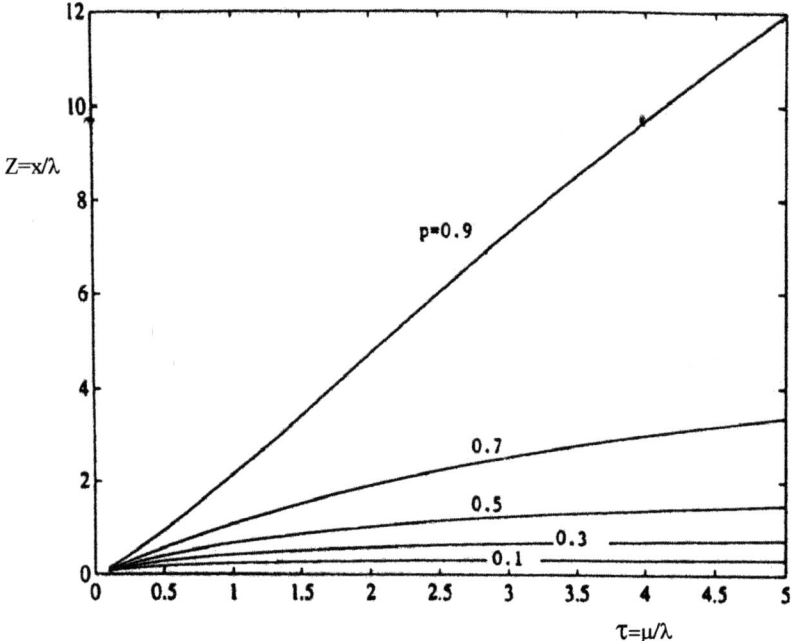

Figure 3.2 The pth percentiles as a fucntion of μ/λ

(c) Tolerance limits when both μ and λ are unknown.
The results from the previous two cases can be combined to derive a conservative one-sided $(1-p)$ content tolerance limit L with confidence coefficient $100(2\gamma - 1)\%$ by appealing to the Bonferroni inequality. Indeed the resulting limit is given by

$$L = \min[\{x \mid F(x, \mu_L, \lambda_L) = p\}, \{x \mid F(x; \mu_L, \lambda_U) = p\}]$$

where $\mu_L = \left(\frac{1}{\overline{X}} + \sqrt{\frac{VF_{1,n-1,\gamma}}{\overline{X}n(n-1)}}\right)^{-1}$ and $(\lambda_L, \lambda_U) = \left(\frac{\chi^2_{1-\gamma}}{V}, \frac{\chi^2_\gamma}{V}\right)$.

To see why, we have from the Bonferroni inequality

$$\begin{aligned} P(x_p \in (L, \infty)) &= P[\mu \in (\mu_L, \infty), \ \lambda \in (\lambda_L^*, \infty)] \\ &\geq 1 - P[\mu \in (\mu_L, \infty)^C] - P[\lambda \in (\lambda_L, \infty)^C \\ &= 1 - (1-\gamma) - (1-\gamma) = 2\gamma - 1 \end{aligned}$$

(λ_L^* is the value of λ corresponding to $p \leq 0.7$).

3.8 Tests of separate families

Definition 3.2 Suppose there are two families of density functions $\{f(x \mid \alpha) \mid \alpha \in \Theta_f\}$ and $\{g(x \mid \beta \in \Theta_g\}$; according to Cox (1961,1962), they are said to be separate, for any $\alpha_0[\beta_0]$, if $f(x \mid \alpha_0)[g(x \mid \beta_0)]$ cannot be approximated arbitrarily closely by any $g(x \mid \beta)[f(x \mid \alpha)]$.

Examples of such separate families are the log-normal and exponential families, the lognormal and inverse Gaussian families.

Let $X_1, \ldots, X_n$ be a random sample from $f(x \mid \alpha)$ under H_f and $g(x \mid \beta)$ under H_g where α, β are parameters such that $\dim \Theta_f = d_1$ and $\dim \Theta_g = d_2$. Further let

$$F_k = \log f(x_k \mid \alpha), \quad F_{k,\alpha_j} = \frac{\partial}{\partial \alpha_j} F_k, \quad F_{k,\alpha_i,\alpha_j} = \frac{\partial^2}{\partial \alpha_i \partial \alpha_j} F_k$$

for $1 \leq k \leq n$, $1 \leq i, j \leq d_1$. Moreover, let G_R, G_{k,β_j}, G_{k,β_i,β_j} be similarly defined for $1 \leq i, j \leq d_2$. Cox (1961,1962) based his statistic on L_{fg} defined by

$$L_{fg} = \sum_{i=1}^n \log \frac{f(X_i \mid \hat{\alpha})}{g(X_i \mid \hat{\beta})}$$

where $\hat{\alpha}, \hat{\beta}$ are the maximum likelihood estimates of α and β under H_f and H_g for testing H_f, the null hypothesis against the alternative H_g. The test statistic is

$$nT_f = L_{fg} - \mathbb{E}_{\hat{\alpha}}(L_{fg})$$

where $\mathbb{E}_{\hat{\alpha}}(L_{fg})$ is the expectation taken under H_f when α is replaced by $\hat{\alpha}$. In many situations $T_f = h(\hat{\alpha}, \hat{\beta})$ and $h(\alpha, \beta_\alpha) = 0$ where β_α is a value such that under f: $\hat{\beta} \xrightarrow{P} \beta_\alpha$. If $h(\hat{\alpha}, \hat{\beta})$ is approximated by a linear function of $(\hat{\alpha} - \alpha, \hat{\beta} - \beta)$ so that

$$h(\hat{\alpha}, \hat{\beta}) = (\hat{\alpha} - \alpha, \hat{\beta} - \beta) l_\alpha$$

where l_α is gradient vector, the variance of T_f is

$$V_\alpha(T_f) = l_\alpha^t C_\alpha l_\alpha$$

where C_α is the covariance matrix of $(\hat{\alpha}, \hat{\beta})$ under H_f. The test of separate families is based on the asymptotic (standard) normality of the statistic

$$T'_f = \frac{T_f}{\sqrt{V_\alpha(T_f)}}.$$

It is at once evident that the roles of f and g can be switched to yield

$$T'_g = \frac{T_g}{\sqrt{V_\beta(T_g)}}.$$

Now in using each of T'_f or T'_g there are three possible conclusions, namely:
(a) consistency with H_f (H_g),
(b) evidence of departure in the direction of H_g (H_f),
(c) evidence of departure away from H_g (H_f).
An application of T'_f and T'_g then leads to 3^2 qualitatively different conclusions.

Test for lognormal vs the inverse Gaussian

Let us consider first the test of H_f: lognormal vs H_g: inverse Gaussian. Under the null hypothesis that H_f is lognormal, $\alpha = (\alpha_1, \alpha_2; \alpha_2 > 0)$

$$f(x \mid \alpha) = \frac{1}{x\sqrt{2\pi\alpha_2}} \exp\left(-\frac{(\log x - \alpha_i)^2}{2\alpha_2}\right) 1_{\mathbb{R}^+}(x).$$

Under the alternative that H_g is $IG(\beta_1, \beta_2)$ where $\beta = (\beta_1, \beta_2; \beta_1 > 0, \beta_2 > 0)$ then

$$g(x \mid \beta) = \sqrt{\frac{\beta_2}{2\pi x^3}} \exp\left(-\frac{\beta_2^2(x - \beta_1)^2}{2\beta_1^2 x}\right) 1_{\mathbb{R}^+}(x).$$

The maximum likelihood estimates of α and β are respectively

$$\hat{\alpha}_1 = \frac{\sum_{i=1}^n \log X_i}{n}, \quad \hat{\alpha}_2 = \frac{\sum_{i=1}^n (\log X_i - \hat{\alpha}_1)^2}{n}$$

$$\hat{\beta}_1 = \frac{\sum_{i=1}^n X_i}{n} = \overline{X},$$

$$\hat{\beta}_2 = \frac{\overline{X}}{\overline{X}\,\overline{X_{-1}} - 1}.$$

Thus we can show that
$$L_{fg} = \frac{n}{2}(\hat{\alpha}_1 - \log \hat{\alpha}_2 - \log \hat{\beta}_2)$$

and
$$\mathbb{E}_{\hat{\alpha}}\left(\log \frac{f(x \mid \hat{\alpha})}{g(x \mid \hat{\beta})}\right) = \frac{1}{2}(\alpha_1 - \log \alpha_2 - \log \beta_{2,\alpha})$$

so that
$$T'_f = \frac{1}{2} \log \left(\frac{\beta_{2,\hat{\alpha}}}{\hat{\beta}_2}\right).$$

Now
$$\hat{\beta}_2 \xrightarrow{a.s.} \beta_{2,\alpha} = \frac{e^{(\alpha_1 + \frac{\alpha_2}{2})}}{e^{\alpha_2} - 1}$$

and
$$\log \frac{\beta_{2,\alpha}}{\hat{\beta}_2} \approx \frac{1}{\beta_{2,\alpha}} \left(\frac{\partial \beta_{2,\alpha}}{\partial \alpha_1}(\hat{\alpha}_1 - \alpha_1) + \frac{\partial \beta_{2,\alpha}}{\partial \alpha_2}(\hat{\alpha}_2 - \alpha_2) + (\hat{\beta}_2 - \hat{\beta}_1)\right).$$

Finally for testing H_f against H_g we use

$$\frac{\sqrt{n} \log \frac{\beta_{2,\hat{\alpha}}}{\hat{\beta}_2}}{\left(e^{\hat{\alpha}_2} + 1 - \hat{\alpha}_2 - \frac{\hat{\alpha}_2^2}{2}\left(\frac{e^{\hat{\alpha}_2}+1}{e^{\hat{\alpha}_2}-1}\right)^{\frac{1}{2}}\right)} \qquad (3.8)$$

and for large negative values we conclude that there is evidence of departure from H_f in the direction of H_g.

Test of IG vs the lognormal

Under H_g as null hypothesis we have
$$\hat{\alpha}_1 \xrightarrow{a.s.} \alpha_{1\beta} = \log \beta_1 + A_\theta\left(-\frac{1}{2}\right)$$
$$\hat{\alpha}_2 \xrightarrow{a.s.} A'_\theta\left(-\frac{1}{2}\right)$$

where
$$\theta = \left(\frac{\beta_2}{\beta_1}\right), \quad A_\theta(\lambda) = \frac{\partial}{\partial \lambda} \log K_\lambda(\theta), \text{ and } A'_\theta = \frac{\partial}{\partial \lambda} A_\theta(\lambda),$$

$K_\lambda(\theta)$ being the modified Bessel function of the third kind.
$$(A_\theta(\lambda) = -A_\theta(-\lambda), \; A'_\theta(\lambda) = A'_\theta(-\lambda)).$$

Now
$$\mathbb{E}_\beta\left(\frac{g(x\mid\beta)}{f(x\mid\alpha_\beta)}\right) = \frac{1}{2}(\log\alpha_{2,\beta} + \log\beta_2 - \alpha_{1,\beta})$$

and
$$nT_g = L_{gf} - \mathbb{E}_{\hat\beta}\left(\frac{g(x\mid\beta)}{f(x\mid\alpha_\beta)}\right)$$
$$= \frac{n}{2}\left(\log\frac{\hat\alpha_2}{\alpha_{2\hat\beta}} + (\alpha_{1\hat\beta} - \hat\alpha_1)\right).$$

Using only the linear term in the expansion of T_g, Liao (1995) has shown that
$$2T_g \approx l_\beta^t(\hat\alpha_1 - \alpha_{1\beta}, \hat\alpha_2 - \alpha_{2,\beta}, \hat\beta_1 - \beta, \hat\beta_2 - \beta_2)^t$$

where
$$l_\beta^t = \left(-1, \frac{1}{A'}, -2\frac{\theta}{\beta_1}\left(A + \frac{2A}{A'}\right), \frac{1}{\beta_2}\left(1 + 2\theta A + \frac{4\theta A^2}{A'}\right)\right),$$

and the asymptotic covariance of $(\hat\alpha_1, \hat\alpha_2, \hat\beta_1, \hat\beta_2)$ under H_g is given by

$$nC_\beta = \begin{pmatrix} A' & A'' & -2\beta_1 A & 2\beta_2(1+2\theta A) \\ A'' & A''' + 2A'^2 & 4\beta_1 A^2 & -8\beta_2\theta A2 \\ -2\beta_1 A & 4\beta_1 A^2 & \frac{\beta_1^3}{\beta_2} & 0 \\ 2\beta_2(1+2\theta A) & -8\beta_2\theta A2 & 0 & 2\beta_2^2 \end{pmatrix}$$

Replacing θ by $\hat\theta = \frac{\hat\beta_2}{\hat\beta_1}$ one rejects for large negative values of

$$\frac{(\alpha_{1\hat\beta} - \hat\alpha_1) + \log\frac{\hat\alpha_2}{\alpha_{2\hat\beta}}}{\sqrt{V}} \qquad (3.9)$$

where
$$(V = l_\beta^t C_\beta l_\beta),$$

as evidence of departure from H_g in the direction of H_f.

Applications to simulated and physiological data

In this section, we consider the above results as applied to the simulated data and real data by Liao. He conducted the following tests: (1) H_0: lognormal (LN) against H_1: exponential (EX); (2) H_0: exponential (EX) against H_1: lognormal (LN); (3) H_0: lognormal (LN) against H_1: inverse Gaussian (IG); (4) H_0: inverse Gaussian (IG) against H_1: lognormal (LN). Since sample sizes in the data were large enough (about 1000),

Tests of separate families

high power could be expected for all tests. Using (3.8), and (3.9), we have presented the test statistics in Table 3.6 for all the data sets. IMSL was used by Liao to evaluate the modified Bessel function of the third kind and its derivatives involved in the test for the inverse Gaussian against lognormal, and MINITAB package for the rest of the computations. He used $\alpha = 0.05$, for which $c_\alpha = -1.645$. For the inverse Gaussian against lognormal, all statistics are greater than -1.64, so we cannot reject the null hypothesis that the model is the inverse Gaussian; meanwhile, all statistics are less than -1.64 for the lognormal against inverse Gaussian, and so we reject the null hypothesis in favor of the alternative that the model is the inverse Gaussian.

The results of Liao (1995) on separate tests applied to simulated and physiological data, for tests relating to H_0: lognormal (LN) against H_1: inverse Gaussian (IG) as well as H_0: inverse Gaussian (IG) against H_1: lognormal (LN), are summarized in Table 3.6 for all the data sets.

Table 3.6 *The test statistics for lognormal against inverse Gaussian (LNvsIG) and inverse Gaussian against lognormal (IGvsLN) are based on (3.8) and (3.9) respectively.*

Noise distribution or data	LNvsIG	IGvsLN
Normal	-3.360	0.589
Gamma	-1.741	-0.159
Uniform	-3.245	0.672
f68.dat	-5.727	4.305
f72.dat	-5.129	3.239
f75.dat	-5.689	4.291
f78.dat	-7.374	7.022
f80.dat	-7.508	7.072
f82.dat	-6.934	5.492
f83.dat	-6.864	5.531
f87.dat	-6.377	4.530
f90.dat	-6.629	5.294
f98.dat	-4.795	3.184

At a level $\alpha = 0.05$ with $Z_\alpha = 1.645$ all statistics are greater than -1.64 for testing IG against LN implying that we cannot reject the null hypothesis that the model is IG: On the other hand all statistics for testing LN against IG are less than -1.64 and hence the null hypothesis of lognormality is rejected in favour of the inverse Gaussian.

The inverse Gaussian therefore seems to be the best model among the candidates. (Liao also has tested exponentiality vs lognormality and vice versa and concluded that lognormality was more plausible than exponentiality).

3.9 Bahadur-efficient tests

Let us recall that Bahadur efficiency considers the relative rates at which the attained significance levels for two tests converge to zero for a fixed alternative and power as the sample size becomes large.

Under the regularity conditions (necessary) for the existence of the Cramer-Rao lower bound, if the maximum likelihood estimators are consistent, then their asymptotic effective variance is equal to the Cramer-Rao lower bound; when this happens these two estimators are said to be Bahadur-efficient. Durairajan (1985) proposed a test for the parameters of the inverse Gaussian law based on Fisher's method of combining independent tests and showed that the test is Bahadur-efficient. We now outline this procedure in the ensuing section.

We assume that $X = (X_1, \cdots, X_n)$ is a random sample from $IG(\mu, \lambda)$ and let

$$\Theta = \{\theta = (\mu, \lambda) \mid 0 < \mu < \infty, \ 0 < \lambda \leq \lambda_0 \ \text{(known)}\}$$

and

$$\Theta_0 = \{\theta \mid \mu = \mu_0, \ \lambda = \lambda_0 \ \text{where } \mu_0, \lambda_0 \text{ are both known}\}.$$

We wish to test

$$H_0 : \theta \in \Theta_0$$

against

$$H_A : \theta \in \Theta \backslash \Theta_0.$$

From section 3.1 (ii) we know that for testing

$$H_0' : \mu = \mu_0 \ (\lambda \text{ unknown}),$$

against

$$H_A' : \mu \neq \mu_0 \ (\lambda \text{ unknown}),$$

the uniformly most powerful unbiased test is based on T (equivalently T^2) where

$$T^2 = \frac{U_n}{V_n}$$

and where U_n and V_n are defined by

$$U_n = \frac{n\lambda(\overline{X} - \mu_0)^2}{\overline{X}\mu_0^2}, \quad V_n = \lambda_0 \sum_{i=1}^{n}\left(\frac{1}{X_i} - \frac{1}{\overline{X}}\right).$$

Bahadur efficient tests

Moreover, from section 3.1 (iv) it is known that the uniformly most powerful unbiased test of

$$H_0'' : \lambda = \lambda_0 \quad (\mu \text{ unknown}),$$

against

$$H_A'' : \lambda < \lambda_0, \quad \mu \text{ unknown}$$

is based on V_n.

Now define Q_n by

$$Q_n = -2 \log P(U_n > u \mid H_0) - 2 \log P(V_n > v \mid H_0)$$

where u and v are the observed values of U_n and V_n respectively.

Fisher's method of combining independent tests for testing H_0 against H_A (it being assumed that U_n is the statistic used for testing H_0' against H_A'). is based on Q_n and the critical region of the test is given by $Q_n \geq c$, c being chosen to attain the desired level α. Since U_n and V_n are independently distributed according to the chisquare law indeed then $\mathcal{L}(Q_n) = \chi_4^2$.

The following propositions due to Durairajan lead to the proof of the Bahadur efficiency of the test based on Q_n.

Proposition 3.3 *The exact slope of the sequence of tests $\{U_n\}$, $\{V_n\}$ and $\{Q_n\}$ are given respectively by $C_U(\theta)$, $C_V(\theta)$ and $C_Q(\theta)$ where for every $\theta \in \Theta \setminus \Theta_0$*

(i) $C_U(\theta) = \dfrac{\lambda_0}{\mu} \left(\dfrac{\mu_0 - \mu}{\mu_0} \right)^2$

(ii) $C_V(\theta) = \dfrac{\lambda_0 - \lambda}{\lambda} - \log \left(\dfrac{\lambda_0}{\lambda} \right)$

and

(iii) $C_Q(\theta) = C_U(\theta) + C_V(\theta)$.

Proof We first recall that the exact slope $C_A(\theta)$ of a sequence of tests $\{A_n\}$ is given by

$$\mathbb{E}_\theta \left\{ \log \dfrac{f(x \mid \theta)}{f(x \mid \theta_0)} \right\} \quad \text{for each } \theta \in \Theta \setminus \Theta_0,$$

where $f(x \mid \theta)$ is the inverse Gaussian density $IG(\mu, \lambda)$.

Let $U_1(n) = \dfrac{U_n}{\sqrt{n}}$. Since $\mathcal{L}(U_n) = \chi_1^2$, it can be shown that for fixed $t > 0$ and large n

$$P(U_n \geq nt) = \sqrt{\dfrac{2}{\pi}} (nt)^{-\frac{1}{2}} \exp\left(-\dfrac{nt}{2}\right) \left[1 + O\left(\dfrac{1}{n}\right)\right].$$

Hence

$$-\frac{1}{n}\log P(U_n \geq nt) = \frac{c}{n} + \frac{\log n}{2n} + \frac{t}{2} - \frac{1}{n}\log\left[1 + O\left(\frac{1}{n}\right)\right]$$

c being a constant independent of n. Thus

$$\lim_{n\to\infty}\left\{-\frac{1}{n}\log P(U_n \geq nt)\right\} = \frac{t}{2}$$

and it follows from the definition of $U_1(n)$ that

$$\lim_{n\to\infty}\left\{-\frac{1}{n}\log P(U_1(n) \geq \sqrt{n}t \mid H_0)\right\} = \frac{t}{2}.$$

Since $\frac{U_1(n)}{\sqrt{n}}$ tends almost surely (θ) for every $\theta \in \Theta\backslash\Theta_0$ to $\frac{\lambda_0}{\mu}\left(\frac{\mu_0-\mu}{\mu_0}\right)^2$, $C_U(\theta)$, the exact slope of $U_1(n)$, and $U_1(n)$ are increasing in U_n, both U_n and $U_1(n)$ have the same slope.

In like manner, since $\mathcal{L}(V_n) = \chi^2_{n-1}$, we can define $V_1(n) = \frac{V_n}{\sqrt{n}}$ and show that $\frac{V_1(n)}{\sqrt{n}}$ tends almost surely (θ) to $\frac{\lambda_0}{\lambda}$ for every $\theta \in \Theta\backslash\Theta_0$. Furthermore

$$\lim_{n\to\infty}\left\{-\frac{1}{n}\log P(V_1(n) \geq \sqrt{n}t \mid H_0)\right\} = \frac{1}{2}(t-1)\log t.$$

Therefore the exact slope of $V_1(n)$ is

$$\frac{\lambda_0}{\lambda} - 1 - \log\left(\frac{\lambda_0}{\lambda}\right)$$

which is the (same) exact slope of V_n.

Finally the exact slope of Q_n is obtained by adding $C_U(\theta)$ and $C_V(\theta)$. Note that for every $\theta \in \Theta\backslash\Theta_0$

$$\log\left[\frac{f(x\mid\theta)}{f(x\mid\theta_0)}\right] = \frac{1}{2}\log\left(\frac{\lambda}{\lambda_0}\right) + \frac{1}{2}\left[\frac{\lambda_0(X-\mu_0)^2}{X\mu_0^2} - \frac{\lambda(X-\mu)^2}{X\mu^2}\right]$$

so that

$$\mathbb{E}_\theta\left[\log\left\{\frac{f(x\mid\theta)}{f(x\mid\theta_0)}\right\}\right] = \frac{\lambda_0}{\mu}\left(\frac{\mu_0-\mu}{\mu_0}\right)^2 + \frac{\lambda_0}{\lambda} - 1 - \log\left(\frac{\lambda_0}{\lambda}\right) = C_Q(\theta).$$

Thus the test based on Q_n is Bahadur-efficient. ♣

CHAPTER 4

SEQUENTIAL METHODS

4.0 Introduction

One of the recommendations of a sequential testing procedure is that when both the type I and type II errors are specified, it requires a smaller sample on the average than when a fixed sample scheme is used. Thus in reliability testing and acceptance sampling where cost consideration and time constraints are of the essence sequential sampling plans have a distinct edge over fixed sampling schemes. We therefore describe some sequential methods which have been considered by Wasan (1969), Edgeman and Salzburg (1991), Joshi and Shah (1990) and Chaturvedi (1991).

4.1 Sequential probability ratio test

We assume that the random variables $X_1, \ldots, X_n$ are drawn sequentially from $IG(\mu, \lambda)$ where we further assume that λ is known. In the language of sequential analysis we take α (type I error) as the producer's risk and β (the type II error) as the consumer's risk to be fixed numbers. The hypotheses concern the mean of a process to be at a level μ_0 as opposed to a level $\mu_1 > \mu_0$. At any stage as samples are drawn one by one the probability ratio

$$L_n = \frac{\prod_{i=1}^n f(x_i : \mu_1)}{\prod_{i=1}^n f(x_i : \mu_0)}$$

is computed. We then select two numbers A and B depending on α and β satisfying the requirement

$$A \leq \frac{1-\beta}{\alpha}, \quad B \geq \frac{\beta}{1-\alpha}$$

where $0 < \beta < 0.5 < 1 - \alpha < 1$. Then by Wald's sequential probability ratio test, as long as $B < L_n < A$, we keep drawing additional samples. But at the first instance $L_n \geq A$, the process is terminated with the rejection of H_0 (accepting H_A) and at the first instance $L_n \leq B$, the process is again terminated and H_0 is accepted. It is well-known that such a test terminates with probability one. Denote by Z_i the ratio

$$Z_i = \log\left(\frac{f(x_i; \mu, \lambda)}{f(x_i; \mu_0, \lambda)}\right).$$

In terms of Z_i, a more convenient form for the sequential rule is to terminate sampling if either $\sum_{i=1}^{n} Z_i \geq \log A$ or $\sum_{i=1}^{n} Z_i \leq \log B$. Thus for the $IG(\mu, \lambda)$ situation we continue sampling if

$$2\log B + \frac{2n(\mu_1 - \mu_0)}{\mu_0 \mu_1} \leq \frac{(\mu_1^2 - \mu_0^2)}{\mu_0^2 \mu_1^2} \sum_{i=1}^{n} X_i \leq 2\log A + \frac{2n(\mu_1 - \mu_0)}{\mu_0 \mu_1}$$

that is, if

$$\log \frac{\beta}{1-\alpha} \left(\frac{2\mu_1^2 \mu_0^2}{\mu_1^2 - \mu_0^2}\right) + \frac{2n\mu_1 \mu_0}{\mu_1 - \mu_0} \leq \sum_{i=1}^{n} X_i \leq \log\left(\frac{1-\beta}{\alpha}\right) \frac{2\mu_1^2 \mu_0^2}{\mu_1^2 - \mu_0^2}$$

$$+ 2n \frac{\mu_0 \mu_1}{\mu_1 - \mu_0}.$$

Therefore we accept if

$$\sum_{i=1}^{n} X_i \leq \left(\log \frac{\beta}{1-\alpha}\right) \frac{2\mu_1^2 \mu_0^2}{\mu_1^2 - \mu_0^2} + \frac{2n\mu_1 \mu_0}{\mu_1 - \mu_0}$$

and reject if

$$\sum_{i=1}^{n} X_i \geq \left(\log \frac{1-\beta}{\alpha}\right) \frac{2\mu_1^2 \mu_0^2}{\mu_1^2 - \mu_0^2} + \frac{2n\mu_1 \mu_0}{\mu_1 - \mu_0}. \qquad (4.1)$$

It should be borne in mind that since the probabilities of the excess over the boundaries are neglected the test is an approximate test.

Operating Characteristic function

Let $h(\mu)$ be a function of μ and consider

$$L^{h(\mu)} = \left\{\frac{f(x; \mu_1)}{f(x; \mu_0)}\right\}^{h(\mu)}.$$

Then

$$\mathbb{E}_\mu\left(L^{h(\mu)}\right) = \int_0^\infty \left(\frac{f(x; \mu_1)}{f(x; \mu_0)}\right)^{h(\mu)} f(x; \mu) dx = 1. \qquad (4.2)$$

Thus $L^{h(\mu)}f(x;\mu) = g(x;\mu)$ is a frequency function for all values of μ if (4.2) holds. We now will determine $h(\mu)$. Consider the integral equation

$$\int_0^\infty \sqrt{\frac{\lambda}{2\pi x^3}} \exp\left(-\frac{\lambda}{2x}\left(\frac{(x-\mu_1)^2}{\mu_1^2} - \frac{(x-\mu_0)^2}{\mu_0^2}\right)h(\mu) - \frac{(x-\mu)^2}{\mu^2}\right) dx$$

$= 1$. According to Wald (1947) the integral equation has at least one non-zero value of $h(\mu)$ as solution. Let

$$a = h(\mu)\left(\frac{1}{\mu_1^2} - \frac{1}{\mu_0^2}\right) + \frac{1}{\mu^2}$$

$$b = h(\mu)\left(\frac{1}{\mu_1} - \frac{1}{\mu_0}\right) + \frac{1}{\mu}.$$

The exponent when expanded in powers of x is

$$-\frac{\lambda}{2x}(ax^2 - 2bx + 1) = -\frac{\lambda}{2x}\left(a\left(x - \frac{b}{a}\right)^2 + \left(1 - \frac{b^2}{a}\right)\right).$$

Suppose that $a \neq 0$ and $\frac{b^2}{a} = 1$. The integral reduces to the familiar integral of an $IG\left(\frac{b}{a}, \lambda\right)$ law which gives us the value of $h(\mu)$ as the solutions to $b^2 = a$, or

$$h^2(\mu)\left(\frac{1}{\mu_1} - \frac{1}{\mu_0}\right)^2 + \frac{2h(\mu)}{\mu}\left(\frac{1}{\mu_1} - \frac{1}{\mu_0}\right) + \frac{1}{\mu^2} = h(\mu)\left(\frac{1}{\mu_1^2} - \frac{1}{\mu_0^2}\right) + \frac{1}{\mu^2}.$$

Simplifying we have

$$h(\mu)\left(\frac{1}{\mu_1} - \frac{1}{\mu_0}\right)\left[h(\mu)\left(\frac{1}{\mu_1} - \frac{1}{\mu_0}\right) - \left(\frac{1}{\mu_1} + \frac{1}{\mu_0}\right) + \frac{2}{\mu}\right] = 0.$$

Thus if $\mu_1 \neq \mu_0$, $h(\mu)$ satisfies

$$h(\mu) = \frac{\left(\frac{1}{\mu_1} - \frac{1}{\mu_0}\right) - \frac{2}{\mu}}{\left(\frac{1}{\mu_1} - \frac{1}{\mu_0}\right)} = \frac{\frac{2\mu_1\mu_0}{\mu} - (\mu_0 + \mu_1)}{\mu_0 - \mu_1} \quad (4.3).$$

It is clear $h(\mu_0) = 1$ and $h(\mu_1) = -1$ satisfy the above equation (4.3). This result is due to Wasan (1969) and has been obtained independently by Edgeman and Salzberg. The operating characteristic function is given by

$$L(\mu) \approx \frac{\left(\frac{1-\beta}{\alpha}\right)^{h(\mu)} - 1}{\left(\frac{1-\beta}{\alpha}\right)^{h(\mu)} - \left(\frac{\beta}{1-\alpha}\right)^{h(\mu)}}$$

where $h(\mu) = \frac{\frac{1}{\mu_1}+\frac{1}{\mu_0}-\frac{2}{\mu}}{\frac{1}{\mu_1}-\frac{1}{\mu_0}}$. One could plot $L(\mu)$ against μ or regard $h(\mu)$ as another parameter and plot $L(\mu)$ versus $h(\mu)$. Note that $\mu = \frac{2\mu_0\mu_1}{\mu_0+\mu_1}$ is also a point on the operating characteristic curve. The famous Wald's identity gives the expected sample size as

$$\mathbb{E}(N) = \frac{\mathbb{E}(Z_n)}{\mathbb{E}(Z)}$$

where Z_n is a random variable that approximately takes the value $\log A$ with probability $1 - L(\mu)$ and the value $\log B$ with probability $L(\mu)$. (These are the only two values for the sampling to terminate.) The variable $Z = \log\left(\frac{f(x;\mu_1)}{f(x;\mu_0)}\right)$ and the expectation is taken with respect to $IG(\mu, \lambda)$. Thus

$$\mathbb{E}(N(\mu)) = \frac{L(\mu)\log B + (1 - L(\mu))\log A}{\mathbb{E}(Z)}. \qquad (4.4)$$

It is easy to find $\mathbb{E}(Z)$. Indeed we have

$$\begin{aligned}
\mathbb{E}(Z) &= \mathbb{E}\left(-\frac{\lambda}{2X}\frac{(X-\mu_1)^2}{\mu_1^2} + \frac{\lambda}{2X}\frac{(X-\mu_0)^2}{\mu_0^2}\right) \\
&= \frac{\lambda}{2}\left(\frac{\mu_1^2 - \mu_0^2}{\mu_0^2\mu_1^2}\mu - \frac{2(\mu_1-\mu_0)}{\mu_0\mu_1}\right) \qquad (4.5) \\
&= \frac{\lambda\mu}{2}\left(\frac{1}{\mu_0^2} - \frac{1}{\mu_1^2}\right) + \lambda\left(\frac{1}{\mu_1} - \frac{1}{\mu_0}\right).
\end{aligned}$$

Finally $\mathbb{E}(N(\mu_0))$ and $\mathbb{E}(N(\mu_1))$ are easily calculated using (4.4) and (4.5). Edgeman and Salzberg consider an example of times to failure of air conditioning equipment first examined by Jørgensen (1982) in relation to the inverse Gaussian law. Specifically they are concerned with a decrease in mean times to failure which implies an increase in the reciprocals of the mean times to failure. Assuming that a preliminary test of fit is deemed satisfactory one wishes to test if $\mu_0 = 20$ against $\mu_1 = 15$ with $\alpha = 0.05$, $\beta = 0.10$ and $\lambda = 1$. In terms of reciprocals $\frac{1}{\mu_0} = 0.05$ and $\frac{1}{\mu_1} = 0.0667$. The sequential rule then recommends acceptance of H_0 if the sum of the reciprocals of the life test times $\leq -0.0257 + 0.0571$, and rejection if this sum $\geq 0.0330 + 0.0571$ using (4.1). If $\mu = \frac{1}{16}$, $h(\mu) = -0.60$ and $L(\mu) \approx 0.2235$. Finally $\mathbb{E}(Z) = 0.476$ so that the average sample number is 3.66.

The second sequential probability ratio test concerns a test of $H_0 : \lambda = \lambda_0$, $\mu = 1$ against $H_A : \lambda = \lambda_1 > \lambda_0$, $\mu = 1$. Using arguments similar to the test of the mean we can arrive at a rule: Reject H_0 if

$$\frac{n}{2}\log\frac{\lambda_1}{\lambda_0} - (\lambda_1 - \lambda_0)\sum_{i=1}^{n}\frac{(X_i-1)^2}{2X_i} \geq \log A$$

and accept H_0 if

$$\frac{n}{2}\log\frac{\lambda_1}{\lambda_0} - (\lambda_1 - \lambda_0)\sum_{i=1}^{n}\frac{(X_i-1)^2}{2X_i} \leq \log B.$$

As for the operating characteristic function we have to determine $h(\lambda)$ from the integral equation

$$\int_0^\infty \left(\sqrt{\frac{\lambda_1}{\lambda_0}}\exp\left(-(\lambda_1-\lambda_0)\frac{(x-1)^2}{2x}\right)\right)^{h(\lambda)} \sqrt{\frac{\lambda}{2\pi x^3}}\exp\left(-\frac{\lambda(x-1)^2}{2x}\right)dx = 1.$$

Therefore we have

$$\left(\frac{\lambda_1}{\lambda_0}\right)^{\frac{h(\lambda)}{2}}\sqrt{\frac{\lambda}{2\pi}}\int_0^\infty x^{-\frac{3}{2}}\exp\left(-((\lambda_1-\lambda_0)h(\lambda))\frac{(x-1)^2}{2x}\right)dx = 1.$$

Integrating the now familar $IG(1, \lambda + (\lambda_1 - \lambda_0)h(\lambda))$ law we have

$$\left(\frac{\lambda_1}{\lambda_0}\right)^{h(\lambda)}\frac{\sqrt{\lambda}}{\sqrt{\lambda + (\lambda_1-\lambda_0)h(\lambda)}} = 1.$$

Therefore $h(\lambda)$ satisfies the equation

$$\lambda\left(\frac{\lambda_1}{\lambda_0}\right)^{h(\lambda)} = \lambda + (\lambda_1 - \lambda_0)h(\lambda). \tag{4.6}$$

Taking the logarithmic derivative of (4.6) gives

$$\frac{1}{\lambda} + \log\left(\frac{\lambda_1}{\lambda_0}\right)h'(\lambda) = \frac{1 + (\lambda_1-\lambda_0)h'(\lambda)}{\lambda + (\lambda_1-\lambda_0)h(\lambda)}.$$

Hence, noting that $h(\lambda_0) = 1$ and $h(\lambda_1) = -1$ satisfy (4.6) we obtain

$$h'(\lambda_0) = \frac{\frac{1}{\lambda_1} - \frac{1}{\lambda_0}}{\frac{\lambda_0}{\lambda_1} - 1 - \log\frac{\lambda_0}{\lambda_1}} < 0$$

and

$$h'(\lambda_1) = \frac{\frac{1}{\lambda_0} - \frac{1}{\lambda_1}}{\log\frac{\lambda_1}{\lambda_0} + 1 - \frac{\lambda_1}{\lambda_0}} < 0.$$

This shows that $L(\lambda) = \frac{\left(\frac{1-\beta}{\alpha}\right)^{h(\lambda)} - 1}{\left(\frac{1-\beta}{\alpha}\right)^{h(\lambda)} - \left(\frac{\beta}{1-\alpha}\right)^{h(\lambda)}}$ is increasing in $h(\lambda)$ and that when $\lambda = \lambda_0$ and $\lambda = \lambda_1$ $(\lambda_1 > \lambda_0)$ $L(\lambda)$ is decreasing in λ. One can show that the average sample number is

$$\mathbb{E}(N(\lambda)) = \frac{L(\lambda) \log B + (1 - L(\lambda)) \log A}{\frac{1}{2}\left(\log \frac{\lambda_1}{\lambda_0} - \frac{(\lambda_1 - \lambda_0)}{\lambda}\right)}.$$

4.2 Sequential test for the mean and asymptotics

In the case of a fixed sample size n we have seen that the likelihood ratio test of $\mu = \infty$ versus $\mu < \infty$ has optimum properties. Formulated in terms of the underlying Brownian motion the hypothesis tests for the drift ν namely $H_0 : \nu = 0$ against $H_A : \nu > 0$, for the density function (see Section 3.2)

$$f(t : a, \nu, \sigma) = \frac{a}{\sigma\sqrt{2\pi t^3}} \exp\left\{-\frac{(a - \nu t)^2}{2\sigma^2 t}\right\} 1_{(0,\infty)}(t)$$

the likelihood ratio test rejects H_0 if

$$\sum_{i=1}^{n} X_i < \frac{n^2 \lambda}{Z_{\alpha/2}^2}$$

when $\lambda = \frac{a^2}{\sigma^2}$ is known. Here $Z_{\alpha/2}$ is the upper $100\frac{\alpha}{2}\%$ point of the standard normal law. Lombard (1978) observes that, writing $S_n = \sum_{i=1}^{n} X_i$,

$$S_n < \frac{n^2 \lambda}{Z_{\alpha/2}^2}$$

if and only if, for some $k = 1, 2, \cdots, n$

$$S_k < \frac{n^2 \lambda}{Z_{\alpha/2}^2}.$$

Thus if the n observations are taken one by one and H_0 is accepted at stage $k(\leq n)$ if and only if

$$S_k \geq \frac{n^2 \lambda}{\lambda Z_{\alpha/2}^2}$$

then the sequential procedure will result in the same power as the fixed sample rule and in many cases result in a smaller sample necessary to reach the decision.

Lombard shows that when λ is unknown the sequential procedure corresponding to the likelihood ratio test based on the rule reject H_0 if

$$\frac{V}{(n-1)} S_n < \frac{n^2}{t^2_{n-1,\alpha/2}}$$

still retains the power asymptotically and results in a saving of observations, when H_0 is true. ($t_{n-1,\alpha/2}$ being the upper $100\alpha/2$ percent point of Student's t based on $(n-1)$ degrees of freedom).

We now describe the procedure as outlined by Lombard.
Recall the following facts

(a) $P[T \le t] = F(t; a, \nu, \sigma)$

$$= \Phi\left(\frac{a - \nu t}{\sigma\sqrt{t}}\right) + \exp\left(\frac{2\nu a}{\sigma^2}\right) \Phi\left(-\frac{a + \nu t}{\sigma\sqrt{t}}\right)$$

(b) $\mathbb{E}[\exp(-\theta T)] = \exp\left[\frac{a\nu}{\sigma^2}\left\{1 - \left(1 + \frac{2\sigma^2\theta}{\nu^2}\right)^{\frac{1}{2}}\right\}\right]$

(c) $\lim_{n \to \infty} P\left[\frac{S_{[nz]}}{n^2} \le t_n \mid \nu = \frac{y}{n}\right] \to F(t; az, y, \sigma)$

for any fixed y and a sequence $\{t_n\} \to t > 0$, and where $[nz]$ denotes the integer part of nz.

Define for $n \ge 2$

(i) $U_n = \lambda V_n$ with $V_n = \sum_{i=1}^{n}\left(\frac{1}{X_i} - \frac{1}{\overline{X}_n}\right)$

where $\mathcal{L}(X_i) = IG\left(\mu = \frac{a}{\nu}, \lambda = \frac{a^2}{\sigma^2}\right)$

(d) $\mathcal{L}(U_n) = \chi^2_{n-1}$

(e) $P\left[S_k \ge b \text{ for some } [nd] \le k \le [nz] \mid \nu = \frac{y}{n}\right]$

$= P\left[S_{[nz]} \ge b \mid \nu = \frac{y}{n}\right]$ whenever $b, y > 0$ and $0 < d < z$.

(ii) for $y \ge 0$ and $\psi \ge 0$

$$G(y; \psi, \sigma) = 1 - \Phi\left(\frac{\psi}{Z_{\alpha/2}} - yZ_{\alpha/2}\right) - \exp(2\psi y)\Phi\left(-\frac{\psi}{Z_{\alpha/2}} - yZ_{\alpha/2}\right).$$

Lombard states and proves the following two theorems which provide results on the asymptotic power and lead to the assertions made at the outset.

Theorem 4.1 *For testing $H_0 : \nu = 0$ against $H_A : \nu > 0$ at a level α where $0 < \alpha < 1$*

(i) $P(\text{reject } H_0 \mid \nu = 0) = \alpha$

(ii) $\lim_{n \to \infty} P\left(\text{reject } H_0 \mid \nu = \dfrac{\psi \sigma^2}{na}\right) = 1 - G(1; \psi, \alpha).$

Proof Rejection of H_0 given $\nu = 0$ implies that

$$\frac{S_n V_n}{(n-1)} \leq \frac{n^2}{t^2_{n-1,\alpha/2}}.$$

When $\nu = 0$, $\mathcal{L}\left(\dfrac{n^2 a^2}{\sigma^2 S_n}\right) = \chi^2_1$ and moreover, from the independence of S_n and V_n and the fact that $\mathcal{L}\left(\dfrac{a^2}{\sigma^2} V_n\right) = \mathcal{L}(\lambda V_n) = \chi^2_{n-1}$, it follows that

$$P\left[\frac{S_n V_n}{(n-1)} < \frac{n^2}{t^2_{n-1,\alpha/2}} \,\Big|\, H_0\right] = \alpha.$$

To prove (b) one has to obtain an estimate of $P\left[\dfrac{S_n V_n}{(n-1)} < \dfrac{n^2}{t^2_{n-1,\alpha/2}} \,\Big|\, H_A\right]$.
Thus for $0 < \varepsilon < \frac{1}{2}$

$$P\left[\frac{S_n V_n}{(n-1)} < \frac{n^2}{t^2_{n-1,\alpha/2}} \,\Big|\, H_A\right]$$

$$= P\left[\frac{S_n U_n}{n-1} < \frac{n^2}{t^2_{n-1,\alpha/2}} \,\Big|\, X \sim IG\left(\frac{na^2}{\psi \sigma^2}, \frac{a^2}{\sigma^2}\right)\right]$$

$$= P\left[\left(\frac{S_n}{\lambda}\right)\left(\frac{V_n}{n-1}\right) < \frac{n^2}{t^2_{n-1,\alpha/2}} \,\Big|\, S_n \sim IG\left(\frac{n^2\lambda}{\psi}, n^2\lambda\right), \lambda V_n \sim \chi^2\right]$$

$$\leq P\left[\frac{S_n}{\lambda} < \frac{n^2}{t^2_{n-1,\alpha/2}} \cdot \frac{1}{(1-\varepsilon)} \,\Big|\, S_n \sim IG\left(\frac{n^2}{\psi}, n^2]\lambda\right)\right]$$

$$+ P\left[\frac{\lambda V_n}{(n-1)} < (1-\varepsilon) \,\Big|\, \lambda V_n \sim \chi^2_{n-1}\right].$$

From (c) we know that:

$$\lim_{\varepsilon \to 0} \lim_{n \to \infty} P\left[\frac{S_n}{\lambda} < \frac{n^2}{(1-\varepsilon) t^2_{n-1,\alpha/2}} \,\Big|\, S_n \sim IG\left(\frac{n^2\lambda}{\psi}, n^2\lambda\right)\right]$$

$$\to \Phi\left(\sqrt{\frac{(n^2\lambda)(n^2\lambda)\psi^2}{\chi^2_{\alpha/2}(n^2\lambda)^2}} - \sqrt{\frac{n^2\lambda Z^2_{\alpha/2}}{n^2\lambda}}\right)$$

$$+ \exp\left(2\frac{n^2\lambda\psi}{n^2\lambda}\right) \Phi\left(-\sqrt{\frac{(n^2\lambda)^2\psi^2}{(n^2\lambda)^2}} - \sqrt{\frac{n^2\lambda Z_{\alpha/2}^2}{n^2\lambda}}\right)$$

$$= 1 - G(1; \psi, \alpha).$$

Using Chebychev's inequality it can be shown that

$$\lim_{n\to\infty} P(\lambda V_n < (n-1)(1-\varepsilon) \mid \lambda V_n \sim \chi_{n-1}^2) \to 0.$$

Thus

$$\limsup_{n\to\infty} P\left(\text{reject } H_0 \mid \nu = \frac{\psi\sigma^2}{na}\right) \leq 1 - G(1; \psi, a).$$

Likewise it can be established that

$$\liminf_{n\to\infty} P\left(\text{reject } H_0 \mid \nu = \frac{\psi\sigma^2}{na}\right) \geq 1 - G(1; \psi, a)$$

and hence that

$$\lim_{n\to\infty} P\left(\text{reject } H_0 \mid \nu = \frac{\psi\sigma^2}{na}\right) = 1 - G(1; \psi, a).$$

The sequential version of the fixed sample rule of the likelihood ratio test recommends that we stop taking observations when

$$N(n) = \min(N, n) \text{ observations are taken}$$

where N is the first $k \geq 2$ such that

$$\frac{V_k}{k-1} \geq \frac{n^2}{t_{n-1,\alpha/2}^2}.$$

The sequential rule says that H_0 is rejected if $N > n$. ♣

The next theorem due to Lombard provides the asymptotic distribution of $\frac{N(n)}{n}$.

Theorem 4.2 *For all $\psi \geq 0$*

$$\lim_{n\to\infty} P\left[\frac{N(n)}{n} \leq y \mid \nu = \frac{\psi\sigma^2}{na}\right] = G(y; \psi, \sigma)$$

for $0 \leq y \leq 1$.

The proof consists in showing that

$$\lim_{n\to\infty} P\left[N \leq [ny] \mid \nu = \frac{\psi\sigma^2}{na}\right] = G(y; \psi, \sigma) \text{ for all } y \geq 0.$$

A lower bound $K_1(n,d)$ and an upper bound $K_1(n,d) + K_2(n,d)$ are used for fixed $0 < d < y$ and it is shown that

$$\lim_{n \to \infty} K_1(n,d) = G(y; \psi, \sigma),$$

and

$$\lim_{d \to 0} \lim_{n \to \infty} K_2(n,d) = 0$$

Here

$$K_1(n,d) = P\left[\frac{S_k V_k}{k-1} \geq \frac{n^2}{t^2_{n-1,\alpha/2}} \text{ for some } [nd] \leq k \leq [ny] \mid \nu = \frac{\psi \sigma^2}{na}\right]$$

and

$$K_2(n,d) = P\left[\frac{S_k V_k}{k-1} \geq \frac{n^2}{t^2_{n-1,\alpha/2}} \text{ for some } 2 \leq k \leq [nd] \mid \nu = \frac{\psi \sigma^2}{na}\right].$$

The local asymptotic power of the sequential procedure is the same as that of the fixed sample procedure and equals $1 - G(1; \psi, \alpha)$ for $\nu = \frac{\psi \sigma^2}{an}$.

From Theorem 4.2 it is clear that $\frac{N(n)}{n}$ converges in distribution to a mixed type random variable Y with distribution function

$$P[Y \leq y] = \begin{cases} 0 & \text{for } y \leq 0 \\ G(y; \psi, \alpha) & \text{for } 0 < y < 1 \\ 1 & y \geq 1 \end{cases}$$

There is thus an atom $1 - G(1; \psi, \alpha)$ at $y = 1$.

The local relative efficiency (in the Pitman sense) of the fixed sample procedure to that of the sequential procedure is

$$e(\psi, \alpha) = \lim_{n \to \infty} n^{-1} \mathbb{E}\left(N(n) \mid \nu = \frac{\psi \sigma^2}{an}\right).$$

When $\psi = 0$ we have

$$e(\psi, \alpha) = 2 Z_{\alpha/2} \int_0^1 y \phi(-y Z_{\alpha/2}) dy$$

where $\phi(\cdot)$ is the standard normal density.

Letting $X = \dfrac{y^2 Z_{\alpha/2}}{2}$

$$e(0,\alpha) = \dfrac{1}{Z_{\alpha/2}}\sqrt{\dfrac{2}{\pi}}\int_0^{\frac{Z_{\alpha/2}^2}{2}} e^{-X}\,dx$$

$$= \sqrt{\dfrac{2}{\pi}}\dfrac{\left(1 - e^{-\frac{z_{\alpha/2}^2}{2}}\right)}{Z_{\alpha/2}}$$

For $\psi > 0$

$$e(\psi,\alpha) = 1 - G(1;\psi,\alpha)$$

since

$$\mathbb{E}\dfrac{N(n)}{n} = \int_0^1 [1 - P(Y \le y)]\,dy$$

$$= 1 - \int_0^1 dG(y;\psi,\alpha)$$

$$= 1 - G(1;\psi,\alpha).$$

When $\psi > 0$ and $\alpha \to 0$, $G(1;\psi,\alpha) \to 0$ while when $\psi = 0$ it can be shown that $\lim_{\alpha \to 0} e(0,\alpha) = (\pi \log \alpha^{-1})^{-\frac{1}{2}}$, thus establishing the asymptotic superiority of the sequential procedure under H_0.

Lombard has done some Monte Carlo studies to verify the speed of the approximation for the power when $\nu = 0$. Since the joint law of $\dfrac{S_k V_k}{k-1}$ does not depend on $\lambda \left(= \dfrac{a^2}{\sigma^2}\right)$, $P(\text{reject } H_0 \mid \nu = 0)$ and $\mathbb{E}(N(n) \mid \nu = 0)$ were computed by taking $\lambda = 1$. On the basis of 5000 independent runs he reports the results given in the Table 4.1.

Table 4.1 *Error probability estimates and expected sample sizes (for $\nu = 0$)*

n	α	$\hat{\alpha}$	$\hat{\mathbb{E}}(N(n))$
5	0.10	0.0922	3.02
	0.05	0.0492	2.81
	0.01	0.0076	2.44
10	0.10	0.0960	5.07
	0.05	0.0500	4.36
	0.01	0.0140	3.74

4.3 Tests with known coefficient of variation

Joshi and Shah (1990) have studied a sequential probability-ratio test of the inverse Gaussian mean where $\mathcal{L}(X_i) = IG\left(\mu, \frac{\mu}{a^2}\right)$, a being a known coefficient of variation. For testing $H_0 : \mu = \mu_0$ against $H_A : \mu = \mu_1$ it turns out that the maximum of the expected sample number occurs when the mean μ is approximately equal to the geometric mean of μ_0 and μ_1 and depends on μ_0 and μ_1 only through the ratio $\left(\frac{\mu_1}{\mu_0}\right)$ and not on a. We now describe the steps leading to this sequential analysis. Suppose that $\mathcal{L}(X_i) = IG\left(\mu, \frac{\mu}{a^2}\right)$ and $\mu_1 > \mu_0$. Then

$$Z = \log\left(\frac{f(X; \mu_1)}{f(X; \mu_0)}\right)$$
$$= \log\left[\frac{\sqrt{\mu_1}}{\sqrt{\mu_0}} \exp\left(\frac{(X-\mu_0)^2}{2a^2 X \mu_0} - \frac{(X-\mu_1)^2}{2a^2 X \mu_1}\right)\right]$$
$$= \frac{1}{2}\log\frac{\mu_1}{\mu_0} - \frac{1}{2a^2}\left(X\frac{\mu_0 - \mu_1}{\mu_0 \mu_1} + \frac{\mu_1 - \mu_0}{X}\right). \tag{4.7}$$

The sequential probability-ratio test gives rise to the following stopping rule for $0 < \beta < 1 < A$. Stop taking observations when N is such that $\sum_{i=1}^{N} Z_i \leq \log B$ or $\sum_{i=1}^{N} \geq \log A$. We can express this by saying that

$$N = \inf\left\{n \geq 1 \;\middle|\; \sum_{i=1}^{n} Z_i \leq \log B \text{ or } \sum_{i=1}^{n} Z_i \geq \log A\right\}.$$

By the Wald approximation rule A and B satisfy $A \leq \frac{1-\beta}{\alpha}$ and $B \geq \frac{\beta}{1-\alpha}$. Now $\sum_{i=1}^{n} Z_i \leq \log B$ implies that

$$\frac{n}{2}\log\frac{\mu_1}{\mu_0} + \frac{(\mu_1 - \mu_0)}{2a^2\mu_1\mu_0}\sum_{i=1}^{n} X_i - \frac{(\mu_1-\mu_0)}{2a^2}\sum\frac{1}{X_i} \leq \log B$$

or

$$\frac{1}{\mu_1\mu_0}\sum_{i=1}^{n} X_i - \sum_{i=1}^{n}\frac{1}{X_i} \leq \frac{2a^2}{\mu_1 - \mu_0}\left(\log B - \frac{n}{2}\log\frac{\mu_1}{\mu_0}\right).$$

Likewise, $\sum_{i=1}^{n} Z_i \geq \log A$ implies that

$$\frac{1}{\mu_1\mu_0}\sum_{i=1}^{n} X_i - \sum_{i=1}^{n}\frac{1}{X_i} \geq \frac{2a^2}{\mu_1 - \mu_0}\left(\log A - \frac{n}{2}\log\frac{\mu_1}{\mu_0}\right).$$

To obtain the operating characteristic function we note that $\mathbb{E}_\mu(L^{h(\mu)}) = \mathbb{E}_\mu(\exp h(\mu)Z)$ and we then have to solve

$$e^{\frac{1}{a^2}}\int_0^\infty \left(\frac{\mu_1}{\mu_0}\right)^{\frac{h(\mu)}{2}} \left(\frac{\mu}{2\pi a^2 x^3}\right)^{\frac{1}{2}} \exp\left(-\frac{Px}{2a^2} - \frac{Q}{2a^2 x}\right) dx = 1$$

where $P(\mu) = \frac{h(\mu)}{\mu_1} - \frac{h(\mu)}{\mu_0} + \frac{1}{\mu}$, and $Q(\mu) = h(\mu)(\mu_1 - \mu_0) + \mu$. When $P(\mu) \geq 0$ the integral can be evaluated and we obtain

$$\left(\frac{\mu_1}{\mu_0}\right)^{\frac{h(\mu)}{2}} \left(\frac{\mu}{Q(\mu)}\right)^{\frac{1}{2}} \exp\left(\frac{1}{a^2}\left(1 - \sqrt{P(\mu)Q(\mu)}\right)\right) = 1. \quad (4.8)$$

We note that when $\mu = \mu_0$, $h(\mu_0) = 1$ and when $\mu = \mu_1$, $h(\mu_1) = -1$. Taking logarithms of both sides of (4.8) and rearranging we have

$$\frac{a^2}{2} \log\left(\frac{\mu_1}{\mu_0}\right) h(\mu) + \frac{a^2}{2} \log \mu - \frac{a^2}{2} \log\{\mu + (\mu_1 - \mu_0)h(\mu)\}$$

$$= \left(\frac{1}{\mu} + \left(\frac{1}{\mu_1} - \frac{1}{\mu_0}\right) h(\mu)\right)^{\frac{1}{2}} \{(\mu + (\mu_1 - \mu_0)h(\mu)\}^{\frac{1}{2}}.$$

Differentiating both sides with respect to μ and substituting $h(\mu_0) = 1$ or $h(\mu_1) = -1$ we get

$$\left(\frac{a^2}{2}\left(\log\left(\frac{\mu_1}{\mu_0}\right) - \frac{\mu_1 - \mu_0}{\mu_0}\right) - \frac{a^2}{2} \frac{(\mu_1 - \mu_0)}{\mu_1 \mu_0}\right) h'(\mu_1)$$

$$= \left\{\frac{d}{d\mu}\sqrt{P(\mu)Q(\mu)}\right\}_{\mu=\mu_1}$$

$$= \frac{h'(\mu_1)(\mu_1 - \mu_0)}{2} \left(\frac{\sqrt{\frac{1}{\mu_0}}}{\sqrt{\mu_0}} - \frac{\sqrt{\mu_0}}{\sqrt{\frac{1}{\mu_0}}} \frac{1}{\mu_0 \mu_1}\right)$$

$$+ \frac{1}{2}\left(\frac{\sqrt{\frac{1}{\mu_0}}}{\sqrt{\mu_0}} - \frac{\sqrt{\mu_0}}{\sqrt{\frac{1}{\mu_0}}} \frac{1}{\mu_1^2}\right)$$

$$= \frac{(\mu_1 - \mu_0)^2}{2\mu_0 \mu_1} h'(\mu_1) + \frac{\mu_1^2 - \mu_0^2}{2\mu_0 \mu_1^2}.$$

Hence

$$h'(\mu_1) = \frac{a^2 \frac{\mu_1 - \mu_0}{\mu_0 \mu_1} + \frac{\mu_1^2 - \mu_0^2}{\mu_0 \mu_1^2}}{a^2 \left(\log\left(\frac{\mu_1}{\mu_0}\right) - \frac{\mu_1 - \mu_0}{\mu_0}\right) - \frac{(\mu_1 - \mu_0)^2}{\mu_0 \mu_1}} \quad (4.9)$$

and

$$h'(\mu_0) = \frac{a^2 \frac{\mu_1 - \mu_0}{\mu_0 \mu_1} + \frac{\mu_0^2 - \mu_1^2}{\mu_0^2 \mu_1}}{a^2 \left(\log\left(\frac{\mu_1}{\mu_0}\right) - \frac{\mu_1 - \mu_0}{\mu_0}\right) + \frac{(\mu_1 - \mu_0)^2}{\mu_0 \mu_1}}. \quad (4.10)$$

To see that $h'(\mu_1) < 0$ we note that the numerator > 0 and the coefficient of a^2 in the denominator is always < 0 for $\frac{\mu_1}{\mu_0} > 0$ ($\log \frac{\mu_1}{\mu_0} < \frac{\mu_1}{\mu_0} - 1$).

(We have assumed here that $\mu_1 > \mu_0$.) In a similar manner it can be argued that $h'(\mu_0) < 0$. This proves that $L(\mu)$ is strictly increasing in $h(\mu)$ and that when $\mu = \mu_0$ and $\mu = \mu_1$, $L(\mu)$ is decreasing. Thus the sequential test can also be extended to testing one-sided composite hypothesis $\mu \leq \mu_0$ against $\mu > \mu_0$.

Average sample number

To calculate $\mathbb{E}(N(\mu))$ we first compute $\mathbb{E}(Z)$. From (4.7) it follows that

$$\mathbb{E}(Z) = \frac{1}{2}\log\left(\frac{\mu_1}{\mu_0}\right) - \frac{(\mu_0 - \mu_1)}{2a^2\mu_0\mu_1}\mu - \frac{(\mu_1 - \mu_0)}{2a^2}\left(\frac{1}{\mu} + \frac{a^2}{\mu}\right).$$

Hence

$$\mathbb{E}(N(\mu)) = \frac{2(L(\mu)\log B + (1 - L(\mu))\log A)}{\log\left(\frac{\mu_1}{\mu_0}\right) + \left(1 - \frac{\mu_1}{\mu_0}\right) - \frac{1}{a^2}\left(\frac{\mu_0}{\mu_1} + \frac{\mu_1}{\mu_0} - 2\right)}.$$

Under H_0, $h(\mu_0) = 1$ and $\alpha = 1 - L(\mu)$ while under H_A, $h(\mu_1) = 1$ and $L(\mu) = 1 - \beta$. Therefore $\mathbb{E}_{H_0}(N(\mu))$ and $\mathbb{E}_{H_A}(N(\mu))$ are easily obtainable. Joshi and Shah show that the value of μ for which $\mathbb{E}(Z) = 0$, say μ_S is given by

$$\frac{\mu_0\mu_1}{2(\mu_0 - \mu_1)}\left(a^2\log\left(\frac{\mu_1}{\mu_0}\right) - \left\{a^4\left(\log\frac{\mu_1}{\mu_0}\right) + 4\frac{(\mu_1 - \mu_0)^2}{\mu_0\mu_1}(1 + a^2)\right\}^{\frac{1}{2}}\right)$$

which is approximately equal to $\sqrt{\mu_0\mu_1} = \tilde{\mu}$ for $\frac{\mu_0}{\mu_1}$ near unity. In practice, they claim that $\frac{\mu_0}{\mu_1}$ can deviate significantly from unity and still have $\mathbb{E}(Z \mid \tilde{\mu})$ close to zero, as shown in Table 4.2.

Table 4.2 Values of $\frac{\mu_0}{\mu_1}$ versus $\mathbb{E}(Z \mid \tilde{\mu})$

$\frac{\mu_0}{\mu_1}$	$\mathbb{E}(Z \mid \tilde{\mu})$
0.50	-0.006980
0.60	-0.002786
0.70	-0.000947
0.80	-0.000231
0.90	-0.000024
0.95	-0.000003

The maximum value of the average sample number occurs in the vicinity of μ_s where $\mathbb{E}(Z) = 0$ (Ghosh 1970, p. 125). Then

$$\mathbb{E}(N(\mu_s)) \approx \frac{\log A \log B}{\mathbb{E}(Z^2 \mid \mu_s)}.$$

The denominator when estimated at $\mu_s = \tilde{\mu}$ has the value (with $r = \frac{\mu_0}{\mu_1}$)

$$\mathbb{E}(Z^2 \mid \tilde{\mu}) = \frac{1}{4}(\log r)^2 + \frac{(4 + 3a^2)(r-1)^2}{4a^2 r} - \frac{(r-1)\log r}{2\sqrt{r}}.$$

The maximum of the average sample number can then be calculated for selected values of α, β, a and r. Joshi and Shah claim that $\mathbb{E}(N(\mu))$ calculated for $\mu = \mu_s$ and $\mu = \tilde{\mu}$ are in close agreement. Table 4.3 shows the values of $\mathbb{E}(N(\mu))$ for $\mu = \mu_0$, $\mu = \tilde{\mu}$ and $\mu = \mu_1$ for $\alpha = \beta = 0.5$ and $\mu_0 = 1$ (μ_1 is determined from $r = \frac{\mu_0}{\mu_1}$).

Table 4.3 *Values of the Average Sample Number*

		Values	of		ASN	
a	r	0.5	0.6	0.7	0.8	0.9
0.25	E_0	0.63	1.19	2.48	6.40	28.87
	$E_{\tilde{\mu}}$	1.05	1.97	4.08	10.50	47.28
	E_1	0.64	1.21	2.50	6.43	28.93
0.50	E_0	2.29	4.33	9.04	23.36	105.58
	$E_{\tilde{\mu}}$	3.85	7.82	14.98	38.53	173.39
	E_1	2.41	4.50	9.28	23.75	106.41
1.00	E_0	6.56	12.54	26.43	68.95	314.31
	$E_{\tilde{\mu}}$	11.55	21.67	44.95	115.59	520.18
	E_1	7.64	14.04	28.61	72.46	321.76
2.00	E_0	12.27	23.81	50.94	134.66	621.44
	$E_{\tilde{\mu}}$	23.11	43.34	89.90	231.19	1040.36
	E_1	16.65	29.86	59.57	148.69	651.24

4.4 Asymptotically risk-efficient sequential estimation

Chaturvedi et al., (1991) have developed a class of sequential procedures for point estimation of the parameter of a population under a family of loss functions together with a cost function. The method is applicable in the presence of an unknown nuisance parameter. A condition on the initial sample size necessary to ensure the asymptotic risk-efficiency is also given. We will not prove the general theorem which guarantees this procedure but will state it together with the preliminary assumptions made by the authors and then show it can be applied to the $IG(\mu, \lambda)$ model.

Assumptions

(a) Let $\{X_i\}_{i=1}$, $i = 1, \ldots$ be a sequence of independent identically distributed random variable from a d-variate ($d \geq 1$) absolutely continuous population $f(X; \Theta, \psi)$, where Θ is a $(d \times 1)$ vector of unknown parameters of interest, ψ is a scalar nuisance parameter assumed unknown. Thus $(\Theta, \psi) \in \mathbb{R}^d \times \mathbb{R}^+$,

(b) $X_1, \ldots, X_n$ is a random sample of size $n \ (\geq d+1)$.

(c) $\hat{\Theta}_n = \hat{\Theta}(X_1, \ldots, X_n)$ and $\hat{\psi}_n = \hat{\psi}(X_1, \ldots, X_n)$ are estimates of Θ and ψ respectively.

(d) There exist a known $(d \times d)$ positive definite matrix Q, a number $\delta \in (0, 1]$ and an integer $r \ (\geq 1)$ such that

$$n\psi^{-1}((\hat{\Theta}_n - \hat{\Theta})^t Q(\hat{\Theta}_n - \hat{\Theta}))^\delta \sim \chi_r^2.$$

(e) For all $n \geq d+1$, $\hat{\Theta}_n \perp\!\!\!\perp \hat{\psi}_n$.

(f) There exist integers $p \geq 1$, $q \geq 1$ such that for all $n \geq q+1$,

$$\frac{p(n-q)\hat{\psi}_n}{\psi} = \sum_{j=1}^{n-q} Z_j^{(p)}$$

where $Z_j^{(p)}$ are independent identically distributed random variables such that $\mathcal{L}(Z_g^{(p)}) = \chi_p^2$.

(g) $\hat{\psi}_n$ is a consistent estimator of ψ.

(h) The loss function is given by

$$L(\Theta, \hat{\Theta}) = A((\hat{\Theta}_n - \Theta)^t Q(\hat{\Theta}_n - \Theta))^\alpha + C_n^d$$

where A, α, C and t are known positive constants. Using (h) and (d) the risk can be shown to be

$$R_n(C) = A \left(\frac{\psi}{n}\right)^{\frac{\alpha}{\delta}} \mathbb{E}(\chi_r^2)^{\frac{\alpha}{\delta}} + C_n^d.$$

The value of n_0 which minimizes $R_n(C)$ is

$$n_0 = \left(\frac{\alpha}{\delta} \frac{A\Gamma\left(\frac{\alpha}{\delta} + \frac{\gamma}{2}\right)}{cd2^{\frac{\gamma}{2}}}\right)^{\frac{\delta}{(\alpha+\delta d)}} \psi^{\frac{d}{\alpha}+\delta d}$$

and the minimum risk is

$$R_{n_0}(C) = \left(1 + \frac{\delta t}{\alpha}\right) Cn_0^t.$$

The sequential procedure begins with a sample of size $m \geq \max(d+1, q+1)$. A stopping variable $N \equiv N(c)$ is introduced as the smallest-positive integer $n \geq m$ such that

$$n \geq \left(\frac{K}{Cd}\right)^{\frac{\delta}{a+\delta d}} \hat{\psi}_n^{\frac{\alpha}{a+\delta d}}, \quad \text{where } K = \frac{\alpha}{\delta} A \frac{\Gamma\left(\frac{\alpha}{\delta} + \frac{r}{2}\right)}{2^{\frac{r}{2}}}.$$

After stopping Θ is estimated by $\hat{\Theta}_N$. Chaturvedi et al., now estabilish several lemmas used to prove the following theorem.

Theorem 4.3

$$\lim_{c \to 0} \frac{R_N(c)}{R_{n_0}(C)} = \begin{cases} 1 & \text{if } m > q + \frac{2\alpha^2}{\delta p(\alpha+\delta d)} \\ 1+K & \text{if } m = q + \frac{2\alpha^2}{\delta p(\alpha+\delta d)} \\ \infty & \text{if } m < q + \frac{2\alpha^2}{\delta p(\alpha+\delta d)} \end{cases}$$

where $\frac{R_N(C)}{R_{n_0}(C)}$ is defined as the risk-efficiency of the procedure.

Application to $IG(\mu, \lambda)$

To apply Theorem 4.3 we first observe that $d = 1$, $\Theta = \mu$, $\psi = \frac{1}{\lambda}$, $\hat{\Theta}_n = \overline{X}_n$, $\tilde{\psi}_n = \frac{V}{n-1}$. We take $Q = \mu^{-3}$, $\delta = 1$ and verify at once that assumptions (d), (e), (f) and (g) hold when we have $r = p = q = 1$. The loss is

$$L(\mu, \overline{X}) = A \left(\frac{(X - mu)^2}{\mu^3} \right)^s + Cn^d$$

for some positive constants A, C and s. Then

$$R_n(C) = \left(\frac{K}{s}\right)(\lambda n)^{-s} + Cn^d$$

$$n_0 = \left(\frac{K}{Cd}\right)^{\frac{1}{s+d}} \lambda^{-\frac{s}{s+d}},$$

$$R_{n_0}(C) = \left(1 + \frac{d}{s}\right) Cn_0{}^d,$$

and

$$R_N(C) = \left(\frac{Cd}{s}\right) n_0^d \mathbb{E}\left(\frac{n_0}{N}\right) + C\mathbb{E}(N^d).$$

Two key lemmas used in the proof of Theorem 4.3 state that $\lim_{c \to 0} \mathbb{E}\left(\frac{N}{n_0}\right)^{\lambda} \equiv 1$ for fixed $\lambda > 0$ and that

$$\lim_{c \to 0} \mathbb{E}\left(\frac{n_0}{N}\right)^{\lambda} = \begin{cases} 1 & \text{if } m < q + \frac{2\alpha\lambda}{p(\alpha+\delta d)}, \\ 1+K & \text{if } m = q + \frac{2\alpha\lambda}{p(\alpha+\delta d)}, \\ \infty & \text{if } m > q + \frac{2\alpha\lambda}{p(\alpha+\delta d)}, \end{cases}$$

$$\lim_{c \to 0} \frac{R_N(C)}{R_{n_0}(C)} = \begin{cases} 1 & \text{if } m > 1 + \frac{2s^2}{s+d}, \\ 1+K & \text{if } m = 1 + \frac{2s^2}{s+d}, \\ \infty & \text{if } m < 1 + \frac{2s^2}{s+d}. \end{cases}$$

4.5 Control charts

Control charts are a device for monitoring the stability of process output. They provide a visual display of the changes in the process mean and dispersion as measured by the sample estimate of the population parameters. A control chart is typically based on normality assumption of the process output or often when the Central Limit Theorem is applicable. For small samples, however, the Central Limit Theorem is no longer valid. Edgeman (1989) presents inverse Gaussian control charts and discuses their sensitivity through simulation studies. His findings are discussed in the next section. We assume that the process output follows an $IG(\mu, \lambda)$ law and that M samples of n items each have been selected from the process when it is in control. Denote by $\overline{X}_j$ the jth sample mean and by V_j the value of V for the jth sample. Further let

$$\overline{\overline{X}} = \frac{1}{M} \sum_{i=1}^{M} \overline{X}_i, \quad \overline{V} = \frac{1}{M} \sum_{i=1}^{M} V_i.$$

M is chosen between 20 and 25. The idea now is to use $\overline{\overline{X}}$ in the expression for the confidence intervals for μ. Writing LCL and UCL for the lower and upper control limits respectively we have

$$\overline{\overline{X}} \left[\max \left\{ 0, 1 - Z_{1-\frac{\alpha}{2}} \sqrt{\overline{\overline{X}}/n\lambda} \right\} \right]^{-1} = \text{UCL}$$

and

$$\overline{\overline{X}} \left[\max \left\{ 0, 1 + Z_{1-\frac{\alpha}{2}} \sqrt{\overline{\overline{X}}/n\lambda} \right\} \right]^{-1} = \text{LCL}$$

when λ is known, and

$$\overline{\overline{X}} \left[\max \left\{ 0, 1 - t_{1-\frac{\alpha}{2}} \sqrt{\overline{\overline{X}}\, \overline{V}/n(n-1)} \right\} \right]^{-1} = \text{UCL}$$

and

$$\overline{\overline{X}} \left[\max \left\{ 0, 1 + t_{1-\frac{\alpha}{2}} \sqrt{\overline{\overline{X}}\, \overline{V}/n(n-1)} \right\} \right]^{-1} = \text{LCL}$$

when λ is unknown. This is allowed since $\overline{\overline{X}}$ is still uniformly minimum variance unbiased for μ and $\overline{V}$ is a linear unbiased estimator of λ^{-1}.

The centre line on each chart is given by $\overline{\overline{X}}$ and the charts are then plotted in standard fashion. Edgeman cautions that when the LCL becomes infinite, one should construct charts for the reciprocal process centrality. The centre line for this chart will be $\left(\frac{1}{\overline{\overline{X}}}\right)$, the UCL will be the reciprocal of the LCL and the new LCL the reciprocal of the old UCL. Thus for example the UCL for the reciprocal process centrality is

$$\frac{1}{\overline{\overline{X}}} + \frac{Z_{1-\alpha/2}}{\sqrt{n\lambda \overline{\overline{X}}}}.$$

These limits can be obtained directly by examining the confidence limits on $\frac{1}{\mu}$ control charts for the process dispersion is based on substituting $\overline{V}$ for V and using the fact that $\mathcal{L}(\lambda \overline{V}) = \chi^2_{n-1}$. The charts that monitor the reciprocal process dispersion will have (see Table 4.1) confidence interval limits for λ, μ (unknown) as

$$\text{LCL} = \overline{V}\chi^2_{\frac{\alpha}{2}}, \quad \text{UCL} = \overline{V}\chi^2_{1-\frac{\alpha}{2}}. \tag{4.11}$$

Designating this chart as the V chart, one observes that, it has the blemish that the 'center line' is not a true center line because of the asymmetry of the χ^2 law. If the process is in control, the values of V will fluctuate around $\overline{V}$. The control chart is constructed by positioning horizontal lines at the control limits given by (4.11) and a center line at $\overline{V}$. Observed values from sucessive samples are then plotted on the vertical axis of the chart corresponding to the sample number on the horizontal axis. Kappenman (1979) has shown that $\overline{X}$ and $\overline{V}$ are independent and hence the process should be monitored by simultaneous maintenance of an $\overline{X}$ ($\frac{1}{\overline{\overline{X}}}$) chart for centrality and a V chart for process dispersion. A Monte Carlo study done by Edgeman examines the sensitivity of the $\frac{1}{\overline{X}}$ and V charts instead of the $\overline{X}$ and R charts. His results indicate that for small λ/μ the $\overline{X}$ chart gives progressively better (yet still poor) results with increasing n. The trend for the R chart is in the opposite direction. The $\frac{1}{\overline{X}}$ and V charts are generally recommended for small $\phi = \frac{\lambda}{\mu}$.

CHAPTER 5

RELIABILITY AND SURVIVAL ANALYSIS

5.0 Introduction

We now consider the $IG(\mu, \lambda)$ law from the point of view of modelling for reliability and survival analysis. The reliability of a system at time t is defined as the probability of the system lasting at least until a time t. Thus if X represents failure time, then symbolically $R(t)$ the reliability is given by $P(X \geq t)$. Since the random variable X has a distribution indexed by a parameter θ it is more convenient to write $R(t; \theta)$ for the reliability function. For the IG distribution we will write from now on $R(t; \mu, \lambda)$ to denote this function. Since

$$F(t) = \Phi\left[\sqrt{\frac{\lambda}{t}}\left(\frac{t}{\mu} - 1\right)\right] + \exp\left(\frac{2\lambda}{\mu}\right)\Phi\left[-\sqrt{\frac{\lambda}{t}}\left(1 + \frac{t}{\mu}\right)\right],$$

$$R(t; \mu, \lambda) = \Phi\left[\sqrt{\frac{\lambda}{t}}\left(1 - \frac{t}{\mu}\right)\right] - \exp\left(\frac{2\lambda}{\mu}\right)$$

$$\Phi\left[-\sqrt{\frac{\lambda}{t}}\left(1 + \frac{t}{\mu}\right)\right]. \qquad (5.1)$$

The estimation of $R(t; \mu, \lambda)$ depends on the status of (μ, λ). One method which guarantees an unbiased estimate with smallest variance is based on the Rao-Blackwell principle.

5.1 Estimation of Reliability

(i) Point estimate of $R(t; \mu, \lambda)$: (λ known, μ unknown).

In this case the minimum variance unbiased estimate of $R(t; \mu, \lambda)$ is based on the conditional law of X_i (say X_1) given the compelete sufficient statistic for μ. We use Theorem 1.8 to obtain $\hat{R}(t; \mu, \lambda)$ as

$$\hat{R}(t; \mu, \lambda) = \int_t^{n\bar{x}} p(x_1 \mid \bar{x}) dx_1.$$

Estimation of Reliability

A change of variable $Y = \dfrac{\sqrt{n\lambda}(x_1 - \bar{x})}{\sqrt{x_1 \bar{x}(n\bar{x} - x_1)}}$ transforms the integral into the range (y^*, ∞) and permits the evaluation of the transformed integral in terms of the standard normal distribution. Thus

$$\hat{R}(y^*; \mu, \lambda) = \int_{y^*}^{\infty} \left(1 - \frac{(n-2)y\sqrt{\bar{x}}}{\sqrt{4n(n-1)\lambda + n^2 \bar{x} y^2}}\right) \phi(y) \, dy$$

$\phi(y)$ being the standard normal density, and $y^* = \dfrac{\sqrt{n\lambda}(t - \bar{x})}{\sqrt{t\bar{x}(n\bar{x} - t)}}$. If we let $z = \sqrt{y^2 + \dfrac{4n(n-1)\lambda}{n\bar{x}}}$

$$\int_{y^*}^{\infty} \frac{y\sqrt{\bar{x}}\phi(y)}{\sqrt{n(n-1)\lambda + n^2 \bar{x} y}} dy = e^{\frac{2(n-1)\lambda}{n\bar{x}}} \Phi\left(-\sqrt{\frac{4n(n-1)\lambda}{n\bar{x}} + y^{*2}}\right).$$

Then we obtain

$$\hat{R}(x; \mu, \lambda) = \mathbb{E}\left(P(X_1 > x \mid \bar{X})\right)$$

$$\bar{R}(t; \mu, \lambda) = \Phi(-y^*) - \frac{n-2}{n} \exp\left(\frac{2(n-1)\lambda}{n\bar{x}}\right)$$

$$\times \Phi\left(-\sqrt{\frac{4(n-1)\lambda}{n\bar{x}} + y^{*2}}\right)$$

$$= R_1(y^*), \quad (\text{say})$$

for $y^* \in \mathbb{R}$. Thus we have the desired estimate as

$$\hat{R}(t; \mu, \lambda) = \begin{cases} 1 & t < 0 \\ R_1(y^*) & 0 < t < n\bar{x} \\ 0 & t > n\bar{x}. \end{cases}$$

(ii) Point estimate of $R(t; \mu, \lambda)$: (λ unknown, μ known)

We use Theorem 1.9 and the density $f(x_1 \mid z)$. The clever idea is to convert the integral to one involving a weighted Student's t. This is achieved by writing

$$Y = \frac{\sqrt{n-1}(X_1 - \mu)}{\sqrt{ZX_1}} \left(1 - \frac{(X_1 - \mu)^2}{ZX_1}\right)^{-\frac{1}{2}}$$

where $Y \in \mathbb{R}$. It turns out that $X_1 \in (L, \mu)$ when $Y \in (-\infty, 0)$ and $X_1 \in (\mu, U)$ when $Y \in (0, \infty)$ where

$$L = \frac{1}{2}\left(2\mu + z - \sqrt{z^2 + 4\mu z}\right), \quad U = \frac{1}{2}\left(2\mu + z + \sqrt{z^2 + 4\mu z}\right).$$

After a tedious computation one obtains

$$\hat{R}(t;\mu,\lambda) = \begin{cases} 1 & t < L \\ S_{n-1}(a(t)) - \left(\frac{z+4\mu}{z}\right)^{\frac{n}{2}-1} S_{n-1}(b(t)) & L < t < U \\ 0 & t > U \end{cases}$$

where

$$a(t) = \frac{\sqrt{n-1}(t-m)}{\sqrt{zt-(t-m)^2}}, \quad b(t) = \frac{\sqrt{n-1}(t+m)}{\sqrt{zt-(t-m)^2}}.$$

(iii) Point estimate of $R(t;\mu,\lambda)$: (μ, λ unknown)

In this case we use Theorem 1.10 and an appropriate transformation to obtain tail areas of Student's t as in the previous case. We omit the cumbersome details since they are of mere academic interest. All these estimates are due to Chhikara and Folks (1974).

(iv) Bayes estimate

Padgett (1979) has presented Bayes estimators of $R(t;\mu,\lambda)$ for the case when μ is known. He employs vague priors as well as a conjugate family of priors. However when both μ and λ are unknown an estimator is proposed which, besides its ease of computation, seems to perform well for large t. The likelihood based on the data $X = (x_1, \ldots, x_n)$ is preportional to

$$l(\mu,\lambda) \propto \lambda^{\frac{n}{2}} \exp\left[\left(\frac{n\lambda}{\mu}\right) - \frac{n\lambda}{2\mu^2}\bar{x} - \frac{\lambda}{2}\sum x_i^{-1}\right].$$

Now the Fisher information matrix has determinant $(\lambda\mu^3)^{-1}$ so that Jeffrey's prior is given by

$$\pi(\mu,\lambda) = (\lambda\mu^3)^{-\frac{1}{2}}.$$

Since this prior does not yield tractable results one resorts to the non-informative prior (improper) $\pi(\lambda) \propto \lambda^{-1}$ for known μ to obtain the following posterior distribution.

$$f(\lambda \mid x,\mu) = C\lambda^{\frac{n}{2}-1} \exp\left[-\frac{\lambda}{2\mu^2}\sum_{i=1}^{n}\frac{(x_i-\mu)^2}{x_i}\right]$$

where

$$C = \frac{1}{\Gamma\left(\frac{n}{2}\right)}\left[\frac{1}{2\mu^2}\sum_{i=1}^{n}\frac{(x_i-\mu)^2}{x_i}\right]^{\frac{n}{2}}.$$

It is clear that $f(\lambda \mid x,\mu)$ is a gamma law of the form

$$\frac{\beta^{-\alpha}\lambda^{\alpha-1}\exp(-\lambda/\beta)}{\Gamma(\alpha)} 1_{\mathbf{R}^+}(\lambda)$$

where $\alpha = \frac{n}{2}$, $\beta = \frac{2\mu^2}{\sum_{i=1}^{n}\frac{(x_i-\mu)^2}{x_i}}$. For a squared error loss, routine calculations give the Bayes estimator of λ as

$$\hat{\lambda}_B = n\mu^2 \left(\sum_{i=1}^{n} \frac{(x_i - \mu)^2}{x_i} \right)^{-1}.$$

The Bayes estimator of $R(t; \mu, \lambda)$ with respect to the improper prior and a squared error is then

$$\hat{R}_B(t) = \mathbb{E}_\Lambda \left(R(t; \mu, \Lambda) \mid \underline{x} \right).$$

Thus we need to compute

$$\mathbb{E}_\Lambda \left[\Phi \left\{ \left(\frac{\Lambda}{t} \right)^{\frac{1}{2}} \left(1 - \frac{t}{\mu} \right) \right\} \right] -$$

$$\mathbb{E}_\Lambda \left[\exp\left(\frac{2\Lambda}{\mu} \right) \Phi \left\{ \left(-\frac{\Lambda}{t} \right)^{\frac{1}{2}} \left(1 - \frac{t}{\mu} \right) \right\} \right].$$

The first term (from Lemma 1 of Padgett and Wei (1977) has the value

$$P\left[T_{2\alpha} < \left(1 - \frac{t}{\mu} \right) \left(\frac{\alpha\beta}{t} \right)^{\frac{1}{2}} \right]$$

where α and β are as defined previously and $T_{2\alpha}$ is a Student 't' random variable with $2\alpha = n$ degrees of freedom. Likewise the second term yields

$$\left(1 - \frac{2\beta}{\mu} \right)^{-\alpha} P\left[T_{2\alpha} < -\left(1 - \frac{t}{\mu} \right) \left(\frac{\alpha\beta\mu}{t(\mu - 2\beta)} \right)^{\frac{1}{2}} \right].$$

These two probabilities are easily evaluated from the tables of a 't_n' and $\hat{R}_B(t)$ can be obtained. One could use a more general gamma family as the prior for λ and obtain parallel results. When λ and μ are unknown one can obtain a modified Bayes estimator by using $\overline{X}$ as the estimator of μ. Padgett provides results based on Monte Carlo simulation which give the average estimated reliability, the average squared error for the estimators based on the maximum likelihood, the minimum variance unbiased estimator, and the modified estimator using $\overline{X}$ in place of μ. The simulated results suggest that the last estimator behaves conservatively with a larger average squared error for small t but improves considerably for large t.

Example 3.1 (continued) For the von Alven data the estimates of reliablility are given in the following table.

Table 5.1 *Estimates of Reliability*

t	1	2	3	5	10	15
$\tilde{R}_B(t)$	0.6934	0.4578	0.33305	0.1984	0.0789	0.0388
MLE	0.6986	0.4607	0.3325	0.1996	0.0791	0.0386
MVUE	0.6951	0.4618	0.3368	0.1057	0.0829	0.0396

5.2 Confidence bounds and tolerance limits

The problem of confidence intervals has been investigated by Padgett (1979) and more recently by Goh et al., (1989) and Tang and Chang (1993). We will follow the trail of the latter authors where the development is based on optimization methods. When the reliability function is monotone in the parameter, confidence bounds are obtained by using the idea of equivariant confidence sets as proposed in Lehmann (1986). In the case of $IG(\mu, \lambda)$, unfortunately, $R(t; \mu, \lambda)$ is not monotone with respect to λ, although it is monotone with respect to μ. Nevertheless, in realistic situations it is highly improbable that any one of the parameters is known to the experimenter. Goh et al., have proposed lower confidence bounds for $R(t; \mu, \lambda)$ when the version of the IG law is parametrized in the following form

$$f(t; \delta, \sigma) = \frac{1}{\sqrt{2\pi\sigma^2 t^3}} \exp\left(-\frac{(\delta t - 1)^2}{2\sigma^2 t}\right) 1_{\mathbf{R}^+}(t) \qquad (5.2)$$

where $\delta = \frac{1}{\mu}$ and $\sigma^2 = \frac{1}{\lambda}$, $\sigma > 0, \delta > 0$. In the model (5.2) the parameters δ and σ^2 carry meaningful interpretations with respect to the Weiner process assumptions underlying the IG law. Thus if we are dealing with wear-out models or fatigue failure models, $X(t)$ is a Weiner process with drift ν and variance parameter σ'^2 so that $X(t) \sim N(\nu t, \sigma'^2 t)$. If $X(t)$ is measured at fixed time intervals Δt we are dealing with a Wiener process with stationary independent increments and $X_i(\Delta t) = X(t_i) - X(t_{i-1})$ where $\Delta t = t_i - t_{i-1}$, $i = 1, \ldots, n$. (One can take $\Delta t = 1$.) As $X(t)$ changes with time, failure will be defined as the instant when the process $X(t) > a$ called a barrier or threshold value. In this case the reliability can be shown to be (Goh et al., 1989)

$$\Phi\left(\frac{a}{\sigma'\sqrt{t}}\left(1 - \frac{\nu t}{a}\right)\right) - \exp\left(\frac{2a\nu}{\sigma'^2}\right) \Phi\left(-\frac{a}{\sigma'\sqrt{t}}\left(1 + \frac{\nu t}{a}\right)\right).$$

Writing $\delta = \frac{\nu}{a}, \sigma = \frac{\sigma'}{a}$

$$R(t; \delta, \sigma) = \Phi\left(\frac{1 - \delta t}{\sigma\sqrt{t}}\right) - \exp\left(\frac{2\delta}{\sigma^2}\right) \Phi\left(-\frac{1 + \delta t}{\sigma\sqrt{t}}\right). \qquad (5.3)$$

We state the following theorem due to Tang and Chang which describes a method of construction of confidence intervals for $R(t; \delta, \sigma)$ as well as $R(t; \mu, \lambda)$.

Theorem 5.1 (a) A $100(1-\alpha_1)(1-\alpha_2)\%$ confidence interval for $R(t; \delta, \sigma)$ is given by the interval

$$\left[\inf_{\underline{\sigma}\leq\sigma\leq\overline{\sigma}} R(t; \overline{\delta}, \sigma), \sup_{\underline{\sigma}\leq\sigma\leq\overline{\sigma}} R(t; \underline{\delta}, \sigma)\right]$$

for each t, where

$$(\underline{\delta}, \overline{\delta}) = \left(\hat{\delta} - \frac{s}{\sqrt{n}}t_{n-1,1-\frac{\alpha_1}{2}}, \hat{\delta} + \frac{s}{\sqrt{n}}t_{n-1,1-\frac{\alpha_1}{2}}\right)$$

$$(\underline{\sigma}, \overline{\sigma}) = \left(\sqrt{\frac{(n-1)s^2}{\chi^2_{n-1,1-\frac{\alpha_2}{2}}}}, \sqrt{\frac{(n-1)s^2}{\chi^2_{n-1,\frac{\alpha_2}{2}}}}\right)$$

and

$$s^2 = \frac{1}{n-1}\sum(X_i - \mu)^2.$$

(b) A $(1-\alpha_1-\alpha_2+\alpha_1\alpha_2)$ level confidence interval for $R(t; \mu, \lambda)$ in (5.1) is given by the interval

$$\left[\inf_{\underline{\lambda}\leq\lambda\leq\overline{\lambda}} R(t; \mu, \lambda), \sup_{\underline{\lambda}\leq\lambda\leq\overline{\lambda}} R(t; \overline{\mu}, \lambda)\right]$$

for each t, where

$$(\underline{\mu}, \overline{\mu}) = \left[\left\{\frac{1}{\overline{X}} + \sqrt{\frac{V}{n(n-1)\overline{X}}}t_{1-\frac{\alpha_1}{2}}\right\}^{-1}, \left\{\frac{1}{\overline{X}} - \sqrt{\frac{V}{n(n-1)\overline{X}}}t_{1-\frac{\alpha_1}{2}}\right\}^{-1}\right]$$

and

$$(\underline{\lambda}, \overline{\lambda}) = \left[\frac{\chi^2_{1-\frac{\alpha_2}{2}}}{V}, \frac{\chi^2_{\frac{\alpha_2}{2}}}{V}\right].$$

Proof The confidence coefficient $(1-\alpha_1)(1-\alpha_2)$ of an interval $(\underline{T}, \overline{T})$ for $q(\theta)$ is defined as the largest possible confidence level, that is to say

$$\inf_\theta P\left(\underline{T}(X) \leq q(\theta) \leq \overline{T}(X)\right)$$

giving the minimum probability of coverage (see Bickel and Doksum, 1977). Therefore, one has to show that

$$P\left(\inf_{\underline{\sigma}\leq\sigma\leq\overline{\sigma}} R(t;\overline{\delta},\sigma) \leq R(t;\delta,\sigma) \leq \sup_{\underline{\sigma}\leq\sigma\leq\overline{\sigma}} R(t;\underline{\delta},\sigma)\right) \geq (1-\alpha_1)(1-\alpha_2).$$

To do so we first observe that $R(t;\delta,\sigma)$ is monotone decreasing in δ, since

$$\frac{\partial R}{\partial \delta} = -\frac{2}{\sigma^2}\exp\left(\frac{2\delta}{\sigma^2}\right)\Phi\left(-\frac{1+\delta t}{\sigma\sqrt{t}}\right) \leq 0, \quad \forall \sigma, t.$$

Now we apply the **Kuhn-Tucker** optimality criterion, which we state below. Let x^* be a relative minimum point for the problem –minimize $f(x)$ subject to $h(x) = 0$, $g(x) \leq 0$ and suppose that x^* is a regular point for the constraints (i.e. $\nabla h_i(x^*)$, $\nabla g_j(x^*)$, $1 \leq i \leq m$, $j \in J$ a set of indices for which $g_j(x^*) = 0$, are linearly independent). Then there exists a vector $\lambda \in E$ and a vector $\mu \in E^p$ with $\mu \geq 0$ such that

$$\nabla f(x^*) + \lambda^t \nabla h(x^*) + \mu^t \nabla g(x^*) = 0$$
$$\mu^t \nabla g(x^*) = 0.$$

(Here g is a p-dimensional function.) We then have

$$\inf_{\substack{\underline{\delta}\leq\delta\leq\overline{\delta}\\\underline{\sigma}\leq\sigma\leq\overline{\sigma}}} R(t;\delta,\sigma) = \inf_{\underline{\sigma}\leq\sigma\leq\overline{\sigma}} R(t;\overline{\delta},\sigma)$$

and

$$\sup_{\substack{\underline{\delta}\leq\delta\leq\overline{\delta}\\\underline{\sigma}\leq\sigma\leq\overline{\sigma}}} R(t;\delta,\sigma) = \sup_{\underline{\sigma}\leq\sigma\leq\overline{\sigma}} R(t;\underline{\delta},\sigma).$$

Denoting by $A(\delta,\sigma) = \{\underline{\delta}\leq\delta\leq\overline{\delta},\ \underline{\sigma}\leq\sigma\leq\overline{\sigma}\}$, we get

$$\inf_{A(\delta,\sigma)} R(t;\delta,\sigma) \leq R(t;\delta,\sigma) \leq \sup_{A(\delta,\sigma)} R(t;\delta,\sigma).$$

An equivariant set of $A(\delta,\sigma)$ is $\bar{A}(\delta,\sigma)$ given by

$$\left\{\left(\hat{\delta}\frac{\sigma}{\sqrt{n}}Z_{1-\frac{\alpha_1}{2}} < \hat{\delta} < \delta - \frac{\sigma}{\sqrt{n}}Z_{1-\frac{\alpha_1}{2}}\right),\right.$$

$$\left.\left(\sqrt{\frac{\sigma^2}{n}\chi^2_{n-1,\frac{\alpha_2}{2}}} \leq \hat{\sigma} \leq \sqrt{\sigma^2\chi^2_{n-1,1-\frac{\alpha_2}{2}}}\right)\right\}.$$

Therefore

$$P\left(\inf_{A(\delta,\sigma)} R(t;\delta,\sigma) \le R(t;\delta,\sigma) \le \sup_{A(\delta,\sigma)} R(t;\delta,\sigma)\right)$$
$$\ge P(A(\hat\delta,\mu)) = P(\bar A(\hat\delta,\mu)) = (1-\alpha_1)(1-\alpha_2)$$

since $\hat\delta \perp\!\!\!\perp \hat\sigma$. The proof of (b) is analogous to the above. First we verify that $\frac{\partial R}{\partial \mu} \ge 0$ for all $\lambda > 0$ and $t \ge 0$ and then use the fact $\hat\mu \perp\!\!\!\perp \hat\lambda$.

As observed by Tang and Chang, over the region where $R(t;\delta,\sigma)$ ($R(t;\mu,\lambda)$) is monotone with respect to both parameters

$$\inf_{\delta,\sigma} P\left(\inf_{A(\delta,\sigma)} R(t;\delta,\sigma) \le R(t;\delta,\sigma) \le \sup_{A(\delta,\sigma)} R(t;\delta,\sigma)\right)$$
$$= P\left(A(\delta,\sigma)\right) = (1-\alpha_1)(1-\alpha_2)$$

thus reinforcing the efficiency of the procedure. For the density (5.2) leading to $R(t;\delta,\sigma)$ note that $R(t;\delta,\sigma)$ is monotone decreasing in σ for all $t \le \frac{1}{\delta}$ just as $R(t;\mu,\lambda)$ is monotone increasing in λ for all $t \le \mu$. Therefore we have

$$t \le \frac{1}{\delta} \Rightarrow \inf_{\underline\sigma \le \sigma \le \overline\sigma} R(t;\overline\delta,\sigma) = R(t;\overline\delta,\overline\sigma)$$
$$\text{and} \quad \sup_{\underline\sigma \le \sigma \le \overline\sigma} R(t;\underline\delta,\sigma) = R(t;\underline\delta,\underline\sigma).$$

Likewise

$$t \le \mu \Rightarrow \inf_{\underline\lambda \le \lambda \le \overline\lambda} R(t;\underline\mu,\lambda) = R(t;\underline\mu,\underline\lambda)$$
$$\text{and} \quad \sup_{\underline\lambda \le \lambda \le \overline\lambda} R(t;\overline\mu,\lambda) = R(t;\overline\mu,\overline\lambda). \quad \clubsuit$$

When $t > \mu$ the authors recommend a graphical approach. Either we plot $R(t;\delta,\sigma)$ against $\sigma/\sqrt{\delta}$ for several values of δt or we plot $R(t;\mu,\lambda)$ against λ/μ for different values of t/μ (Figure 5.1, Figure 5.2). Padgett gives an approximate point of change from increasing $R(t;\mu,\lambda)$ to decreasing $R(t;\mu,\lambda)$ when μ is known as $\lambda_0(t,\mu) = 2t^2 \Big/ \left(1+\frac{t}{\mu}\right)^2 (t-\mu)$. Using the graphical method, how does one obtain the lower $(\lambda_L^*(t))$ and upper $(\lambda_U^*(t))$ bounds? In Figure 5.2 concentrate on the range $(\underline\lambda/\mu, \overline\lambda/\mu)$ and choose that one which gives the smallest $R(t;\mu,\lambda)$. This determines $\lambda_L^*(t)$. Similarly considering the range $(\underline\lambda/\overline\mu, \overline\lambda/\overline\mu)$ choose that one giving the maximum $R(t;\mu,\lambda)$, to obtain $\lambda_U^*(t)$. A similar strategy is employed to determine $\sigma_L^*(t)$ and $\sigma_U^*(t)$.

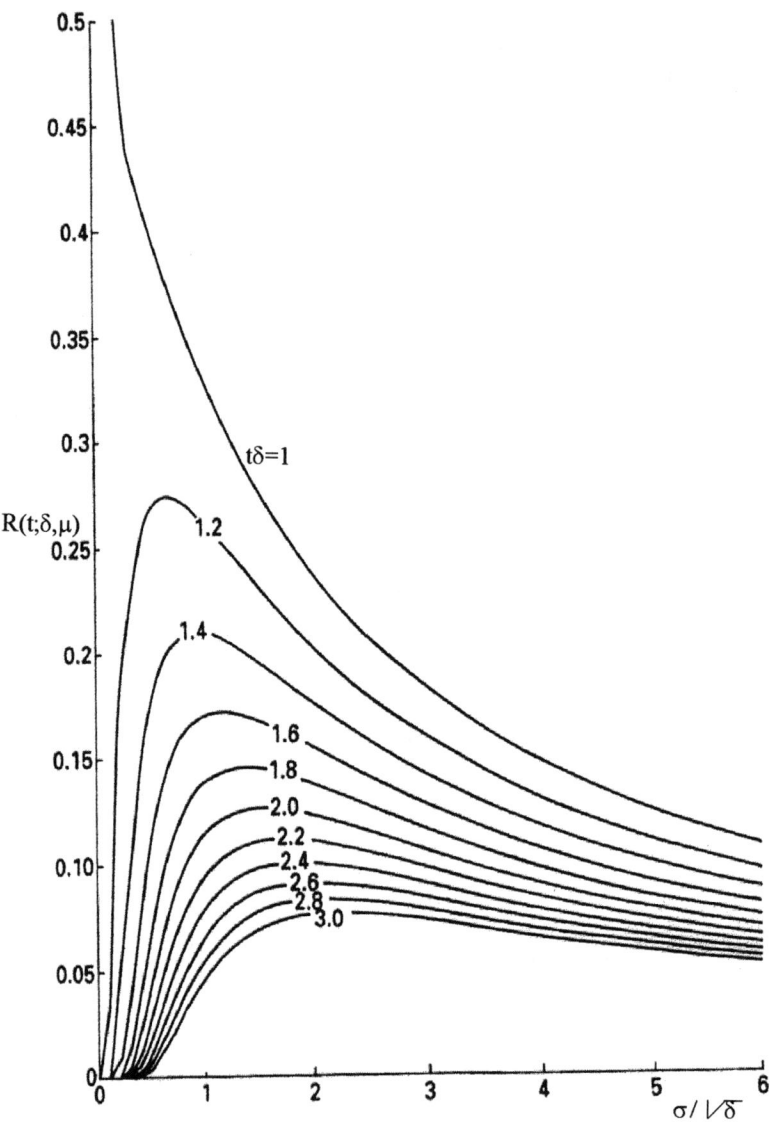

Figure 5.1 $R(t)$ as a fucntion of $\sigma/\sqrt{\delta}$

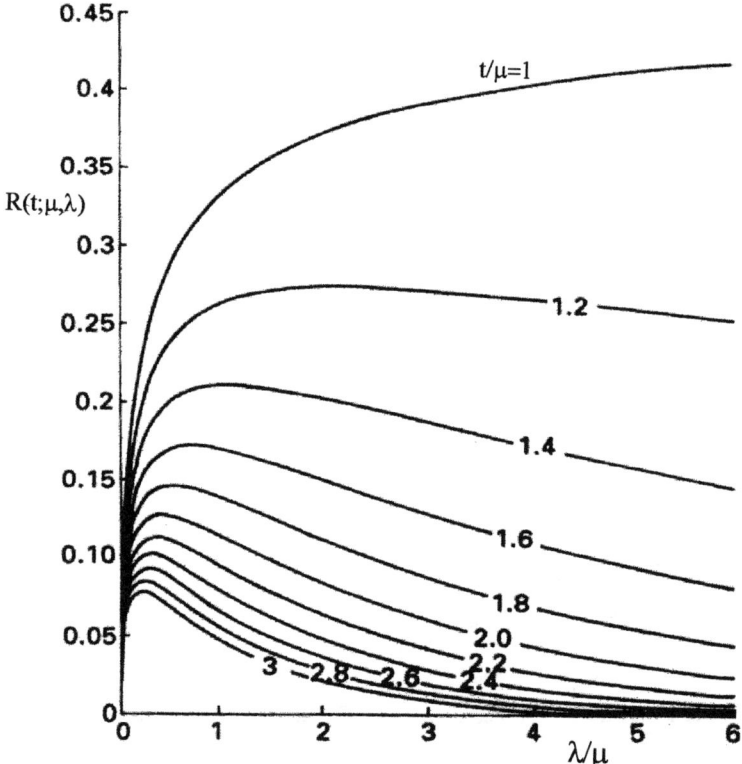

Figure 5.2 $R(t)$ as a fucntion of λ/μ

Example 3.1 (continued)

For the von Alven data, $n = 46$, $\bar{\mu} = 3.61$ and $\bar{\lambda} = 1.66$. Taking $\alpha_1 = \alpha_2 = 0.05$, 95% confidence limits for μ and λ are respectively

$$\mu \in [2.500; 6.471], \quad \lambda \in [1.023, 2.359].$$

To determine 90.25% confidence bounds for $R(t; \mu, \lambda)$ one has to minimize $R(t; 2.500, \lambda)$ subject to $\lambda \in [1.023, 2.359]$ iteratively for some fixed t, thus obtaining the lower bound; and then maximizing $R(t; 6.471, \lambda)$ subject to the same constraint obtain the upper bound (see Figure 5.3). The same procedure can be adapted to obtain tolerance limits as in the case of the distribution function.

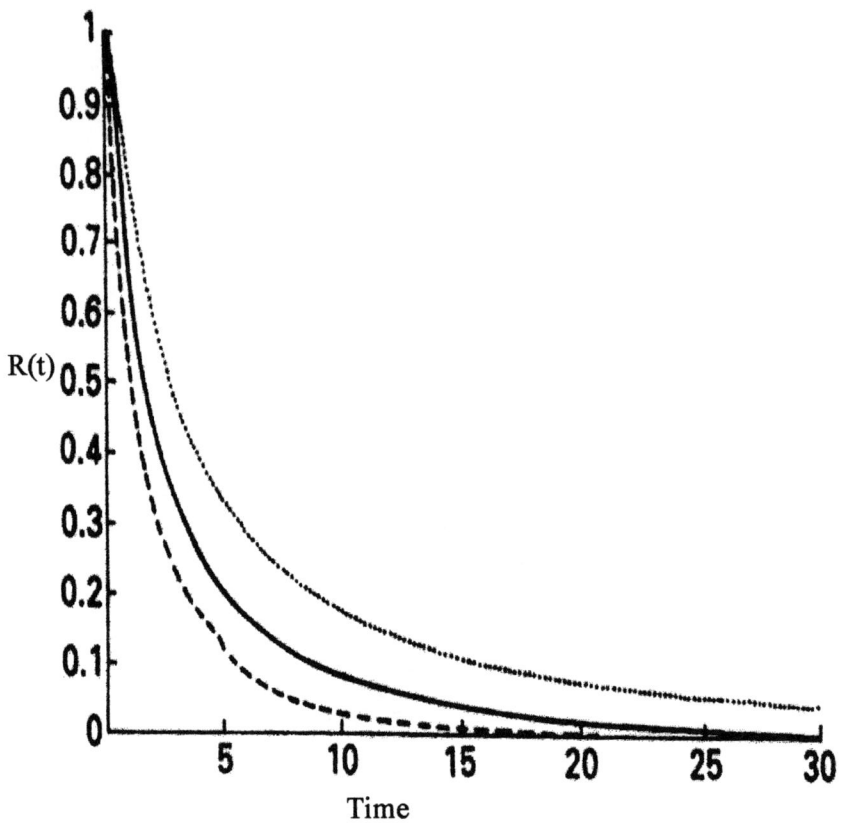

Figure 5.3 *90.25% confidence bound for $R(t)$*

5.3 Hazard rate

Definition 5.1 The hazard rate of a system is defined as

$$h(t) = \frac{f(t)}{R(t)} = \frac{f(t)}{\overline{F}(t)}.$$

Strictly speaking $h(t)dt$ is the probability that a system of age t will fail in the interval $(t, t + dt)$ i.e.,

$$h(t) = \lim_{\Delta t \to 0} \left\{ \frac{P(\text{system of age } t \text{ fails in } (t, t + \Delta t))}{\Delta t} \right\}.$$

Other names for the hazard rate are intensity function, Mill's ratio and force of mortality. For a good discussion of the various types of failure see Mann et al., (1974). For $IG(\mu, \lambda)$

we have

$$h(t;\mu,\lambda) = \left(\frac{\lambda}{2\pi t^3}\right)^{\frac{1}{2}} \frac{\exp\left(-\frac{\lambda(t-\mu)^2}{2\mu^2 t}\right)}{\Phi\left(\frac{\sqrt{\lambda}(\mu-t)}{\mu\sqrt{t}}\right) + e^{\frac{2\lambda}{\mu}}\Phi\left(-\frac{\sqrt{\lambda}(\mu+t)}{\mu\sqrt{t}}\right)}.$$

Chhikara and Folks show that $h(t;\mu,\lambda)$ is not a monotone function of t in $(0,\infty)$ and that it increases from zero at time $t=0$ till it attains a maximum at some critical time t^* and then decreases to a non-zero asymptotic value. A phenomenon of this nature makes the IG law a good candidate for modelling lifetimes with high early failure rates. The following proposition relates to the non-monotonic aspect of $h(t;\mu,\lambda)$.

Proposition 5.1 *The hazard rate $h(t;\mu,\lambda)$ is increasing in $(0,t_m)$ where t_m is the mode of $IG(\mu,\lambda)$. Moreover there exists a value t^* at which $h(t;\mu,\lambda)$ attains its maximum and $t^* \in (t_m,t_0)$ where t_0 is a value such that the hazard rate is decreasing for $t > t_0$.*

Proof The mode t_m of $IG(\mu,\lambda)$ is given by

$$t_m = \mu\left(1 + \frac{9\mu^2}{4\lambda^2}\right)^{\frac{1}{2}} - \frac{3\mu^2}{2\lambda}$$

Clearly $f(t)$ increases in $(0,t_m)$ while $\bar{F}(t)$ is decreasing in $(0,t_m)$. Thus $h(t)$ is increasing in $(0,t_m)$. Consider the logarithmic derivative of $h(t)$.

$$\frac{h'(t)}{h(t)} = \frac{f'(t)}{f(t)} + h(t).$$

Now denote by $p(t)$

$$p(t) = -\frac{f'(t)}{f(t)} = \frac{3}{2t} + \frac{\lambda}{2\mu^2} - \frac{\lambda}{2t^2}.$$

Observe that
a) $p'(t) = -\frac{3}{2t^2} + \frac{\lambda}{t^3} < 0$ if $t > \frac{2\lambda}{3} = t_0$ (say),
b) $p'(t) > 0$ if $t < t_0$,
c) $t_m < t_0$,
and
d) $p(t)$ is decreasing in t if $t > t_0$.
  Now let

$$\frac{h'(t)}{h(t)} = \frac{p(t)}{\bar{F}(t)}\left(\int_t^\infty \frac{f'(x)}{p(x)}dx + \frac{f(t)}{p(t)}\right).$$

From this representation it follows that

$$\frac{h'(t)}{h(t)} < \frac{p(t)}{\bar{F}(t)}\left(\frac{1}{p(t)}\int_t^\infty f'(x)dx + \frac{f(t)}{p(t)}\right) = 0.$$

Thus $\log h(t)$ and therefore $h(t)$ is decreasing in t for $t > t_0$. But $h(t)$ is increasing in t for $t < t_m$. Hence there exists a root of $\frac{h'(t)}{h(t)}$ in $t \in (t_m, t_0)$. We now show that there is at most one such root. To do so define

$$g(t) = \int_t^\infty \frac{f'(x)}{p(x)}dx + \frac{f(t)}{p(t)}.$$

Then

$$g'(t) = -\frac{f(t)p'(t)}{p^2(t)}.$$

Moreover

$$g'(t) = \begin{cases} < 0 & t_m < t < t_0, \\ = 0 & t = t_0, \\ > 0 & t > t_0. \end{cases}$$

Now $g(t)$ is decreasing in $t \in (t_m, t_0)$ and $g(t)$ is increasing for $t > t_0$ whereas $p(t) > 0$ for $t > t_m$ and $\frac{h'(t)}{h(t)} = \frac{p(t)g(t)}{\bar{F}(t)} < 0$ for $t > t_0$. Therefore $g(t)$ vanishes at one point in $t \in (t_m, t_0)$ implying that $\frac{h'(t)}{h(t)}$ can have at most one root for $t > t_m$. Thus $h(t)$ attains its maximum at $t^* \in (t_m, t_0)$. Since

$$\frac{d}{dt}\log h(t) = -\frac{\lambda}{2\mu^2} - \frac{3}{2t} + \frac{\lambda}{2t^2} + h(t)$$

t^* is a solution of $\frac{d}{dt}\log h(t) = 0$ so that $h(t^*) = \frac{\lambda}{2\mu^2} + \frac{3}{2t^*} - \frac{\lambda}{2t^{*2}}$. Clearly as $t \to \infty$

$$\lim_{t\to\infty} h(t) = \lim_{t\to\infty} -\frac{f'(t)}{f(t)} = \lim_{t\to\infty} p(t) = \frac{\lambda}{2\mu^2}. \quad \clubsuit$$

5.4 Estimation of critical time

Case (a) μ, λ known. The problem of estimation of the critical time is useful in reliability studies to determine the duration of a burn-in. Hsieh (1990) examines several methods of estimation and finds that there is no estimator among those investigated that is uniformly better in terms of root mean square error and bias. However, he recommends just two estimators and develops an algorithm for one of them. We describe below his technique.

Note that the critical time t^* is a root of

$$h(t; \mu, \lambda) - \frac{\lambda}{2\mu^2} - \frac{3}{2t} + \frac{\lambda}{2t^2} = 0 \tag{5.4}$$

Estimation of critical time

where $t^* \in \left(t_m, \frac{2\lambda}{3}\right)$. When μ and λ are known one has to construct tables to solve the equation iteratively. To reduce the number of parameters we let

$$X = \frac{T}{\lambda}, \quad \phi = \frac{\lambda}{\mu}, \quad \text{where } \mathcal{L}(T) = IG(\mu, \lambda).$$

Multiplying (5.4) by λ and substituting we have the equation

$$g(x, \varphi) = \lambda h(t; \mu, \lambda) - \frac{\phi^2}{2} - \frac{3}{2x} + \frac{1}{2x^2} = 0.$$

Since $h(t; \mu, \lambda) = \frac{f(t;\mu,\lambda)}{R(t;\mu,\lambda)}$ and $f(t;\mu,\lambda)$ becomes $f(x;\varphi^{-1},1)$ under the transformation $X = \frac{T}{\lambda}, \phi = \frac{\lambda}{\mu}$, we have

$$g(x, \phi) = \frac{f(x;\phi^{-1},1)}{R(x;\phi^{-1},1)} - \frac{\varphi^2}{2} - \frac{3}{2x} + \frac{1}{2x^2} = 0$$

$$= p(x;\phi) - \frac{\phi^2}{2} - \frac{3}{2x} + \frac{1}{2x^2} = 0.$$

Now we solve $g(x; \phi)$ by the Newton-Raphson method using the modal value x_m as the initial value. Iteration is based on the equation

$$x_{k+1} = x_k - \frac{g(x_k, \phi)}{g'(x_k, \phi)} \tag{5.5}$$

and $x_m = \frac{1}{2}\left(\frac{2}{3} + \frac{1}{\phi}\left(1 + \frac{9}{2\phi^2}\right)^{\frac{1}{2}} - \frac{3}{2\phi^2}\right)$. If x^* is the solution obtained by iteration of (5.5) then $\lambda x^* = t^*$ is the desired estimate of the critical time.

Case (b) μ, λ unknown. Hsieh remarks that of six estimators investigated by him two estimators of the critical time stand out as desirable. Estimators of the critical time are obtained as solutions to an equation containing maximum likelihood estimators of λ and ϕ. These are biased and have large root mean square errors. He is led to the choice of two estimators of λ, ϕ namely $(\overline{\lambda}, \overline{\phi})$ and $(\overline{\lambda}, \hat{\phi})$ where

$$\left(\overline{\lambda} = \frac{n-3}{V}, \quad \overline{\phi} = \frac{n-3}{\overline{X}V} - \frac{1}{n}\right)$$

and

$$\left(\overline{\lambda} = \frac{n-3}{V}, \quad \hat{\phi} = \frac{\overline{\lambda}}{\overline{x}}\right).$$

With sample sizes of $n = 10$, 20 and 5000 repetitions Hsieh concludes that an estimator based on $(\overline{\lambda}, \overline{\phi})$ performs better for values of $\overline{\phi} < 5$ in terms of smaller bias and root mean square error while for $\overline{\phi} \geq 5$ Hsieh recommends an estimator based on $(\overline{\lambda}, \hat{\phi})$. Using the Jackknife method Hsieh proposes construction of confidence intervals for t^*. We will see how the technique of Chang (1994) and Tang and Chang (1994) can be used to obtain confidence intervals for the hazard rate as well as for the critical time. Their method depends on the solution of a non-linear programming problem when both parameters are unknown.

5.5 Confidence intervals for hazard rate

The following theorem is due to Chang (1994).

Theorem 5.2 *The $100(1-\alpha_1)(1-\alpha_2)\%$ confidence interval for $h(t; \mu, \lambda)$ is given by the interval*

$$\left[\underline{h}(t; \mu, \lambda), \overline{h}(t; \mu, \lambda)\right] = \left[\inf_{\underline{\lambda} \leq \lambda \leq \overline{\lambda}} h(t; \overline{\mu}, \lambda), \sup_{\underline{\lambda} \leq \lambda \leq \overline{\lambda}} h(t; \underline{\mu}, \lambda)\right]$$

for each t where $(\underline{\mu}, \overline{\mu})$ and $(\underline{\lambda}, \overline{\lambda})$ are the $100(1-\alpha_1)\%$ and $100(1-\alpha_2)\%$ confidence intervals for μ and λ respectively.

Proof Define $\Omega = \{(\mu, \lambda) \mid \underline{\mu} \leq \mu \leq \overline{\mu}, \underline{\lambda} \leq \lambda \leq \overline{\lambda}\}$. Then for $(\mu, \lambda) \in \Omega$

$$\inf_{\Omega} h(t; \mu, \lambda) \leq h(t; \mu, \lambda) \leq \sup_{\Omega} h(t; \mu, \lambda).$$

Hence

$$P\left(\inf_{\Omega} h(t; \mu, \lambda) \leq h(t; \mu, \lambda) \leq \sup_{\Omega} h(t; \mu, \lambda)\right)$$
$$\geq P\left(\underline{\mu} \leq \mu \leq \overline{\mu}, \underline{\lambda} \leq \lambda \leq \overline{\lambda}\right) = 1 - \alpha_1 - \alpha_2 + \alpha_1 \alpha_2$$

since $\hat{\mu} \perp\!\!\!\perp \hat{\lambda}$. ($\underline{\mu}, \overline{\mu}, \underline{\lambda}, \overline{\lambda}$ are defined in Table 4.2.) Observe that $h(t; \mu, \lambda)$ is monotone decreasing in μ. Therefore

$$\inf_{\Omega} h(t; \mu, \lambda) = \inf_{\underline{\lambda} \leq \lambda \leq \overline{\lambda}} h(t; \overline{\mu}, \lambda)$$

and

$$\sup_{\Omega} h(t; \mu, \lambda) = \sup_{\underline{\lambda} \leq \lambda \leq \overline{\lambda}} h(t; \underline{\mu}, \lambda).$$

This then yields the desired reuslt. ♣

Example 3.1 (continued) For the von Alven data $\hat{\mu} = 3.6065$ and $\hat{\lambda} = 1.6589$, and $n = 46$. The 97.5% confidence intervals for μ and λ

Confidence interval for hazard rate

are respectively $[2.3911, 7.3639]$ and $[0.9548, 2.5411]$. We can then plot $h(t; \mu, \lambda)$ as well as the trajectories of $\lambda_L^*(t)$ and $\lambda_U^*(t)$. Therefore, it turns out that the failure rate for the inverse Gaussian distribution is monotone decreasing in μ. Hence, it follows that

$$\inf_{\Omega} h(t; \mu, \lambda) = \inf_{\underline{\lambda} \leq \lambda \leq \overline{\lambda}} h(t; \overline{\mu}, \lambda)$$

and

$$\sup_{\Omega} h(t; \mu, \lambda) = \sup_{\underline{\lambda} \leq \lambda \leq \overline{\lambda}} h(t; \underline{\mu}, \lambda).$$

Furthermore

$$\hat{\mu} = 3.6065 \text{ and } \hat{\lambda} = 1.6589.$$

The 95.0625% confidence bounds for $h(t; \mu, \lambda)$ are shown in Figure 5.4 and the $\lambda_{\text{LB}}^*(t)$ and $\lambda_{\text{UB}}^*(t)$ that give these bounds can be plotted likewise.

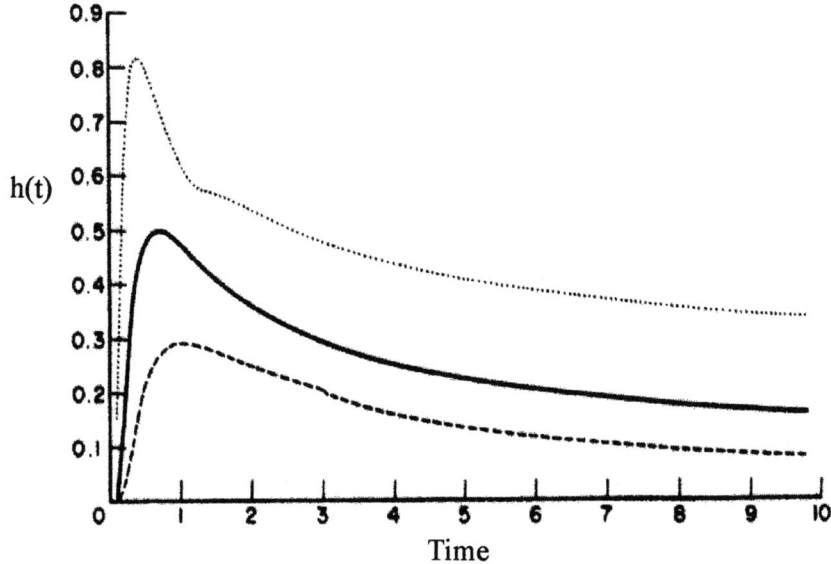

Figure 5.4 *The 95.0625% confidence bounds of $h(t)$*

Confidence interval for critical time

Case (a) μ known, λ unknown.

We make the substitution $X = \frac{T}{\mu}$ where $\mathcal{L}(T) = IG(\mu, \lambda)$ and $\phi = \frac{\lambda}{\mu}$ in both the density and the reliability function to obtain the hazard rate as $h_1(x, \phi)$. An estimate of the critical time is now $\mu x^* = t^*$ where x^* is the root of

$$h_1(x, \phi) + \frac{\hat{\phi}}{2x^2} - \frac{3}{2x} - \frac{\hat{\phi}}{2} = 0$$

(see equation 5.4). (Here $\hat{\phi} = \frac{\hat{\lambda}}{\mu}$.) Using a graph where x^* is plotted against ϕ, one can see that x^* is increasing monotonically in ϕ so that for fixed μ, x^* is increasing in λ. Under the transformation $X = \frac{T}{\mu}$, a complete sufficient statistic for ϕ is $\sum_{i=1}^{n} \frac{(X_i - 1)^2}{X_i} = S$, while the maximum likelihood estimate is $\frac{n}{S}$. Since $\mathcal{L}\left(\frac{n\phi}{Z}\right) = \chi_n^2$, a $100(1-\alpha)\%$ confidence interval for ϕ is $\left(\phi_L = \frac{\chi_{n, \frac{\alpha}{2}}^2}{S}, \phi_U = \frac{\chi_{n, 1-\frac{\alpha}{2}}^2}{S}\right)$. The monotonicaity of x^* implies that

$$\phi \in (\phi_L, \phi_U) \text{ if and only if } x^* \in (x_L^*, x_U^*) \text{ say.}$$

Hence

$$P(t^* \in (t_L^*, t_U^*)) = P(\mu x^* \in (\mu x_L^*, \mu x_U^*))$$
$$= P(\phi \in (\phi_L, \phi_U))$$
$$= 1 - \alpha.$$

Therefore from the graph (see Figure 5.5) one can obtain bounds for x^* from the bounds ϕ_L, ϕ_U for ϕ and then construct $\mu x_L^*, \mu x_U^*$.

Case (b) μ unknown, λ known.

We now use the transformation (cf. Hsieh's method) $Y = \frac{T}{\lambda}, \gamma = \frac{\mu}{\lambda}$ where $\mathcal{L}(T) = IG(\mu, \lambda)$. Proceeding exactly as in (a) we substitute μ by its maximum likelihood estimate and obtain $t^* = \lambda y$ where y^* is the root of

$$h_2(y, \gamma) + \frac{1}{2y^2} - \frac{3}{2y} - \frac{1}{2\hat{\gamma}^2} = 0$$

where $\hat{\gamma} = \frac{\hat{\mu}}{\lambda}$ and $h_2(y, \gamma)$ the transformed hazard rate. We plot y^* against $\gamma = \frac{\mu}{\lambda}$. Tang and Chang point out that the range of interest is at the higher extremity indicating a large variance for the IG variate as burn-in has proven useful in reducing the variability of the life of a device. The graph (Figure 5.6) shows that y^* is monotone decreasing in $\gamma(\mu)$ and approaches the value 0.383328 asymptotically, being the root of

$$h_2(y, \infty) + \frac{1}{2y^2} - \frac{3}{2y} = 0.$$

Confidence interval for hazard rate

The maximum likelihood of γ is $\overline{Y} = \hat{\gamma}$ and $\mathcal{L}\left(\frac{n(\hat{\gamma}-\gamma)^2}{\hat{\gamma}\gamma^2}\right) = \chi_1^2$ so that (see Table 3.2)

$$(\gamma_L, \gamma_U) = \left[\left(\frac{1}{\overline{Y}} + \sqrt{\frac{\chi_{1,1-\alpha}^2}{n\overline{Y}}}\right)^{-1}, \left(\frac{1}{\overline{Y}} - \sqrt{\frac{\chi_{1,1-\alpha}^2}{n\overline{Y}}}\right)^{-1}\right].$$

(The usual caution about the upper bound being positive is necessary). Thus y^* being a monotone decreasing function of $\gamma \in (\gamma_L, \gamma_U)$ if and only if $y^* \in (y_L^*, y_U^*)$ and given (γ_L, γ_U) one can construct (y_L^*, y_U^*) and hence $(\lambda y_L^*, \lambda y_U^*)$ from the graph.

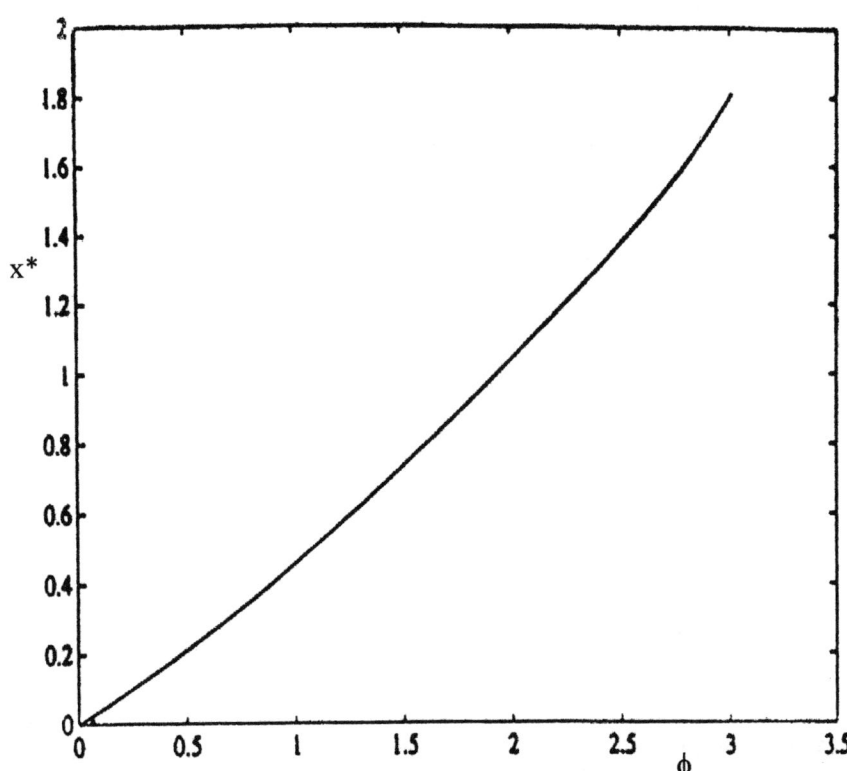

Figure 5.5 $x^* = $ *critical time*$/\mu$ vs $\phi = \lambda/\mu$

Case (c). Both μ and λ unknown.

As observed by Hsieh several choices are available as estimators for the critical time. Tang and Chang select the maximum likelihood estimates of μ and λ and their corresponding $100(1-\alpha)\%$ confidence intervals (μ_L, μ_U) and (λ_L, λ_U) (see Table 3.2). With this choice one then solves for the critical time $t^*(\mu, \lambda)$ from the equation

$$h(t; \mu, \lambda) - \frac{\lambda}{2\mu^2} - \frac{3}{2t} + \frac{\lambda}{2t^2} = 0. \tag{5.6}$$

Now the authors use two nonlinear programs
i) minimize $t^*(\mu, \lambda)$ subject to

$$\bar{\Omega} = \{(\mu_L \leq \mu \leq \mu_L) \cap (\lambda_L \leq \lambda \leq \lambda_U)\}.$$

ii) maximize $t^*(\mu, \lambda)$ subject to $\bar{\Omega}$. From the 2 graphical plots (Figure 5.5, Figure 5.6) it is clear
  a) that $t^*(\mu, \lambda)$ is increasing in λ for every fixed μ and
  b) that $t^*(\mu, \lambda)$ is decreasing in μ for every fixed λ so that $\frac{\partial t^*}{\partial \lambda} > 0$ and $\frac{\partial t^*}{\partial \mu} < 0$.

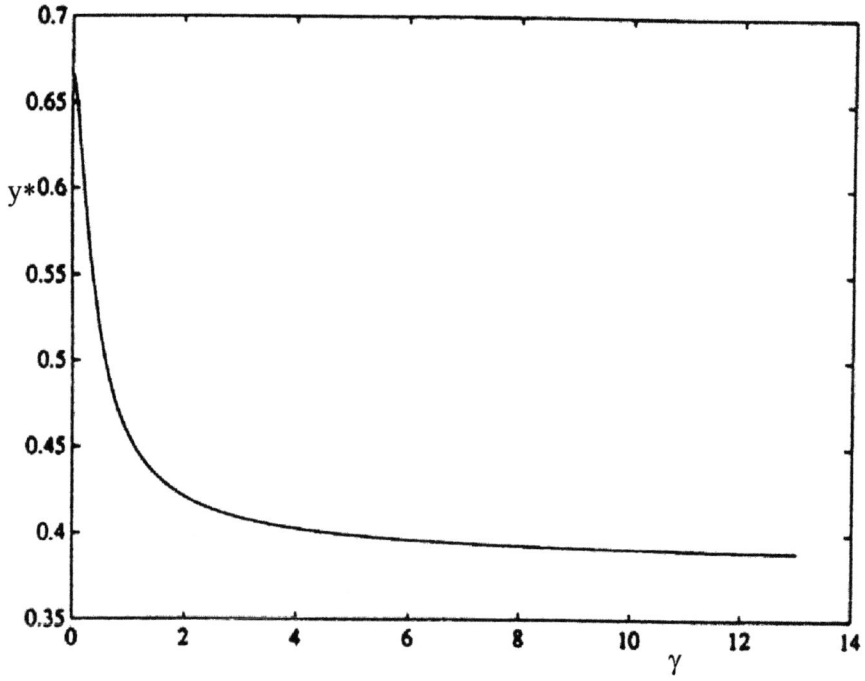

Figure 5.6 $y^* = \text{critical time}/\lambda$ vs $\gamma = \mu/\lambda$

The **Kuhn-Tucker** optimality condition then implies that $\mu \leq \bar{\mu}$ and $\underline{\lambda} \leq \lambda$ must be tight in $\bar{\Omega}$ so that $\mu = \bar{\mu}$ and $\lambda = \underline{\lambda}$ when the first of the nonlinear programs (i) attains its optimality. In like manner $\mu = \underline{\mu}$ and $\lambda = \bar{\lambda}$ provide optimal solutions for the second of the nonlinear programs. Hence we can solve (5.6) with (μ, λ) replaced by $(\bar{\mu}, \underline{\lambda})$ for the first program and (μ, λ) replaced by $(\underline{\mu}, \bar{\lambda})$ for the second program. Thus

$$t^* \in (\text{solution under program (i)}, \text{solution under program (ii)})$$

whenever $(\mu, \lambda) \in \bar{\Omega}$. This then implies that

$$(\mu, \lambda) \in \bar{\Omega} \subseteq \{t^* \mid t^* \in (\text{solution under (i)}, \text{solution under (ii)})\}.$$

Finally

$$P(t^* \in (\text{solution under program (i)}, \text{solution under program (ii)}))$$
$$\geq P(\underline{\mu} \leq \mu \leq \bar{\mu}, \underline{\lambda} \leq \lambda \leq \bar{\lambda})$$
$$\geq 1 - P(\{\underline{\mu} \leq \mu \leq \bar{\mu}\}^c) - P(\{\underline{\lambda} \leq \lambda \leq \bar{\lambda}\}^c) \text{ (Bonferroni)}$$
$$= 1 - 2\alpha.$$

This solution scheme suggests that in practice one can obtain t_L^* by using (μ_U, λ_L) and t_U^* by using (μ_L, λ_U). Moreover, using the Bonferroni inequality an improvement in the minimum coverage probability can be obtained by using $(1 - \alpha)^2$ instead of $1 - 2\alpha$. Lastly this methodology works in the case of censored data also. Tang and Chang point out that despite the fact that the analytic approach is exact when both parameters are unknown the width of the confidence intervals may be too wide and hence too conservative. The figures seems to indicate that when $\mu > \lambda$ (the situation when burn-in is practical and effective) the critical time is more sensitive to variation in λ and hence the estimated critical time also shares the same sensitivity to variability in the estimated value of λ between random samples. They recommend that it may then be reasonable to treat μ as known in order to obtain an interval whose confidence level is identical with the coverage probability as is the case in case (a). They propose an approximate method which is given in the next proposition.

Proposition 5.2 *An approximate $100(1 - \alpha)\%$ confidence interval of the critical time is given by μ replaced by the maximum likelihood estimate $\hat{\mu}$ and χ_n^2 replaced by χ_{n-1}^2 in case (a).*

This proposition is substantiated by a Monte Carlo simulation where they find that the approximate method yields smaller intervals with its coverage probability very close to the desired confidence coefficient.

Example 3.1 (continued) The maximum likelihood estimate of the critical time is obtained from

$$h(t; \mu, \lambda) - \frac{\lambda}{2\mu^2} - \frac{3}{2t} + \frac{\lambda}{2t^2} = 0$$

where μ is replaced by $\hat{\mu} = 3.6065$ and λ is replaced by $\hat{\lambda} = 1.6589$. Then $\hat{t}^* = 0.6934$. At level $\alpha = 0.025$ the 97.5% confidence intervals for λ and μ are respectively $(2.389, 7.357)$ and $(0.9540, 2.484)$. Therefore a confidence interval of t^* with at least 95% coverage is obtained by letting $\mu = 2.484$ and $\lambda = 2.389$ in the equation and solving for t_L^*, and then solving the equation with $\mu = 0.9540$ and $\lambda = 7.357$ to obtain t_U^*. This gives

$$(t_L^*, t_U^*) = (0.3751, 1.142).$$

Using Proposition 5.1 one computes $x_i = \frac{t_i}{\hat{\mu}}$ for $i = 1, \ldots, 46$ and then proceeds as in case (a). This time the 95% confidence interval turns out to be $(0.4141, 1.020)$ with a width of 0.606 as against a width of 0.767 using the analytic approach.

Remarks One may wonder why two different models have been advanced leading to equations (5.1) and (5.3) for the reliability function. The parameters in the modelling of the two have different physical interpretations. In the case of (5.3) the drift and variance parameters are estimated from the increments (wear) of the deterioration process. Moreover, both the time variable and the 'physical wear' have been normalized by the threshold value.

Tang et al., (1988) have studied the reliability of components subject to sliding wear. They have questioned the conventional approach employed in sliding wear theory. Wear factors are derived from the measured wear rates through what is known as Archard's law. Tang et al. view Archard's law as a mere empirical relation that ignores the random effects in the modelling of the reliability of sliding components. They introduce a time-independent random effect of reliability as an extension to Archard's law. The Archard wear theory models the volume V of material removed by a wear process using a normal load L, when the room temperature hardness of the softer material is H_0, u is the sliding velocity and t is the time interval in which the volume V is removed. This relationship is given by

$$\frac{V}{ut} = k\left(\frac{L}{H_0}\right).$$

Here k is a dimensionless coefficient known as the Archard wear factor which characterizes the extent of wear damages. In the modified approach the wear factor is regarded as a stationary discrete random process $k(t,w)$. This leads to a time-averaging process of $k(t,w)$ given by

$$k_{\Delta t}(i,w) = \frac{1}{\Delta t} \int_{i\Delta t}^{(i+1)\Delta t} k(t,w)dt,$$

where Δt is a fixed-time sampling interval. The δ and σ referred to in equation (5.3) can be explained as

$$\delta = \delta_{\Delta t} \text{ and } \sigma = \sigma_{\Delta t} \times \Delta t$$

quantities obtainable experimentally with sampling interval Δt. In reliability studies, say, of a low-alloy steel shaft rotating in its housing under an unlubricated condition if V^* is the critical wear volume beyond which failure occurs as specified by the designer, the threshold value or critical barrier 'a' is then

$$a = \frac{V^* H_0}{uL}.$$

The time to failure T of the sliding component is defined as the first value such that

$$V^* = \frac{uL}{H_0} \int_0^T k(t,w)dt.$$

The discrete nature of k and its implicit relation to T does not permit the evaluation of $R(t) = P(T > t)$. Tang et al., surmise that T fits the inverse Gaussian description as suggested by empirical studies.

CHAPTER 6

GOODNESS-OF-FIT

6.0 Introduction

Goodness-of-fit tests for continuos distributions are generally handled by the Kolmogorov-Smirnov test which in its classical form requires that the distribution is completely specified. In practice this is seldom the case and one then resorts to estimating parameters from the data and then examining Kolmogorov-Smirnov "type" tests. The standard tables can no longer be used and modifications are necessary.

The inverse Gaussian distribution offers good competition to distributions like the Weibull and lognormal and enjoys a similarity in shape befitting long-tailed data. It is quite important in model selection that tests discriminate well between distributions of dissimilar shapes. In the discussion that follow we consider several approaches based on the work of Edgeman et al., (1988, 1992), Edgeman (1990) and O'Reilly and Rueda (1992).

6.1 A modified Kolmogorov-Smirnov-type test

Edgeman et al., begin by scaling the random variable X whose law is $IG(\mu, \lambda)$ by the factor $\frac{1}{\mu}$ to get $Y = \frac{X}{\mu}$ so that $\mathcal{L}(Y) = IG(1, \phi)$. They then scale the observed data $X = (X_1, \ldots, X_n)$ by an estimate of μ, namely $\overline{X}$ and then test the hypothesis

$$H_0 : \mathcal{L}(Y_i) = IG(1, \phi) = F_0 \quad (\text{say})$$

against

$$H_A : H_0 \text{ is false}$$

by employing the statistic $D_n = \max(D_n^+, D_n^-)$ where

$$D_n^+ = \max_i \left\{ \frac{i}{n} - F_0\left(Y_{(i)}\right) \right\}$$

and

$$D_n^- = \max_i \left\{ F_0\left(Y_{(i)}\right) - \frac{i-1}{n} \right\}.$$

Here $F_0(Y_{(i)})$ is the distribution function of $IG(1, \hat{\phi})$ evaluated at the ith ordered value of $(X_1/\bar{X}, \ldots, X_n/\bar{X})$, and $\hat{\phi} = \frac{n}{V\bar{X}}$. Whenever the observed value of D_n exceeds the critical value of D_n given by

$$D_n \left(\sqrt{n} + \frac{b_1}{\sqrt{n}} + \frac{b_2}{n} + \frac{b_3}{n^2} \right)$$

at level α, H_0 is rejected. They present a set of tables of this modified test including values of the coefficients b_1, b_2, b_3 at critical values of the D_n statistic at significance levels 0.20, 0.10, 0.05 and 0.01. Also published are approximate power results against alternatives such as the exponential, lognormal, uniform and Weibull. According to their conclusion the test indicates a good dicriminating ability between the IG and densities of different shapes, but poor ability to distinguish between similarly shaped densities. By way of an example they study the data set analysed by Jorgensen (1982).

Example 6.1 The intervals in operating hours between successive failures of airconditioning equipment in a Boeing 720 aircraft (#13) are: 102, 209, 14, 57, 54, 32, 67, 59, 134, 152, 27, 14, 230, 66, 61 and 34. Are these values from an $IG(\mu, \lambda)$ law?

To use the modified Kolmogorov-Smirnov test we proceed thus.

$$\bar{X} = 82, \hat{\phi} = \frac{16}{V(82)} = 1.0968.$$

The scaled data $\left(\frac{X_i}{\bar{X}}\right)$ are put in ascending order as

0.1707	0.1707	0.3293	0.3902	0.4146	0.6585
0.6951	0.7195	0.7439	0.8049	0.8171	1.2439
1.6341	1.8537	2.5488	2.8409		

D_{16} based on this standardized data is 0.2641. From Table 6.1 D^*_{16} ($\phi = 1.0$) and D^*_{16} ($\phi = 1.5$) are computed and then D^*_{16} ($\phi = 1.0968$) obtained by interpolation as 1.3912. Similarly using interpolation one obtains $D^*_{16,.05}$ from the table. This value 1.421 when compared with the observed value 1.3912 shows that H_0 is not rejected at the 5% level, but the result is significant at the 10% level since $D^*_{16,.10} = 1.294$.

An approximate method

A second method recommended by Edgeman (1990) involves the use of

$$\Phi \left[(X_{(i)} - \bar{X})/\bar{X}\sqrt{\bar{X}V/n - 1} \right]$$

in place of $F_0(Y_{(i)})$ in calculating D_n^+ and D_n^- and then modifying the value $L_n = \max(D_n^+, D_n^-)$ by $L_n^* = L_n(\sqrt{n} - 0.01 + 0.85/\sqrt{n})$. Finally the null hypothesis is rejected at level α if L_n^* exceeds the upper tail percentage points given in Table 6.2. Edgeman's power studies indicate that the approximate powers are uniformly larger and within the margins of error of the results in Edgeman et al (1988).

Example 6.1 (continued) For the aircraft (#3) data we obtain $D_n^+ = 0.2635, D_n^- = 0.2010, L_n = 0.2035$ and $L_n^* = 0.2635[4 - 0.01 + 0.2125] = 1.107$. The 1% table value is 1.035 and this test indicates that the IG assumption is questionable.

6.2 Anderson-Darling statistic

Goodness-of-fit based on the empirical distribution of the Anderson-Darling statistic A^2 (to be made precise) has been advanced by O'Reilly and Rueda (1992) mainly due to its good power properties. This has also been investigated by Pavur et al. (1992) who have examined two other members of the Cramer-von Mises family, namely, the Cramer-von Mises statistic U^2 and the Watson statistic W^2 where

$$U^2 = n \int_{-\infty}^{\infty} \left[(F_n(y) - F(y)) - \int_{-\infty}^{\infty} \{F_n(y) - F(y)\} dF(y) \right]^2 dF(y).$$

For computational purposes we let $Z_i = F(X_i)$, where $X = (X_1, \cdots, X_n)$, the random sample is taken from $F(\cdot)$ and $Z_{(i)}$ the ordered values of Z. Then

$$A^2 = -n - \frac{1}{n} \sum_{i=1}^{n} \left[(2i-1) \ln Z_{(i)} + (2n - 2i + 1) \ln(1 - Z_{(i)}) \right],$$

$$W^2 = \sum_{i=1}^{n} \left[Z_{(i)} - \frac{(2i-1)}{2n} \right]^2 + \frac{1}{12n}, \text{ and}$$

$$U^2 = W^2 - n \left(\frac{\sum_{i=1}^{n} Z_i}{n} - \frac{1}{2} \right)^2.$$

O'Reilly and Rueda have studied the asymptotic distribution of A^2 when the parameters are estimated by maximum likelihood or another asymptotically efficient method — the Rao-Blackwell distribution function estimator method. They conclude that in either case the same asymptotic law holds for A^2. We consider their approach first and then examine the modified approach suggested by Pavur et al.

Proposition 6.1 Let $Z_{(i)} = F(X_{(i)}; \mu, \lambda)$, $i = 1, \ldots, n$ and $\hat{Z}_{(i)} = \hat{F}(X_{(i)}; \hat{\mu}, \hat{\lambda})$. Then the asymptotic distribution of $\hat{A}^2$ when $\hat{Z}_{(i)}$ is used depends only on $\theta = \frac{\lambda}{\mu}$.

Consider the process $\sqrt{n}\{F_n(x) - F(x; \hat{\mu}, \hat{\lambda})\}$ evaluated at $F(x; \hat{\mu}, \hat{\lambda}) = \hat{t}$, where $F_n(\cdot)$ is the sample distribution function. This process converges weakly to the Gaussian process $\{Y(t); t \in (0,1)\}$ with mean zero and covariance

$$\sigma(s,t) = \min(s,t) - st - g^T(t) I^{-1} g^T(s)$$

where I is the Fisher information matrix based on a single observation and

$$g^T(t) = \left(\frac{\partial F(x; \mu, \lambda)}{\partial \mu}, \frac{\partial F(x; \mu, \lambda)}{\partial \lambda} \right)$$

evaluated at $t = F(x; \mu, \lambda)$. Now recall that $\mathcal{L}\left(\frac{\lambda(X-\mu)^2}{\mu^2 X}\right) = \chi_1^2$, so that we can calculate I

$$I = \begin{pmatrix} \frac{\mu^3}{\lambda} & 0 \\ 0 & \frac{1}{2\lambda^2} \end{pmatrix} = \mu^2 \begin{pmatrix} \theta & 0 \\ 0 & \frac{1}{2\theta^2} \end{pmatrix}.$$

Moreover

$$g(t) = \frac{1}{\mu} \begin{bmatrix} -\frac{\sqrt{\lambda x}}{\mu} \phi(R) - \frac{2\lambda}{\mu} e^{2\theta} \Phi(L) + e^{2\theta} \phi(L) \frac{\sqrt{\lambda x}}{\mu} \\ \frac{R\mu}{2\lambda} \phi(L) + 2e^{2\theta} \Phi(L) + e^{2\theta} \phi(L) \frac{L\mu}{2\lambda} \end{bmatrix}$$

where $R = \frac{\sqrt{\lambda x}}{\mu} - \sqrt{\frac{\lambda}{x}}$ and $L = -\frac{\sqrt{\lambda x}}{\mu} - \sqrt{\frac{\lambda}{x}}$, and ϕ is the standard normal density function.

Since

$$F(x) = \Phi\left(-\sqrt{\frac{\lambda}{x}} + \frac{\sqrt{\lambda x}}{\mu}\right) + e^{2\theta} \Phi\left(-\sqrt{\frac{\lambda}{x}} - \frac{\sqrt{\lambda x}}{\mu}\right),$$

letting $z = \frac{x}{\mu}$

$$F_1(z) = \Phi\left(-\sqrt{\frac{\theta}{z}} + \sqrt{\theta z}\right) + e^{2\theta} \Phi\left(-\sqrt{\frac{\theta}{z}} - \sqrt{\theta z}\right).$$

Direct differentiation of $F(x)$ gives

$$\frac{\partial F}{\partial \mu} = -\frac{\sqrt{\lambda x}}{\mu^2} \phi(R) - \frac{2\lambda}{\mu^2} e^{2\theta} \Phi(L) + \frac{\sqrt{\lambda x}}{\mu^2} \phi(L) e^{2\theta}$$

or

$$\mu \frac{\partial F}{\partial \mu} = -\sqrt{\theta z}\, \phi(R) - 2\theta e^{2\theta} \Phi(L) + \sqrt{\theta z} e^{2\theta} \phi(L)$$

Similarly

$$\mu \frac{\partial F}{\partial \lambda} = \frac{R}{2\theta} \phi(R) + 2 e^{2\theta} \Phi(L) + \frac{L}{2\theta} e^{2\theta}\, \phi(L).$$

Hence the expression $-g^T(t)I^{-1}g^T(s)$ is strictly a function of θ only. Here $z = F^{-1}(t; 1, \theta)$.

O'Reilly and Rueda now state and prove the following theorem.

Theorem 6.1 *If $\theta \to \infty$ then $\sigma(s,t) \to \min(s,t) - st - \{\frac{1}{2}\phi(x)x\phi(y)y + \phi(x)\phi(y)\}$ with $x = \Phi^{-1}(t)$ and $y = \Phi^{-1}(s)$, while if $\theta \to 0$ $\sigma(s,t) = \min(s,t) - st - \{2\phi(x)x\phi(y)y\}$ with $x = \Phi^{-1}\left(1 - \frac{t}{2}\right)$ and $y = \Phi^{-1}\left(1 - \frac{s}{2}\right)$.*

These two limiting cases correspond to the process arising in tests of normality or chi-squaredness. The authors provide a table, Table 6.1, of the asymptotic upper tail critical values of A^2 for various values of θ against $\alpha = 0.20, 0.10, 0.05, 0.025$ and 0.01. The second method where the Rao-Blackwell distribution function estimator of $F(x; \mu, \lambda)$ is used leads to a theorem which asserts that the same asymptotic process is obtained. In particular it is shown that

$$\sqrt{n} \sup_x \left\{\hat{F}_n(x) - F(x; \hat{\mu}, \hat{\lambda})\right\} \to 0$$

with probability one. We omit the details of the proof.

Example 6.2 A data set on the endurance of deep groove ball bearings analyzed by Lieblein and Zelen (1956) consists of the number of million revolutions before failure for each of 23 ball bearings used in a life test. The data set is as follows.

17.88	42.12	51.96	68.64	93.12	127.96
28.92	45.60	54.12	68.64	98.64	128.01
33.00	48.48	55.56	68.88	105.12	173.4
41.52	51.84	67.80	84.12	105.84	

The calculated value of A^2 is 0.692 corresponding to the estimated $\theta = 3.21$. The p-value obtained by interpolation from Table 6.1 is close to 0.095.

Pavur et al. (1992) also propose a modified procedure for use with the Anderson-Darling, Cramer-von Mises and Watson tests. Their procedure uses regression relationships between sample size and critical values. In particular if T_n is the test statistic then T_n is modified by

$$T_n^* = T_n \left[\sqrt{n} + \hat{\beta}_1(1/\sqrt{n}) + \hat{\beta}_2 \left(\frac{1}{n}\right)\right].$$

They produce tables of the empirical critical values for the three quadratic statistics corresponding to selected values of ϕ. The critical values $C(\phi)$ are provided for $\alpha = 0.20, 0.10, 0.05$ and 0.01.

Anderson-Darling statistic

Table 6.1 *Asymptotic distribution for A^2. α = significance level (upper tail); $\theta = \frac{\lambda}{\mu}$*

θ		Critical		value	
	$\alpha = 0.20$	0.10	0.05	0.025	0.01
0	0.90	1.19	1.50	1.83	2.27
2^{-10}	0.89	1.18	1.49	1.82	2.26
2^{-9}	0.89	1.18	1.48	1.81	2.24
2^{-8}	0.88	1.17	1.47	1.79	2.22
2^{-7}	0.86	1.15	1.45	1.76	2.19
2^{-6}	0.84	1.12	1.41	1.71	2.13
2^{-5}	0.81	1.07	1.35	1.64	2.03
2^{-4}	0.77	1.01	1.27	1.53	1.90
2^{-3}	0.72	0.93	1.16	1.40	1.73
2^{-2}	0.66	0.85	1.05	1.26	1.54
2^{-1}	0.61	0.78	0.95	1.13	1.37
2^0	0.57	0.72	0.87	1.02	1.23
2^1	0.54	0.68	0.82	0.96	1.14
2^2	0.53	0.66	0.79	0.92	1.09
2^3	0.52	0.65	0.77	0.90	1.07
2^4	0.52	0.64	0.76	0.89	1.05
2^5	0.51	0.64	0.76	0.88	1.05
∞	0.51	0.63	0.76	0.88	1.04

Example 6.2 (continued) For the ball bearings data we have the following table.

Table 6.2 *Critical values of $C(\phi)$ for $\alpha = 0.10, 0.05$.*

ϕ	$\hat{\beta}_1$	$\hat{\beta}_2$	$C_{.10}(\phi)$	$C_{.05}(\phi)$
4.0	33.9452	-43.5218	6.8142	8.0254
3.0	44.9027	-58.1030	7.6985	9.0749

Using the observed value of $A_n^2 = 0.684$

$$A_n^{*2}(4) = 0.684\left[\sqrt{23} + 33.9452\left(\frac{1}{\sqrt{23}}\right) - 43.5218\left(\frac{1}{23}\right)\right] = 6.827$$

$$A_n^{*2}(3) = 0.684\left[\sqrt{23} + 44.9027\left(\frac{1}{\sqrt{23}}\right) - 58.1030\left(\frac{1}{23}\right)\right] = 7.956.$$

By interpolating one obtains

$$A_n^{*2}(3.21) = 7.719, \ C_{.10}(3.21) = 7.512, \ C_{.05}(3.21) = 8.85$$

The p-value is approximately 0.09 which is in conformity with that obtained using the method of O'Reilly and Rueda.

CHAPTER 7

COMPOUND LAWS AND MIXTURES

7.0 Introduction

The versatility of the IG law and its interpretability as a first passage time distribution make it a strong candidate in modelling data in diverse disciplines. Several authors have employed this law as a mixing distribution to generate compound distributions that have many appealing features in fitting long-tailed data. In this chapter we examine some of its ramifications.

7.1 Poisson-inverse Gaussian —$P\text{-}IG(\mu, \lambda)$

One of the first to consider the IG as a mixing tool was Holla (1966). He assumed that the mean parameter θ of a Poisson law followed the $IG(\mu, \lambda)$ distribution and arrived at the compound law which came to be known as the Poisson-inverse Gaussian law, abbreviated at $P\text{-}IG(\mu, \lambda)$. Thus if

$$f(x \mid \theta) = \frac{\theta^x \exp(-\theta)}{x!}, \quad x = 0, 1, 2, \ldots$$

and

$$\pi(\theta) = \sqrt{\frac{\lambda}{2\pi\theta^3}} \exp\left\{-\frac{\lambda}{2\mu^2} \frac{(\theta - \mu)^2}{\theta}\right\} 1_{\mathbf{R}^+}(\theta)$$

then

$$p(x) = \int_0^\infty f(x \mid \theta) \pi(\theta) \, d\theta$$

$$= \sqrt{\frac{\lambda}{2\pi}} \exp\left(\frac{\lambda}{\mu}\right) \left(\frac{\lambda}{2} + \frac{\lambda}{\mu^2}\right)^{\frac{1}{2}(x-\frac{1}{2})} \frac{1}{x!} K_{x-\frac{1}{2}}\left\{\sqrt{\lambda\left(2 + \frac{\lambda}{\mu^2}\right)}\right\}$$

$$x = 0, 1, 2 \ldots$$

(7.1)

is the Poisson-inverse Gaussian law. Its Laplace transform has the simple form

$$\mathbb{E}(\exp(-tX)) = \exp\left[\frac{\lambda}{\mu}\left\{1 - \left(1 + \frac{2\mu^2}{\lambda}(1 - e^{-t})\right)^{\frac{1}{2}}\right\}\right]. \quad (7.2)$$

The probability generating function $P(s)$ is

$$P(s) = \exp\left[\frac{\lambda}{\mu}\left\{1 - \left(1 + \frac{2\mu^2}{\lambda}(1 - s)^{\frac{1}{2}}\right)\right\}\right].$$

From (7.2) it is readily seen that if $X_1, \ldots, X_n$ are independently distributed as $P\text{-}IG(\mu, \lambda)$,

(a) $\mathcal{L}\left(\sum_{i=1}^{n} X_i\right) = P\text{-}IG(n\mu, n\lambda)$ a property inherited from the mixing parent
(b) If μ is fixed and $\lambda \to \infty$ (7.2) converges to $\exp\{\mu(e^t - 1)\}$ indicating a Poisson-like behaviour in the limit
(c) $\mathbb{E}(X_i) = \mu$, $\text{Var}(X_i) = \mu + \frac{\mu^3}{\lambda}$
(d) Pearson's coefficient of skewness β_1 and kurtosis β_2 are given by

$$\beta_1 = \frac{\phi^3}{\mu(\mu + \phi)^3} + \frac{6\phi}{(\mu + \phi)^2} + \frac{9\mu}{\phi(\mu + \phi)}$$

$$\beta_2 = 3\left\{1 + \frac{5\mu^2 + 2\mu\phi(3 + 2\phi)}{\phi(\mu + \phi)^2}\right\}$$

where $\phi = \frac{\lambda}{\mu}$
(e) $P\text{-}IG(\mu, \lambda)$ is unimodal. This is due to a result on the modality of compound laws due to Holgate (1970)

Sankaran (1968) also derives this law and gives a recurrence relation to calculate the probabilities p_x for any x. This relation is

(f) $p_{x+1} = \dfrac{\mu^2}{(1 + \frac{2\mu^2}{\lambda})}\left\{\dfrac{2x - 1}{\lambda(x + 1)}p_x + \dfrac{1}{x(x + 1)}p_{x-1}\right\}$, $x = 1, 2, \ldots$
(g) $P\text{-}IG(\mu, \lambda)$ is infinitely divisible. An asymptotic result for the probability function given by Teugels and Wilmott (1987) is

$$p(x) \sim \sqrt{\frac{\lambda}{2\pi x^3}} \exp\left(\frac{\mu^3}{\lambda}\right) \left(\frac{2\mu^2}{\lambda(1 + \frac{2\mu^2}{\lambda})}\right)^{x - \frac{1}{2}}.$$

Holla's motivation for introducing this law is based on an application in the theory of accident proneness. Suppose that the number of accidents X sustained by an individual in a fixed period of time is Poisson distributed with mean θ, and suppose that this mean varies from individual to individual according to the $IG(\mu, \lambda)$ law; assume that θ retains this law over the successive time period and Y is the number of accidents in this period. Holla shows that (X, Y) has the following distribution

$$\sqrt{\frac{2\lambda}{\pi}} \frac{e^{\frac{\lambda}{\mu}}}{x!y!} \left(\frac{\lambda}{2(2+\frac{\lambda}{\mu^2})}\right)^{\frac{1}{2}(x+y-\frac{1}{2})} K_{x+y-\frac{1}{2}}\left(\sqrt{2\lambda(2+\frac{\lambda}{\mu^2})}\right).$$
$$x = 0, 1, \ldots, \ y = 0, 1, \ldots$$
(7.3)

Holla shows that, given that we consider individuals who are accident-free in the first time period, the mean number of accidents in the second time period tends to decrease. Sichel, in a series of papers (1971, 1973a, 1973b, 1974, 1975, 1982, 1985a, 1985b, 1986) advanced the P-IG law as a viable alternative to the negative binomial law and applied it to examine its fit to distributions involving word frequencies, sentence lengths in texts, theory of repeat-buying by customers of a single brand of a product, and size of diamonds in marine deposits of southwest Africa.

Ord and Whitmore (1986) consider this law to describe species abundance — a subject with a venerable history dating back to the days of Fisher (1943), Willmot (1986, 1988), Teugels and Willmot (1987) have used this law to fit automobile claim data. They indicate that this law can be justified from a physical standpoint to reflect heterogeneity of risk characteristics within an insurance portfolio or as a model to reflect the possibility of multiple claims from a single accident.

Stein et al., (1987) have examined parameter estimation and introduced a new parametrization which induces parameter orthogonality. They have also proposed some multivariate extensions. Atkinson and Yeh (1982) consider inference for a generalized compound distribution using the generalized inverse Gaussian — $GIG(\gamma, \chi, \psi)$ law — and show how to study tests on the shape parameter γ. Surprisingly all these investigators have resorted to different parametrizations to suit their assumptions. We present a synopsis of this in the following table. We write the mixing IG law in the form

$$f(x) = \sqrt{\frac{\chi}{2\pi x^3}} \exp(\sqrt{\chi\psi}) \exp\left\{-\frac{\chi}{2x} - \frac{\psi x}{2}\right\} 1_{\mathbb{R}^+}(x)$$

where $\chi > 0, \psi \geq 0$.

Table 7.1 *Various parametrizations of the generalized inverse Gaussian law*

	χ	ψ	Mean	Variance
A/Y	$\alpha\beta$ $\alpha > 0, 0 < \beta < \frac{\alpha}{2}$	$\frac{\alpha-2\beta}{\beta}$	$\frac{\beta\sqrt{\alpha}}{\sqrt{\alpha-2\beta}}$	$\frac{\beta^2\sqrt{\alpha}}{((\alpha-2\beta)^{3/2}}$
H/S/O/W	λ	$\frac{\lambda}{\mu^2}$	μ	$\mu + \frac{\mu^3}{\lambda}$
Si	$\frac{\alpha^2\theta}{2}$ $\alpha > 0, 0 \leq \theta \leq 1$	$\frac{2(1-\theta)}{\theta}$	$\frac{\alpha\theta}{2\sqrt{1-\theta}}$	$\frac{\alpha\theta^2}{4(1-\theta)^{3/2}}$
St	w $\alpha > 0,$ $w = \sqrt{\mu^2 + \alpha^2} - \mu$	$\frac{w}{\mu}$	μ	$\mu + \frac{\mu^2}{w}$
Wi (a)	$\beta^{-1}\mu^2$ $(\beta > 0)$	β^{-1}	μ	$\mu(1+\beta)$
(b)	$\frac{\mu(\sqrt{\mu^2+\delta^2}-\mu)}{\delta}$ $\delta > 0$	$\frac{\sqrt{\mu^2+\delta^2}-\mu}{\mu\delta}$	μ	$\frac{\mu^2(\sqrt{\mu^2+\delta^2}+\mu)}{\delta}$

A=Atkinson, Y=Yeh, H=Holla, S=Sankaran, O=Ord, W=Whitmore, Si=Sichel, St=Stein et al, Wi=Willmot.

The compound distribution corresponding to the parametrizations are given below

$$f_1(x) = \sqrt{\frac{2\alpha}{\pi}} \exp\left(\sqrt{\alpha^2 - 2\alpha\beta}\right) \frac{\beta^x}{x!} K_{x-\frac{1}{2}}(\alpha)$$

$$f_2(x) = \sqrt{\frac{\lambda}{2\pi}} \exp\left(\frac{\lambda}{\mu}\right) \left(\sqrt{\frac{\mu^2\lambda}{\lambda+2\mu^2}}\right)^{x-\frac{1}{2}} \frac{1}{x!}$$

$$\times K_{x-\frac{1}{2}}\left(\sqrt{\frac{\lambda(\lambda+2\mu^2)}{\mu^2}}\right)$$

Inference

$$f_3(x) = \sqrt{\frac{2\alpha}{\pi}} \exp(\alpha\sqrt{1-\theta}) \left(\frac{\alpha\theta}{2}\right)^x \frac{1}{x!} K_{x-\frac{1}{2}}(\alpha)$$

$$f_4(x) = \sqrt{\frac{2\alpha}{\pi}} \exp(w) \left(\frac{\mu w}{\alpha}\right)^x \frac{1}{x!} K_{x-\frac{1}{2}}(\alpha)$$

$$f_5(x) = \mu\sqrt{\frac{\gamma}{2\pi}} \exp(\mu\gamma) \left(\sqrt{\left(\frac{\gamma\mu^2}{\gamma+2}\right)}\right)^{x-\frac{1}{2}} \frac{1}{x!} K_{x-\frac{1}{2}}(\mu\sqrt{\gamma(\gamma+2)}).$$

Remarks Atkinson and Yeh stick to the original notation used by Barndorff-Nielsen and point out that this is handy for making inference about the shape parameter γ when the mixing law is $GIG(\gamma, \chi, \psi)$. Stein et al., use a parametrization which has an interpretational edge over the Sichel version and also leads to asymptotically uncorrelated maximum likelihood estimates as does the parametrization of Willmot (b). In Sichel's parametrization α characterizes the frequencies for low values of the random variable X while θ is descriptive of the tail behaviour.

7.2 Inference The calculation of probabilities and derivations of the likelihood function is facilitated by the use of the following relations where $\alpha > 0$.

(i) $K_{x+1}(\alpha) = \frac{2x}{\alpha} K_x(\alpha) + K_{x-1}(\alpha)$.

(ii) $K'_{x-\frac{1}{2}}(\alpha) = -\frac{1}{2} K_{x-\frac{3}{2}}(\alpha) - \frac{1}{2} K_{x+\frac{1}{2}}(\alpha)$.

(iii) $R'_x(\alpha) = R^2_{x-\frac{1}{2}}(\alpha) + \frac{2(x-1)}{\alpha} R_{x-\frac{1}{2}}(\alpha) - 1$.

where $R_{x-\frac{1}{2}}(\alpha) = \frac{K_{x+\frac{1}{2}}(\alpha)}{K_{x-\frac{1}{2}}(\alpha)}$.

(iv) $\frac{K'_{x-\frac{1}{2}}(\alpha)}{K_{x-\frac{1}{2}}(\alpha)} = -R_{x-\frac{1}{2}}(\alpha) + \frac{2x-1}{2\alpha}$.

(v) $R_x(\alpha) = \left\{\frac{2x-3}{\alpha} + R_{x-1}(\alpha)\right\}^{-1}$.

To obtain the maximum likelihood estimates of α, β (in parametrization (1)) we see that if the observations are $n_0, n_1, \ldots, n_x, \ldots$, the log-likelihood is for $f_1(x)$

$$\ell(\alpha, \beta) = \sum_{i=0}^{\infty} n_i \log p_i + \text{constant}$$

where

$$\log p_i = \frac{1}{2} \log \alpha + (\alpha^2 - 2\alpha\beta)^{\frac{1}{2}} + i \log \beta + \log K_{i-\frac{1}{2}}(\alpha).$$

Solving $\frac{\partial l}{\partial \beta} = 0$, we obtain

$$\frac{\partial \ell}{\partial \beta} = \sum_{i=0}^{\infty} \frac{in_i}{\beta} - \alpha(\alpha^2 - 2\alpha\beta)^{\frac{1}{2}} = 0$$

$$\frac{1}{\beta}\bar{x} - \frac{\sqrt{\alpha}}{\sqrt{\alpha - 2\beta}} = 0 \Rightarrow \hat{\mu} = \bar{x} = \frac{\beta\sqrt{\alpha}}{\sqrt{\alpha - 2\beta}}.$$

Equating $\frac{\partial \ell}{\partial \alpha}$ to zero, we have

$$\sum_{i=0}^{\infty} \frac{n_i}{2\alpha} + \frac{n(\alpha - \beta)}{\sqrt{\alpha^2 - 2\alpha\beta}} + \sum_{i=0}^{\infty} n_i \frac{K'_{i-\frac{1}{2}}(\alpha)}{K_{i-\frac{1}{2}}(\alpha)} = 0.$$

Thus using (iv)

$$\frac{n}{2\alpha} + \frac{n(\alpha - \beta)}{\sqrt{\alpha^2 - 2\alpha\beta}} - \sum_{i=0}^{\infty} n_i \, R_{i-\frac{1}{2}}(\alpha) + \sum \frac{n_i(2i-1)}{2\alpha} = 0.$$

Hence

$$\frac{\bar{x}}{\alpha} + \frac{\alpha}{\sqrt{\alpha^2 - 2\alpha\beta}} - \frac{\beta\sqrt{\alpha}}{\alpha\sqrt{\alpha - 2\beta}} = \frac{1}{n}\sum_{i=0}^{\infty} n_i \, R_{i-\frac{1}{2}}(\alpha)$$

and we have

$$\frac{\hat{\alpha}}{\sqrt{\hat{\alpha}^2 - 2\hat{\alpha}\hat{\beta}}} = \frac{1}{n}\sum_{i=0}^{\infty} n_i \, R_{i-\frac{1}{2}}(\hat{\alpha}).$$

Since $\hat{\beta} = \frac{\bar{x}}{\sqrt{\hat{\alpha}}}\sqrt{\hat{\alpha} - 2\hat{\beta}}$ we obtain

$$\hat{\beta}^2 + 2\left(\frac{\bar{x}^2}{\alpha}\right)\hat{\beta} - \bar{x}^2 = 0.$$

The admissible root then yields

$$\hat{\beta} = \frac{\bar{x}}{\hat{\alpha}}\left(\sqrt{\bar{x}^2 + \hat{\alpha}^2} - \bar{x}\right)$$

and

$$\hat{\alpha}^2 - 2\hat{\alpha}\hat{\beta} = \hat{\alpha}^2 + 2\bar{x}^2 - 2\bar{x}\sqrt{\bar{x}^2 + \hat{\alpha}^2},$$

so that

$$\sqrt{\hat{\alpha}^2 - 2\hat{\alpha}\hat{\beta}} = \sqrt{\hat{\alpha}^2 + \bar{x}^2} - \bar{x}.$$

Inference

Moreover

$$\frac{\hat{\alpha}}{\sqrt{\hat{\alpha}^2 - 2\hat{\alpha}\hat{\beta}}} = \frac{\hat{\alpha}}{\sqrt{\hat{\alpha}^2 + \bar{x}^2} - \bar{x}} = \frac{\sqrt{\hat{\alpha}^2 + \bar{x}^2} + \bar{x}}{\hat{\alpha}}$$

and we then have

$$\frac{\sqrt{\hat{\alpha}^2 + \bar{x}^2} + \bar{x}}{\hat{\alpha}} = \frac{1}{n} \sum_{i=0}^{\infty} n_i R_{i-\frac{1}{2}}(\hat{\alpha}). \tag{7.4}$$

Since

$$R_{-\frac{1}{2}}(\alpha) = 1$$

and

$$R_\gamma(\alpha) = \frac{2\gamma}{\alpha} + \frac{1}{R_{\gamma-1}(\alpha)}, \quad \gamma = \frac{1}{2}, \frac{3}{2}, \ldots$$

we see that equation (7.4) can be solved numerically for $\hat{\alpha}$ using Newton-Raphson iteration. On the other hand in Sichel's parametrization it can be shown that

$$\frac{\partial \ell}{\partial \theta} = 0 \quad \Rightarrow \quad \bar{x} = \frac{\hat{\alpha}\hat{\theta}}{2\sqrt{1-\hat{\theta}}}$$

so that solving for $\hat{\theta}$ one obtains

$$\hat{\theta} = \frac{2\bar{x}}{\hat{\alpha}^2} \left\{ \sqrt{\bar{x}^2 + \hat{\alpha}^2} - \bar{x} \right\}. \tag{7.5}$$

Similarly solving $\frac{\partial \ell}{\partial \alpha} = 0$ gives

$$\sum_{i=0}^{\infty} n_i \frac{K_{i-\frac{3}{2}}(\alpha)}{K_{i-\frac{1}{2}}(\alpha)} - \frac{1}{\hat{\alpha}} \left\{ \sqrt{\bar{x}^2 + \hat{\alpha}^2} - \bar{x} + 1 \right\} = 0$$

a formula to be solved by iteration for $\hat{\alpha}$ and then substituted in (7.5) to obtain $\hat{\theta}$. In the parametrization of Stein et al. clearly $\hat{\mu} = \bar{x}$ while the maximum likelihood estimate of α is obtained by solving

$$\frac{\hat{\alpha}}{\sqrt{\bar{x}^2 + \hat{\alpha}^2}} \left(1 + \frac{\bar{x}}{\hat{w}}\right) - \frac{1}{n} \sum_{i=0}^{\infty} n_i R_{i-\frac{1}{2}}(\hat{\alpha}) = 0.$$

Proposition 7.1 *The asymptotic variances of $\hat{\mu}$ and $\hat{\alpha}$ are given by $nvar(\hat{\mu}) = \mu \left(1 + \frac{\mu}{w}\right)$ and*

$$nvar(\tilde{\alpha}) = \left[\frac{-\mu\{(w+\mu)(2(w+\mu)+1) + \mu(1-2(w+\mu))\}}{w^2(w+\mu)} - 1 \right.$$
$$\left. + \mathbb{E}\left(R^2_{i-\frac{1}{2}}(\alpha)\right) \right]^{-1}.$$

To see this, consider the derivatives of the log-likelihood $\ell(\alpha, n)$ with respect to α and μ. Using the parametrization of Stein et al., the log-likelihood is proportional to

$$\ell(\alpha, \mu) \propto \frac{1}{2}\log\alpha + \sqrt{\mu^2 + \alpha^2} - \mu + i\log\left(\frac{\mu}{\alpha}\right) + i\log\left(\sqrt{\mu^2+\alpha^2} - \mu\right)$$
$$+ \log K_{i-\frac{1}{2}}(\alpha).$$

$$\frac{\partial \ell}{\partial \mu} = \frac{\mu}{\sqrt{\mu^2+\alpha^2}} - 1 + \frac{i}{\mu} + \frac{i}{\sqrt{\mu^2+\alpha^2}-\mu}\left(\frac{\mu}{\sqrt{\mu^2+\alpha^2}} - 1\right)$$
$$= \frac{\mu}{\sqrt{\mu^2+\alpha^2}} - 1 + \frac{i}{\mu} - \frac{i}{\sqrt{\mu^2+\alpha^2}}$$
$$= \frac{(\mu-i)}{\sqrt{\mu^2+\alpha^2}} - \frac{(\mu-i)}{\mu}$$
$$= (\mu-i)\left[\frac{1}{\sqrt{\mu^2+\alpha^2}} - \frac{1}{\mu}\right].$$

$$\frac{\partial^2 \ell}{\partial \mu^2} = \frac{\sqrt{\mu^2+\alpha^2}}{(\mu^2+\alpha^2)} - \frac{(\mu-i)\mu}{(\mu^2+\alpha^2)^{\frac{3}{2}}} - \frac{i}{\mu^2}$$
$$= \frac{1}{\sqrt{\mu^2+\alpha^2}} - \frac{(\mu-i)\mu}{(\mu^2+\alpha^2)^{\frac{3}{2}}} - \frac{i}{\mu^2}.$$

$$\mathbb{E}\left(\frac{\partial^2 \ell}{\partial \mu^2}\right) = \frac{1}{\sqrt{\mu^2+\alpha^2}} - \frac{1}{\mu} = \frac{\mu - \sqrt{\mu^2+\alpha^2}}{\mu\sqrt{\mu^2+\alpha^2}} = -\frac{w}{\mu(w+\mu)}.$$

$$\frac{\partial \ell}{\partial \alpha} = \frac{\alpha}{\sqrt{\mu^2+\alpha^2}} + \frac{i\alpha}{\mu^2+\alpha^2 - \mu\sqrt{\mu^2+\alpha^2}} - R_{i-\frac{1}{2}}(\alpha).$$

$$\frac{\partial^2 \ell}{\partial \alpha^2} = \frac{1}{\sqrt{\mu^2+\alpha^2}} - \frac{\alpha^2}{(\mu^2+\alpha^2)^{\frac{3}{2}}} + \frac{i}{\mu^2+\alpha^2 - \mu\sqrt{\mu^2+\alpha^2}}$$
$$- \frac{2i\alpha^2}{(\mu^2+\alpha^2 - \mu\sqrt{\mu^2+\alpha^2})} + \frac{i\alpha^2\mu}{(\mu^2+\alpha^2 - \mu\sqrt{\mu^2+\alpha^2})^2\sqrt{\mu^2+\alpha^2}}$$
$$- R'_{i-\frac{1}{2}}(\alpha).$$

Since $R'_{i-\frac{1}{2}}(\alpha) = R^2_{i-\frac{1}{2}}(\alpha) - \frac{2i}{\alpha}R_{i-\frac{1}{2}}(\alpha) - 1$, we can simplify the above expression to obtain

$$\frac{\partial^2 \ell}{\partial \alpha^2} = \frac{\mu^2}{(\mu^2+\alpha^2)^{\frac{3}{2}}} + \frac{i(\mu^2 - \alpha^2 - \mu\sqrt{\mu^2+\alpha^2})}{(\mu^2+\alpha^2 - \mu\sqrt{\mu^2+\alpha^2})^2}$$
$$+ \frac{i\alpha^2\mu}{(\mu^2+\alpha^2 - \mu\sqrt{\mu^2+\alpha^2})^2\sqrt{\mu^2+\alpha^2}} - R^2_{i-\frac{1}{2}}(\alpha) + \frac{2i}{\alpha}R_{i-\frac{1}{2}}(\alpha) + 1.$$

Inference

Now we need to compute $\mathbb{E}\left[iR_{i-\frac{1}{2}}(\alpha)\right]$.

$$\mathbb{E}\left[iR_{i-\frac{1}{2}}(\alpha)\right] = \sum_{i=0}^{\infty} i \frac{K_{i+\frac{1}{2}}(\alpha)}{K_{i-\frac{1}{2}}(\alpha)} p_i$$

$$= \sum_{i=0}^{\infty} \frac{\alpha\theta}{2(1-\theta)} \frac{K_{\frac{3}{2}}(\alpha\sqrt{1-\theta})}{K_{-\frac{1}{2}}(\alpha\sqrt{1-\theta})} \frac{(\sqrt{1-\theta})^{\frac{3}{2}}}{K_{\frac{3}{2}}(\alpha\sqrt{1-\theta})^x}$$

$$\times \frac{(\frac{\alpha\theta}{2})^i}{i!} K_{i+\frac{3}{2}}(\alpha)$$

$$= \frac{\alpha\theta}{2(1-\theta)} \frac{K_{\frac{3}{2}}(\alpha\sqrt{1-\theta})}{K_{-\frac{1}{2}}(\alpha\sqrt{1-\theta})}$$

$$= \frac{\alpha\theta}{2(1-\theta)} \left(1 + \frac{1}{\alpha\sqrt{1-\theta}}\right)$$

$$= \frac{\alpha\mu(\sqrt{\mu^2+\alpha^2} - \mu + 1)}{(\sqrt{\mu^2+\alpha^2} - \mu)^2}.$$

Taking expectations we have

$$\mathbb{E}\left(-\frac{\partial^2 \ell}{\partial \mu^2}\right) = \frac{w}{\mu(\mu+w)},$$

and after some simplification

$$\mathbb{E}\left(-\frac{\partial^2 \ell}{\partial \alpha^2}\right) = -\mu \left[\frac{\sqrt{\mu^2+\alpha^2}\,(1+2\sqrt{\mu^2+\alpha^2}) + \mu(1 - 2\sqrt{\mu^2+\alpha^2})}{(\sqrt{\mu^2+\alpha^2} - \mu)^2 \sqrt{\mu^2+\alpha^2}}\right]$$

$$- 1 + \mathbb{E}\left[R_{i-\frac{1}{2}}^2(\alpha)\right].$$

The maximum likelihood estimates of the parameters can also be obtained from the probability generating function. Let us illustrate this method in the case of $f_5(x)$ following Willmot's arguments. This is also applicable for $f_2(x)$.

First consider the logarithmic derivative of

$$P(s) = \exp\left[\frac{\mu}{\beta}\{1 - 2\beta(s-1)\}^{\frac{1}{2}}\right]$$

$$\frac{P'(s)}{P(s)} = \mu\{1 - 2\beta(s-1)\}^{-\frac{1}{2}}.$$

Then

$$-\frac{P(s)\{1 - 2\beta(s-1)\}}{\mu\beta} = -\frac{P(s)}{\beta}\{1 - 2\beta(s-1)\}^{\frac{1}{2}}.$$

Now
$$\frac{\partial}{\partial \mu} P(s) = \frac{P(s)}{\beta} \left[1 - \{1 - 2\beta(s-1)\}^{\frac{1}{2}} \right]$$
$$= \frac{P(s)}{\beta} - \frac{P'(s)\{1 - 2\beta(s-1)\}}{\mu\beta}$$
$$= \frac{P(s)}{\beta} - \frac{(1+2\beta)}{\mu\beta} P'(s) + \frac{2}{\mu} P'(s).$$

Equating the coefficients of s^k on both sides and then dividing by p_k we have
$$\frac{\partial}{\partial \mu} \log p_k = \frac{1}{\beta} - \frac{(1+2\beta)}{\mu\beta} \frac{(k+1)}{p_k} p_{k+1} + \frac{2k}{\mu}. \qquad (7.6)$$

Hence solving the loglikelihood equation we obtain
$$\sum_{k=0}^{\infty} n_k \left\{ \frac{1}{\beta} - \frac{(1+2\beta)}{\mu\beta} \frac{(k+1)p_{r+1}}{p_r} + \frac{2k}{\mu} \right\} = 0.$$

Writing $t_k = \frac{(k+1)}{p_k} p_{k+1}$,
$$\frac{(1+2\beta)}{\mu\beta} \sum n_k\, t_k = \frac{n}{\beta} + \frac{2n\bar{x}}{\mu} \quad \Rightarrow \quad \bar{x} = \hat{\mu} \qquad (7.7)$$

Likewise
$$\frac{\partial P(s)}{\partial \beta} = P(s)\left[-\frac{\mu}{\beta^2} + \frac{\mu}{\beta^2} \{1 - 2\beta(s-1)\}^{\frac{1}{2}} \right.$$
$$\left. + \frac{\mu(s-1)}{\beta} \{1 - 2\beta(s-1)\}^{-\frac{1}{2}} \right]$$
$$= -\frac{\mu}{\beta^2} P(s) \left[1 - \{1 - 2\beta(s-1)\}^{\frac{1}{2}} \right]$$
$$+ \frac{(s-1)}{\beta} P(s) \mu \{1 - 2\beta(s-1)\}^{-\frac{1}{2}}$$
$$= -\frac{\mu}{\beta} \frac{\partial}{\partial \mu} P(s) + \frac{(s-1)}{\beta} P(s) \frac{P'(s)}{P(s)}$$
$$= -\frac{\mu}{\beta} \frac{\partial}{\partial \mu} P(s) + \frac{(s-1)}{\beta} P'(s).$$

Therefore
$$\frac{\partial \log p_k}{\partial \beta} = -\frac{\mu}{\beta} \left(\frac{\partial}{\partial \mu} \log p_k \right) + \frac{(k - t_k)}{\beta}.$$

Inference

$$\frac{\partial \ell}{\partial \beta} = \sum_{k=0}^{\infty} n_k \frac{\partial \log p_k}{\partial \beta} = 0 \Rightarrow -\frac{\mu}{\beta} \sum n_k \frac{\partial \log p_k}{\partial \mu}$$

$$+ \sum \left(\frac{k-t_k}{\beta}\right) n_k = 0.$$

This in turn implies $\sum \frac{k-t_k}{\beta} n_k = 0$ and we obtain

$$g(\beta) = \sum_{k=0}^{\infty} n_k t_k - n\bar{x} = 0. \qquad (7.8)$$

To solve for $\hat{\beta}$ we start with the relation

$$t_k = \frac{(k+1)p_{k+1}}{p_k}$$

$$\frac{\partial t_k}{\partial \beta} = (k+1) \left\{ \frac{p'_{k+1}}{p_{k+1}} \frac{p_{k+1}}{p_k} - \frac{p_{k+1}}{p_k} \frac{p'_k}{p_k} \right\}$$

$$= (k+1) \frac{p_{k+1}}{p_k} \left\{ \frac{p'_{k+1}}{p_{k+1}} - \frac{p'_k}{p_k} \right\}.$$

To obtain $\frac{p'_k}{p_k}$ we note that

$$\frac{\frac{\partial p_k}{\partial \beta}}{p_k} = -\frac{\mu}{\beta} \left(\frac{\partial}{\partial \mu} \log p_k \right) + \frac{k - t_k}{\beta}.$$

Hence

$$\frac{p'_{k+1}}{p_{k+1}} - \frac{p'_k}{p_k} = -\frac{\mu}{\beta} \left[\frac{\partial}{\partial \mu} \log p_{k+1} - \frac{\partial}{\partial \mu} \log p_k \right] + \frac{1}{\beta} - \frac{(t_{k+1} - t_k)}{\beta}$$

$$= -\frac{\mu}{\beta} \left[\frac{-(1+2\beta)}{\mu\beta} (t_{k+1} - t_k) + \frac{2}{\mu} \right] + \frac{1}{\beta} - \frac{(t_{k+1} - t_k)}{\beta}$$

using (7.6). Hence

$$\frac{\partial t_k}{\partial \beta} = t_k \left[\frac{(1+2\beta)}{\beta^2} (t_{k+1} - t_k) - \frac{1}{\beta} - \frac{(t_{k+1} - t_k)}{\beta} \right]$$

$$= t_k \left[(t_{k+1} - t_k) \frac{(1+\beta)}{\beta^2} - \frac{1}{\beta} \right].$$

Finally, since

$$g'(\beta) = \sum_{k=0}^{\infty} n_k \frac{\partial t_k}{\partial \beta} = 0$$

we have

$$\frac{(1+\beta)}{\beta^2} \sum_{k=0}^{\infty} n_k t_k \left[t_{k+1} - t_k\right] - \frac{1}{\beta} \sum_{k=0}^{\infty} n_k t_k = 0.$$

$\hat{\beta}$ is a solution to $g(\beta) = 0$ (see 7.8) and therefore we can use Newton-Raphson iteration techniques to estimate β. As a starting value we can take the moment estimate of β, $s^2 \bar{x}^{-1} - 1$ where $s^2 = \frac{\sum_{i=1}^{n}(x_i - \bar{x})^2}{n}$, to iterate

$$\hat{\beta}_{k+1} = \hat{\beta}_k - \frac{g(\hat{\beta}_k)}{g'(\hat{\beta}_k)}.$$

A similar technique can be adopted for the density $f_2(x)$. In this case one obtains

$$\frac{\partial p_r}{\partial \mu} = -\frac{\lambda p_r}{\mu^2} + \frac{\lambda(r+1)}{\mu^3} p_{r+1}, \quad r = 0, 1, \ldots$$

and

$$\frac{\partial p_r}{\partial \lambda} = \frac{\lambda + r\mu}{\lambda \mu} p_r - (r+1) p_{r+1} \frac{(\lambda + \mu)^2}{\lambda \mu^2}.$$

If the distribution $f_2(x)$ (in particular) is truncated at $x = 0$, the log-likelihood is

$$\ell(\mu, \lambda) = \sum_{r=1}^{\infty} n_r \log p_r - \sum_{r=1}^{\infty} n_r \log(1 - p_0).$$

Therefore $\frac{\partial \ell}{\partial \mu} = 0$ yields

$$\sum_{r=1}^{\infty} n_r \left(-\frac{\lambda p_r}{\mu^2} + \frac{\lambda(r+1)}{\mu^3} p_{r+1}\right) + \frac{n \frac{\partial p_0}{\partial \mu}}{1 - p_0} = 0.$$

But

$$\frac{\partial p_0}{\partial \mu} = -\frac{\lambda p_0}{\mu^2} + \frac{\lambda}{\mu^3} p_1.$$

Hence

$$\frac{\partial \ell}{\partial \mu} = 0 \Rightarrow \frac{\lambda}{\mu^3} \left\{\sum_{1}^{\infty}(r+1) \frac{p_{r+1}}{p_r} - n\mu + \frac{n(p_1 - \mu p_0)}{1 - p_0}\right\} = 0. \quad (7.9)$$

Likewise

$$\frac{\partial p_0}{\partial \lambda} = \frac{1}{\mu} - p_1 \frac{(\lambda + \mu)^2}{\lambda \mu^2}.$$

Examples

Hence

$$\frac{\partial \ell}{\partial \lambda} = 0 \Rightarrow \frac{1}{\mu^2 \lambda} \left[n\mu\lambda + n\mu^2 \bar{x} - (\mu^2 + \lambda) \sum_{r=1}^{\infty} n_r(r+1) \frac{p_{r+1}}{p_r} \right] \quad (7.10)$$
$$+ \frac{n}{\lambda \mu^2 (1 - p_0)} \left[\lambda \mu p_0 - (\lambda + \mu^2) p_1 \right] = 0.$$

Equations (7.9) and (7.10) have to be solved numerically to obtain the maximum likelihood estimates.

The moment estimates of the parameters are provided in Table 7.2.

Table 7.2 *Moment estimates*

1. $\tilde{\alpha} = \dfrac{\bar{x}\sqrt{\bar{x}(\bar{x} + 2s^2)}}{s^2}$ $\qquad \tilde{\beta} = \dfrac{\bar{x}\sqrt{\bar{x}}}{\sqrt{\bar{x} + 2s^2}}$

2. $\tilde{\lambda}^{-1} = \dfrac{\sum_{i=1}^{n}(x_i - \bar{x})^2}{n}$ $\qquad \tilde{\mu} = \bar{x}$

3. $\tilde{\alpha} = \dfrac{2\bar{x}\sqrt{1 - \hat{\theta}}}{\hat{\theta}}$ $\qquad \tilde{\theta} = 1 - \left(\dfrac{2s^2}{\bar{x}} - 1\right)^{-1}$

4. $\tilde{\alpha} = \bar{x}\sqrt{1 - \left(\dfrac{\bar{x} + s^2}{s^2}\right)^2}$ $\qquad \tilde{\mu} = \bar{x}$

5. $\tilde{\beta} = s^2 \bar{x}^{-1} - 1$ $\qquad \tilde{\mu} = \bar{x}$

Sichel remarks that the moment estimators are extremely inefficient and should be used only if the coefficient of variation is $\leq 35\%$. In the truncated case his recommendation for obtaining parameter estimates involves solving

$$\frac{\sum_{i=1}^{n} i\, n_i}{n} = \frac{\bar{x}}{1 - \hat{p}_0}, \quad \hat{p}_0 = \frac{n_0}{n},$$

$$\frac{n_1}{n} = \frac{\bar{x}}{\sqrt{1 + \frac{2\bar{x}^2}{\lambda}}} \frac{\hat{p}_0}{1 - \hat{p}_0}, \quad \sum_{i=1}^{\infty} n_i = n$$

$$\log\left(\frac{n_0}{n_1}\right) = \frac{\hat{\lambda}}{\bar{x}} - \frac{\hat{\lambda}}{\bar{x}}\sqrt{1 + \frac{2\bar{x}^2}{\hat{\lambda}}}.$$

Sichel also recommends varying the estimates until the total χ^2 is minimized (minimum chi-squared method).

7.3 Examples

Example 7.1 As a first example consider the data(see Table 7.3 on the distribution of European corn beans taken from Kemp and Kemp (1965). Sankaran used this data set to examine the adequacy of the $P - IG(\mu, \lambda)$ fit and compared it with the Hermite law. Sankaran used moment estimates of μ and λ. As can be seen the fit is quite good.

Table 7.3 *Distribution of European corn beans data together with expected frequencies based on the new distribution, and Hermite distribution (M.L.) from Sprott (1958, Table 6) and quoted by Kemp and Kemp (1965, Table 1)*

no. of Larvae per plant	observed frequencies	expected frequenceies	
		P-IG	Hermite (M.L.)
0	423	425.2	427.7
1	414	412.9	389.3
2	253	246.7	262.0
3	117	120.1	131.0
4	53	53.0	55.8
5	22	22.3	20.6
6	4	9.1	6.8
7	5	3.7	2.0
8	3	1.5	0.6
9	2	0.6	0.2
total	1296	1295.1	1296.0

Example 7.2 Willmot (1988) presents six data sets analyzed by Gossiaux and Lemaire (1981) of automobile insurance claims per policy over a fixed time period. We present one data set (# 1) for illustration.

Table 7.4 Auto-insurance claims per policy

No. of claims k	No. of policies F_k	NB	P-IG
0	103 704	103 723.61	103 710.03
1	14 075	13 989.95	13 989.95
2	1 766	1 857.08	1 784.91
3	255	245.19	254.49
4	45	32.29	40.42
5	6	4.24	6.94
6	2	0.56	1.26
Total	119 853	119 852.92	119 852.70

Examples 135

For this example the M.L.E for the NB is -54,615.315 and for the P-IG is -54,609.758 . As for the Pearson goodness-of-fit, the Chi-squared values based on 3 degrees of freedom are 12.37 and 0.78 giving p-values of 0.006 and 0.855 respectively.

Example 7.3 This example is from Sichel (1974) who studied sentence-length distributions from Macaulay, fitted to both the P-IG model and the negative binomial. The data set in Table 7.5 comes from Yule (1939).

Table 7.5 *Senetence-length distributions*

	Ob. no. of sentences	P-IG Ex. no. of sentences	Negative binomial Ex. no. of sentences
No. of words			
1-5	46	57.8	110.3
6-10	204	201.0	185.7
11-15	252	244.6	202.6
16-20	200	209.1	184.2
21-25	186	157.5	152.3
26-30	108	113.0	118.6
31-35	61	79.5	88.6
36-40	68	55.6	64.4
41-45	38	38.9	45.7
46-50	24	27.3	31.9
51-55	20	19.2	32.0
56-60	12	13.6	14.9
61-65	8	9.6	10.1
66-80	14	15.2	14.1
81-90	4	4.3	3.2
91+	6	4.8	2.4

Since the expected frequencies for $x = 1$ were very small they were combined in the first cell, the frequency class corresponding to 1–5 words and zero truncation was unneccesary. The P-IG performs quite well and is far superior to the negative binomial fit. The next example is also from Sichel one involving a market survey of repeat-buyers.

Example 7.4 The observed frequency distribution of the number of bars of toilet soap bought by 614 households during an 8-month period in 1978, for the entire product field, is shown in Table 7.6. Of particular interest is the fact that all households bought at least two bars of toilet soap during this lengthy period. (For the shorter periods a certain number of non-buyers was recorded.)

Table 7.6 *Frequency distribution of toilet soap bars*

Number of units bought	Number of households		
	Observed	Expected	
r	f_0	NBD	IGP
0-3	10	27	13
4-7	69	79	76
8-7	136	106	122
12-15	113	105	117
16-19	19	88	92
20-23	68	68	65
24-27	47	49	44
28-31	23	33	29
32-35	19	22	19
36-39	10	14	13
40-43	10	9	8
44-51	11	9	9
52 and over	6	5	7
Totals	614	614	614
χ^2	-	27	6
d.f.	-	10	10
$P(\chi^2/d.f.)$	-	0.003	0.8

7.4 A compound inverse Gaussian model

In marketing purchase incidence models are needed to obtain an idea of the number of purchases of a low-cost consumer product in a fixed time interval and the waiting period between successive purchases. A popular model for purchasing behaviour of a customer on successive buys has been the Poisson process. This naturally leads to Erlang interpurchase times. The consumer population is quite heterogeneous and the studies of the pioneering investigators in this area like Anscombe (1961), Chatfield (1966, 1973) and Ehrenberg (1959) have slowly shifted to models in which the shape parameter of the gamma was held fixed at a value equal to two and letting the scale parameter account for consumer

A compound inverse Gaussian model 137

heterogeneity. Banerjee and Bhattacharyya (1976) have questioned the validity of this hypothesis. They postulate the $IG(\mu, \lambda)$ law as an interpurchase time distribution and then adopt a natural conjugate family of bivariate laws to justify the heterogenous behaviour of the population of consumer households. Thus they use a two stage scheme in the model formulation. Following are the steps of their procedure.

(1) Check the fit of the IG model, $IG(\frac{1}{\delta}, \lambda)$

(2) If the fit is satisfactory then examine how the estimated parameter values are distributed across a large sample N of households. To do this one really needs data on (δ, λ) which is not observable. However, they are estimated from the interpurchase time data, and a bivariate frequency table is constructed for this sample. The parameters for the distribution of (δ, λ) are estimated using moment estimators. The observed frequencies in this table are then compared with the theoretical frequencies which have to be obtained by numerical integration. From this bivariate frequency table one can obtain the corresponding marginal frequency distributions for δ and λ also and a χ^2 fit is assessed for each.

(3) Lastly the observed frequency distribution of $N(t)$ the number of purchases in a time period t is obtained and checked against the theoretical model.

Example 7.5 Banerjee and Bhattacharyya have analyzed a random sample of 289 households who have made at least 20 purchases of toothpastes during a five year period.

First they considered households who purchased at least 100 times in the 5 year period. They found 12 which fitted this requirement and then fitted an $IG(\frac{1}{\delta}, \lambda)$ model to each of the 12 households by constructing histograms of observed time between successive purchases and then comparing them with the theoretical $IG(\frac{1}{\delta}, \hat{\lambda})$ model (where the parameters are estimated by maximum likelihood). They found that in each case the p-values of the χ^2 goodness of fit ranged from 0.053 to 0.634. Thus at the 5% level the fit was satisfactory. At the next step $\hat{\delta}, \hat{\lambda}$ were estimated for each of the 289 sampled households and a bivariate frequency distribution constructed. This observed distribution was compared to the theoretical law (which is the natural conjugate to the $IG(\frac{1}{\delta}, \lambda)$ law). Denoting this law by $\pi_c(\alpha, \beta, \gamma)$ (see equation 1.21), first the parameters α, β and γ were estimated using the method of moments and then the theoretical probabilities obtained by numerical integration. We reproduce in Table 7.7 the observed and expected frequency distribution of $(\frac{1}{\delta}, \lambda)$ as obtained by these authors. Then a χ^2 goodness of fit was performed for (a) the bivariate model (b) the marginal model for $\frac{1}{\delta}$, and (c) the marginal model for λ. Since none of these were found significant at the 5% level, they proceeded to obtain the observed frequency distribution of $N(t)$, the number of purchases in t weeks and finally compared it with the expected frequencies. The results are given in Table 7.8. The fit is quite good.

Table 7.7 *Observed and expected frequency distribution of* $(1/\delta, \lambda)$

λ	$\leq .43$	.43–.70	.70–.97	.97–1.24	1.24–1.51	≥ 1.51
≤ 1.0	13.0	23.0	23.0	7.0	7.0	11.0
	(20.1)	(24.1)	(22.2)	(12.2)	(6.9)	(8.2)
1.0-1.88	19.0	31.0	17.0	8.0	2.0	2.0
	(23.5)	(25.1)	(16.4)	(7.3)	(1.3)	(1.1)
1.88-2.76	15.0	25.0	15.0	2.0	1.0	
	(15.6)	(19.0)	(17.8)	(2.2)	(0.2)	
2.76-3.64	14.0	15.0	1.0			
	(13.2)	(13.4)	(2.1)			
3.64-4.52	9.0	7.0				
	(7.2)	(8.0)				
4.52-5.40	7.0	3.0				
	(4.5)	(5.2)				
≥ 5.4	9.0	3.0				
	(4.3)	(7.6)				

Expected frequencies are displayed in parentheses.

Table 7.8 *Observed and Expected Frequency Distributions of N(t) for the Toothpaste Data*

n	t=1	t=2	t=3	t=4	t=5
0	143 (155.5)	82 (80.6)	54 (48.6)	28 (32.9)	16 (24.3)
1	119 (92.8)	110 (103.2)	73 (79.2)	71 (57.2)	57 (41.9)
2	20 (25.4)	61 (57.8)	84 (70.8)	74 (66.2)	63 (55.2)
3	5 (8.1)	23 (24.0)	40 (41.3)	47 (52.0)	55 (53.5)
4	2 (3.2)	8 (10.4)	16 (21.1)	27 (32.1)	36 (40.2)
5	(1.4)	3 (5.2)	12 (11.0)	15 (18.2)	17 (26.0)
6	(1.2)	1 (2.9)	5 (6.1)	8 (10.4)	15 (15.9)
7	(0.9)	1 (1.7)	2 (3.5)	11 (6.4)	12 (9.8)
8	(0.3)	(1.2)	2 (2.3)	3 (4.0)	10 (6.4)
9	(0.2)	(0.6)	1 (1.4)	4 (2.6)	1 (4.3)
10		(0.6)	(0.9)	(1.7)	3 (2.9)
11		(0.3)	(0.6)	(1.2)	3 (2.0)
12		(0.3)	(0.6)	(0.9)	(1.4)
13		(0.2)	(0.6)	1 (0.6)	(1.2)
14			(0.3)	(0.6)	1 (0.9)
15			(0.3)	(0.6)	(0.6)
16			(0.2)	(0.3)	(0.6)
17			(0.2)	(0.3)	(0.3)
18				(0.3)	(0.3)
19				(0.3)	(0.3)
20				(0.2)	(0.3)
21					(0.3)
22					(0.2)
23					(0.2)

7.5 A normal-gamma mixture

Whitmore (1986) introduced a family of normal-gamma mixtures based on a similar family studied by Banerjee and Bhattacharyya. The difference between the two approaches lies in the requirement that the drift parameter can be negative under Whitmore's model. Whitmore advances his case by noting that when the drift ν is close to zero, a realistic assumption about the associated Wiener process having either zero drift or negative drift is quite reasonable. In support of this theory he argues that in the case of an employee's level of dissatisfaction with his job, the employee service time $W(t)$ may drift towards zero ($\nu > 0$) or away from the threshold (tolerance) level of dissatisfaction ($\nu < 0$). If $\nu < 0$ he decides to call it quits. It is well known that if $\{X_\nu(t)|t \geq 0\}$ is a Brownian motion process on $\mathbb{R}$ with drift ν and diffusion constant $\frac{1}{\lambda}$ ($\lambda > 0$) the first passage time distribution for the process to reach the level 1 (assuming $X_\nu(0) = 0$ and writing $T_1 = X$ for the first time level 1 is attained) is

$$p(x \mid \nu, \lambda) = \sqrt{\frac{\lambda}{2\pi x^3}} \exp\left\{-\frac{\lambda(\nu x - 1)^2}{2x}\right\} 1_{\mathbb{R}^+}(x). \quad (7.11)$$

When $\nu \geq 0$ $T_1 = X < \infty$ with probability one and when $\nu < 0$ $P[T_1 = X < \infty] = \exp(-2\lambda|\nu|)$ since $T_1 = \infty$ with positive probability. Thus $P(T_1 = \infty) = 1 - \exp(2\lambda\nu)$, $\nu < 0$.

Theorem 7.1 *Under the assumptions stated above if*

$$\mathcal{L}(\lambda) = \Gamma(\alpha, \beta) \text{ and } \mathcal{L}(\nu \mid \lambda) = N\left(\xi, \frac{\sigma}{\lambda}\right)$$

where $\alpha, \beta, \sigma > 0$ and $\xi \in \mathbb{R}$, the unconditional law of X is

$$\frac{\Gamma(\alpha + \frac{1}{2})}{\Gamma(\alpha)} \frac{1}{\{2\pi\beta x^3(\sigma x + 1)\}^{\frac{1}{2}}} \left\{1 + \frac{(x\xi - 1)^2}{2\beta x(x\sigma + 1)}\right\}^{-\alpha - \frac{1}{2}} 1_{\mathbb{R}^+}(x) \quad (7.12)$$

We omit the proof. The density $f(x)$ is improper since $p(x \mid \nu, \lambda)$ is improper for $\nu < 0$. As corollaries we consider the cases $\nu = \xi, \sigma = 0$ and $\lambda = \frac{\alpha}{\beta}$ and take the limit as $\alpha \to \infty$.

Corollary 7.1 *If in Theorem 7.1, we let $\nu = \xi, \sigma = 0$*

$$f_1(x|\xi) = \frac{\Gamma(\alpha + \frac{1}{2})}{\Gamma(\alpha)} \left(\frac{1}{2\pi\beta x^3}\right)^{\frac{1}{2}} \left(1 + \frac{(x-1)^2}{2\beta x}\right)^{-\alpha - \frac{1}{2}} 1_{\mathbb{R}^+}(x). \quad (7.13)$$

Corollary 7.2 *If in Theorem 7.1, we let $\lambda = \frac{\alpha}{\beta}$ and take the limit as $\alpha \to \infty$*

$$f_2(x \mid \lambda) = \left(\frac{\lambda}{2\pi x^3(\sigma x + 1)}\right)^{\frac{1}{2}} \exp\left\{-\frac{\lambda(x\xi - 1)^2}{2x(\sigma x + 1)}\right\} 1_{\mathbb{R}^+}(x). \quad (7.14)$$

Remarks

(1) Both (7.13) and (7.14) are improper densities

A normal-inverse Gaussian mixture 141

(2) If $\nu < 0, F(\infty \mid \nu, \lambda) = \exp(2\nu\lambda)$

(3) $F_1(\infty \mid \xi) = \left(1 - \frac{2\xi}{\beta}\right)^{-\alpha}$ if $\nu < 0$.

Whitmore analyzes three data sets using this normal-gamma mixture. The first study deals with the situation when all the sample values are finite and uncensored — namely the complete sample case. In this context the probability of an infinite first passage time is negligible. The second study has some sample observations censored to the right due to a non-negligible infinite first passage time. The final study contains only finite observations while the number of infinite observations is unknown; so the density is truncated at α. Whitmore refers to this sample as an **incomplete sample**. We present for illustrative purposes an analysis of the von Alven data set.

Example 3.1 (concluded) We have seen in Chapter 6 that a goodness of fit has shown a satisfactory fit. Thus parameter heterogeneity can be checked by fitting the normal-gamma mixture. Table 7.9 provides the maximum likelihood estimates and the sample log-likelihood for both (7.11) and (7.12). It appears that both laws fit equally well which suggests that there is little or no variation in the parameters and that the extra parameters in the mixture model have not contributed to the sample likelihood.

Table 7.9 *Maximum likelihood estimates and sample log-likelihoods*

Model	Estimates		Sample log-likelihood
Normal-gamma	$\hat{d} = 2.773 \times 10^{-1}$	$\hat{a} = 4756$	-99.06
mixture	$\hat{v} = 3.749 \times 10^{-6}$	$\hat{b} = 2865$	
Unmixed	$\hat{\delta} = 2.773 \times 10^{-1}$	$\hat{\lambda} = 1.659$	-99.06

7.6 A normal-inverse Gaussian mixture

Suppose that $\mathcal{L}(X_i) = N(0, \sigma^2)$ and that σ^2 varies from observation to observation according to $IG(\mu, \lambda)$. Then Sankaran (1968) showed that the resulting mixture leads to the following unconditional law for X_i

$$f(x) = \frac{1}{\pi} \exp\left(\frac{\lambda}{\mu}\right) \frac{\lambda}{\mu\sqrt{x^2+\lambda}} K_1\left(\frac{\sqrt{\lambda(x^2+\lambda)}}{\mu}\right)$$

with Laplace transform

$$\exp\left\{\frac{\lambda}{\mu}\left(1 - \sqrt{1 + \frac{t^2\mu^2}{\lambda}}\right)\right\}.$$

The intractability of the Bessel function spelled the doom and no further studies were made. Bhattacharya (1987) on the other hand considered a different line. He assumed that

$$\mathcal{L}(X \mid m, \sigma) = N(m, \sigma^2), \quad \mathcal{L}(m|\sigma) = N(\mu_0, \sigma^2), \quad \sigma > 0$$

and $\mathcal{L}(\sigma^2) = IG(\mu, \alpha)$ so that

$$\pi_2(\sigma) = \sqrt{\frac{2\alpha}{\pi}} \frac{1}{\sigma^2} \exp\left\{\frac{\alpha}{2\mu^2} \frac{(\sigma^2 - \mu)^2}{\sigma^2}\right\} 1_{(0,\infty)}(\sigma) \qquad (7.15)$$

where $\alpha > 0, \mu > 0$.

These considerations proved worthwhile and made Bayesian estimation of μ feasible. The following theorem provides an estimator of μ under quadratic loss.

Theorem 7.2 Let $X = (X_1, \ldots, X_n)$ be a random sample such that $\mathcal{L}(X_i) = N(m, \sigma^2)$. Further let the prior of m given σ and the prior of σ be defined respectively by $\mathcal{L}(m \mid \sigma) = N(\mu_0, \sigma^2)$, and (7.15). Then with respect to quadratic loss the Bayes estimator of μ is

$$\hat{m}_B = \frac{n\overline{X} + \mu_0}{(n+1)}.$$

Proof The likelihood function is proportional to

$$\sigma^{-n} \exp\left\{-\frac{ns^2}{2\sigma^2} - \frac{n(\bar{x} - m)^2}{2\sigma^2}\right\}$$

where $ns^2 = \sum_{i=1}^n (x_i - \bar{x})^2$. The posterior of (m, σ) can be obtained by integrating out m and σ in

$$\sigma^{-n-3} \exp\left\{-\frac{ns^2}{2\sigma^2} - \frac{n(\bar{x} - m)^2}{2\sigma^2} - \frac{(m - \mu_0)^2}{2\sigma^2} - \frac{\alpha(\sigma^2 - \mu^2)}{2\sigma^2 m^2}\right\}.$$

The posterior of (m, σ) will be proportional to

$$\sigma^{-n-3} \exp\left[-\frac{\{ns^2 + n(m - \bar{x})^2 + (m - \mu_0)^2 + \alpha\}}{2\sigma^2} - \frac{\alpha\sigma^2}{2\mu^2}\right] 1_{\mathbf{R} \times \mathbf{R}^+}(m, \sigma).$$

We can now simplify the term in the exponent

$$n(m - \bar{x})^2 + (m - \mu_0)^2 = (n+1)\left[m^2 - 2m\frac{(n\bar{x} + \mu_0)}{n+1}\right]$$

$$= (n+1)\left(m - \frac{n\bar{x} + \mu_0}{n+1}\right)^2 + \frac{(n\bar{x} + \mu_0)^2}{n+1}$$

so that writing

$$a = \frac{n\bar{x} + \mu_0}{n+1}, \quad b = \frac{n(\mu_0 - \bar{x})^2 + n(n+1)s^2 + (n+1)\alpha}{(n+1)^2},$$

the posterior of (m, σ) is proportional to

$$\sigma^{-n-3} \exp\left\{-\frac{\alpha\sigma^2}{2\mu^2} - \frac{(n+1)}{2\sigma^2}(b + (m-a)^2)\right\} 1_{\mathbb{R} \times \mathbb{R}^+}(m, \sigma). \quad (7.16)$$

The marginal posterior of m can be obtained from (7.16). By writing $\sigma^2 = Y$, the posterior of m is proportional to

$$\int_0^\infty y^{-\frac{n+2}{2} - 1} \exp\left\{-\frac{\chi}{2y} - \frac{\psi y}{2}\right\} dy$$

where

$$\chi = (n+1)(b + (m-a)^2), \quad \psi = \frac{\alpha}{m^2}.$$

The integral being the familiar $GIG(-\frac{n+2}{2}, \chi, \psi)$ law we see that the posterior of m, is apart from a normalizing factor c

$$\pi(m|x) = \frac{cK_{\frac{n+2}{2}}\left(\frac{\sqrt{n+1\alpha\{b+(m-a)^2\}}}{\mu^2}\right)}{\{b + (m-a)^2\}^{\frac{n+2}{4}}} 1_{\mathbb{R}}(m).$$

Since in $\int_{\mathbb{R}}(m-a)\pi(m \mid x)dm$, the integrand is an odd function, $\mathbb{E}(m - a \mid x) = 0$ as is $\mathbb{E}(m - a)$. Therefore with respect to squared error loss the Bayes estimator of m is

$$\hat{m}_B = \frac{n\bar{X} + \mu_0}{n+1}. \quad \clubsuit$$

Theorem 7.2 admits of a generalization to an arbitrary prior for σ with the requirement that it be non-negative over $\mathbb{R}^+$ and such that the posterior of (m, σ) exists. Then, as shown by Bhattacharya, the Bayes estimator with respect to quadratic loss has still the same form.

Independently of Bhattacharya, Athreya (1986) has considered a $GIG(\nu, \chi, \psi)$ law for the marginal prior of σ^2 and shown that the marginal posterior of σ^2 given the data is again a GIG law.

7.7 A mixture inverse Gaussian — M-$IG(\mu, \lambda, p)$

Jorgensen et al. (1991) used the distribution function of $IG(\mu, \lambda)$ as a basis for deriving a generalization of the $IG(\mu, \lambda)$. The resultant law is termed the mixture inverse Gaussian law. First we recall that the distribution function of the IG law is expressible as a linear combination of standard normal law. Indeed

$$F(x) = \Phi(\alpha(x)) + \exp\left(\frac{2\lambda}{\mu}\right) \Phi(\bar{\alpha}(x)) \tag{7.17}$$

where

$$\alpha(x) = \frac{\lambda(x-\mu)}{\mu\sqrt{x}}, \quad \bar{\alpha}(x) = \frac{\lambda(x+\mu)}{\mu\sqrt{x}} \tag{7.18}$$

possess an interesting symmetry, namely

$$\alpha(x) = -2x\bar{\alpha}'(x) \text{ and } \bar{\alpha}(x) = -2x\alpha'(x). \tag{7.19}$$

The pair $(\alpha(x), \bar{\alpha}(x))$ constitutes an independent solution set of Euler's differential equation

$$4x^2 y''(x) + 4x y'(x) - y(x) = 0.$$

Consider a pair $(\alpha, \bar{\alpha})$ satisfying (7.19) which generates a distribution function $F(x)$ satisfying (7.17). It can be shown that the distribution function is of the form

$$G_p(x) = \Phi(\alpha(x)) + (1-2p)\Phi(\bar{\alpha}(x))$$

where $0 \leq p \leq 1$. The density corresponding to $G_p(x)$ is

$$\begin{aligned}g_p(x) &= \sqrt{\frac{\lambda}{2\pi x^3}} \left\{(1-p) + \frac{px}{\mu}\right\} \exp\left\{-\frac{\lambda(x-\mu)^2}{2\mu^2 x}\right\} 1_{\mathbf{R}^+}(x) \\ &= \frac{\gamma+x}{\gamma+\mu} f(x)\end{aligned} \tag{7.20}$$

where $f(x)$ is the density of $IG(\mu, \lambda)$. From (7.20) it is clear that $g_p(x)$ is a finite mixture of $IG(\mu, \lambda)$ with its length-biased density $\frac{xf(x)}{\mathbf{E}(x)}$, the weights being $(1-p)$ and p for $0 \leq p \leq 1$ and $p = \frac{\mu}{\gamma+\mu}, \gamma \geq 0$. When $\gamma = \alpha$, g_p reduces to $IG(\mu, \lambda)$ and when $\gamma = 0$ one obtains $RIG(\lambda, \lambda\mu^{-2})$. Finally when $\gamma = \mu$, $g_p(x)$ is the family studied in detail by Birnbaum and Saunders (1969). Desmond (1986) has noted this mixture representation for the case $\gamma = \mu$. When a random variable X has the density (7.20) we say that X has the M-$IG(\mu, \lambda, p)$ law. Following are some salient characteristics of this law (for details see Seshadri 1993).

(A) M-$IG(\mu, \lambda, p)$ is unimodal

(B) For $c > 0$, $\mathcal{L}(x) = M$-$IG(\mu, \lambda, p) \Rightarrow \mathcal{L}(cx) = M$-$IG(c\mu, c\lambda, p)$

(C) $\mathcal{L}(X^{-1}) = M\text{-}IG(\mu^{-1}, (\frac{\mu^2}{\lambda})^{-1}(1-p))$

(D) $\mathcal{L}\left(\frac{\sqrt{\lambda}(X-\mu)}{\mu\sqrt{X}}\right) = N(0,1)$ when $\mathcal{L}(X) = M\text{-}IG(\mu, \lambda, \frac{1}{2})$

(E) For fixed μ, as $\lambda \to \infty$ $M\text{-}IG(\mu, \lambda, p) \to N(\mu, \frac{\mu^3}{\lambda})$ for any $p \in (0,1)$

(F) $M\text{-}IG(\mu, \lambda, p)$ is infinitely divisible

(G) $\mathcal{L}\left(\frac{\lambda(X-\mu)^2}{\mu^2 X}\right) = \chi_1^2$ if $\mathcal{L}(X) = M\text{-}IG(\mu, \lambda, p)$

(H) For fixed γ, $M\text{-}IG\left(\mu, \lambda, \frac{\mu}{\gamma+\mu}\right)$ is a two-parameter exponential family

(I) If $\mathcal{L}(X) = M\text{-}IG(\mu, \lambda, p)$, $\mathbb{E}(X) = \mu + p\frac{\mu^2}{\lambda}$ and $\text{var}(X) = \frac{\mu^3}{\lambda} + (\frac{\mu}{\lambda})^4 p(3-p)$

(J) $M\text{-}IG(\mu, \lambda, p)$ is a convolution of $IG(\mu, \lambda)$ with a compound Bernoulli law

To see (J) let $\mathcal{L}(X_1) = IG(\mu, \lambda)$ and define V as

$$V = \begin{cases} 0 & \text{with probability } (1-p) \\ \frac{\mu^2}{\lambda}\chi_1^2 & \text{with probability } p \end{cases}$$

Further let $X_1 \perp\!\!\!\perp V$. Then $\mathcal{L}(X) = \mathcal{L}(X_1 + V)$.

(K) $M\text{-}IG\left(\mu, \lambda, \frac{\mu}{\gamma+\mu}\right)$ is a special instance of Jørgnesen's concept of mixtures of exponential dispersion models.

Inference - estimation

The subclass of $M\text{-}IG(\mu, \lambda, p)$ with $p = \frac{\mu}{\gamma+\mu}, \gamma \geq 0$ is a full and steep model if γ is known and is regular if $\gamma \in [0, \infty)$. Writing $\theta_1 = -\frac{\lambda}{2\mu^2}, \theta_2 = -\frac{\lambda}{2}$ where $\Theta = \{(\theta_1, \theta_2) \mid -\infty < \theta_1 < 0, -\infty < \theta_2 < 0\}$ the canonical representation of (7.20) is

$$g_\gamma(x) = \sqrt{\frac{1}{2\pi x^3}}(\gamma + x)\exp\left\{\frac{\theta_2}{x} + \theta_1 x - k(\theta_1, \theta_2)\right\} 1_{\mathbb{R}^+}(x)$$

where

$$k(\theta_1, \theta_2) = -2\sqrt{\theta_1\theta_2} - \frac{1}{2}\log(-2\theta_2) + \log\left(\gamma + \sqrt{\frac{\theta_2}{\theta_1}}\right).$$

From exponential family theory, it can be shown that by solving the likelihood equations

$$\mathbb{E}(X) = \frac{\partial k}{\partial \theta_1} = \mu + \frac{\mu^2}{\lambda(\gamma + \mu)} = \bar{X}$$

$$\mathbb{E}\left(\frac{1}{X}\right) = \frac{\partial k}{\partial \theta_2} = \frac{1}{\mu} + \frac{\gamma}{\lambda(\gamma + \mu)} = \bar{X}_-.$$

These relations lead to the two equations

$$f_1(\mu) = \mu^3 \bar{X}_- - \mu^2 + \gamma\mu - \gamma\bar{X} = 0, \text{ and} \qquad (7.21)$$

$$f_2(\mu) = \bar{X}_- + \frac{\bar{X}}{\mu^2} - \frac{2}{\mu} - \frac{1}{\lambda} = 0 \qquad (7.22)$$

which have a unique solution if and only if all $X_i > 0$ and $\bar{X}\bar{X}_- > 1$. Since $\bar{X} \geq \mu$ and $\bar{X}_- \geq \frac{1}{\mu}$, the maximum likelihood estimator of μ is bounded and $\frac{1}{\bar{X}_-} \leq \hat{\mu} \leq \bar{X}$. Noting that $f_1(\frac{1}{\bar{X}_-}) < 0$ and $f_1(\bar{X}) > 0$ we conclude that $f_1(\mu)$ has exactly one root in $(\frac{1}{\bar{X}_-}, \bar{X})$.

If now γ is assumed unknown, we must also solve

$$\frac{n}{\gamma + \mu} = \sum_{i=1}^{n} \frac{1}{\gamma + x_i}. \qquad (7.23)$$

The system of equations (7.21) – (7.23) has multiple solutions. Two of them correspond to $(\gamma = 0, \mu = \frac{1}{\bar{X}_-})$ and $(\gamma = \infty, \mu = \bar{X})$. There is perhaps at least one solution in the interior. For fixed γ one first obtains a solution pair (μ, λ) and then searches numerically for a global optimum of γ in $[0, \infty)$ to determine the maximum likelihood estimator $(\hat{\mu}, \hat{\lambda}, \hat{\gamma})$. Suppose we parametrize by (m, θ_2, γ) where $m = \mathbb{E}X = \mu + \mu^3/\lambda(\mu+\gamma)$, then the maximum likelihood estimators of m and θ_2 are easily obtained from those of μ, λ and γ. Consider

$$g_\gamma(x) = a(x, \gamma) \exp\{\theta_1 x + \theta_2 b(x) - k(\theta_1, \theta_2, \gamma)\}.$$

where

$$a(x, \gamma) = \frac{x + \gamma}{\sqrt{2\pi x^3}}, \quad b(x) = \frac{1}{x} \text{ and}$$

$$k(\theta_1, \theta_2, \gamma) = -2\sqrt{\theta_1 \theta_2} - \frac{1}{2}\log(-2\theta_2) + \log\left(\gamma + \sqrt{\frac{\theta_2}{\theta_1}}\right).$$

When γ is known, m and θ_2 are orthogonal and when θ_2 is known, m and γ are orthogonal. These facts can be verified from Barndorff-Nielsen

A mixture inverse Gaussian

(1978 pages 182–183 and 184). Therefore the observed information matrix for (m, θ_2, γ) has the form

$$-n \begin{pmatrix} A & 0 & 0 \\ 0 & B & C \\ 0 & C & D \end{pmatrix}$$

where the entries are obtained by tedious computation

$$A = \frac{\partial^2 k}{\partial \theta_1^2}, \quad B = \frac{\partial^2 k}{\partial \theta_2^2}, \quad D = \frac{\partial^2 k}{\partial \gamma^2}, \quad C = \frac{\partial^2 k}{\partial \theta_2 \partial \gamma}.$$

We omit the details. The estimated covariance matrix for $(\hat{m}, \hat{\theta}_2)$ when γ is known is the negative inverse of the upper diagonal (2×2) submatrix of the observed information matrix. The negative inverse of the whole matrix gives the asymptotic covariance matrix of $(\hat{m}, \hat{\theta}_2, \hat{\gamma})$.

Example 7.6 Jørgensen et al (1991) considered the aircraft data of Proschan (1963) Table 7.10 which has been analyzed by Cox and Lewis(1966) and Jørgensen (1991). The data in Table 7.10 are intervals in operating hours between successive failures of air-conditioning equipment in 13 Boeing 720 aircraft. A good fit of the RIG law was reported by Jørgensen. Since the M-IG family contains the RIG, the $MIG(\mu, \lambda, p)$ was fitted to the data except for aircraft #11 (only two failures are available) and thus on the whole 211 observations from 12 aircraft are analyzed using the following scheme

(1) For each of the 12 aircraft all three parameters are estimated separately accounting for 36 estimated parameters.

(2) Then a common γ is fitted to all aircraft and μ, λ are estimated for each of the 12 craft so that there are 25 estimated parameters.

(3) In the third stage γ is set at zero ($\gamma = 0 \Rightarrow p = 1 \Rightarrow \mathcal{L}(x) = RIG$) and μ, λ estimated for all 12 aircraft. Thus 24 parameters are estimated.

(4) Lastly γ is set at zero and a common (μ, λ) used for all 12 aircraft so that only 2 parameters are estimated. The fitting procedure involves solving the likelihood equations with γ fixed. In the case where γ has to be estimated the search for the global maximum is used until the maximum likelihood estimate is identified. The results are given in Tables 7.11 and 7.12.

We are concerned here with a sequence of rested hypotheses $\Omega^n = H_0 \supseteq H_1 \supseteq H_2 \ldots$, each H_i being a subset of Ω^n and incorporating an extra constraint on the parameter space. The standard procedure is to go from H_1 by checking its adequacy and then testing H_2 under H_1. If

Table 7.10 *Numbers of operating hours between successive failures of air-conditioning equipment in 13 aircraft.*

						Aircraft						
1	2	3	4	5	6	7	8	9	10	11	12	13
194	413	90	74	55	23	97	50	359	50	130	487	102
15	14	10	57	320	261	51	44	9	254	493	18	209
41	58	60	48	56	87	11	102	12	5		100	14
29	37	186	29	104	7	4	72	270	283		7	57
33	100	61	502	220	120	141	22	603	35		98	54
181	65	49	12	239	14	18	39	3	12		5	32
	9	14	70	47	62	142	3	104			85	67
	169	24	21	246	47	68	15	2			91	59
	447	56	29	176	225	77	197	438			43	134
	184	20	386	182	71	80	188				230	152
	36	79	59	33	246	1	79				3	27
	201	84	27	15	21	16	88				130	14
	118	44	153	104	42	106	46					230
	34	59	326	35	20	206	5					66
	31	29	326		5	82	5					61
	18	118			12	54	36					34
	18	25			120	31	22					
	67	156			11	216	139					
	57	310			3	46	210					
	62	76			14	111	97					
	7	26			71	39	30					
	22	44			11	63	23					
	34	23			14	18	13					
		62			11	191	14					
		130			16	18						
		208			90	163						
		70			1	24						
		101			16							
		208			52							
					95							

H_2 is accepted one then proceeds to test H_3 under H_2 and finally if H_3 is accepted one tests H_4 under H_3. The check is based on the p-value of the associated chi-square statistic (here — twice the difference in log-likelihood $\sim \chi_f^2$ where f is the number of extra parameters estimated i.e, $\dim(H_i) - \dim(H_{i+1})$). When a significant result is obtained at a certain stage, it is customary to retain the last hypothesis accepted as

A mixture inverse Gaussian

the best explanation that the data has to offer. Based on this reasoning we have the following inferences for the aircraft data.

(1) Test of H_2 under H_1 has $(36 - 25) = 11$ d.f., $2(4.14)$ is the observed χ^2_{11}. The p-value corresponding to this is between 0.70 and 0.60 indicating non-rejection of H_2.

(2) $\gamma = 0$ being a boundary value, the asymptotic χ^2 law is strictly not valid for testing H_3 under H_2. Nevertheless the difference in likelihoods is negligible and H_3 does not suffer rejection.

(3) Test of H_4 under H_3 gives a $\chi^2_{22} = 40.92$ yielding a p-value between 0.01 and 0.001, small enough to cause rejection of a common model for all 12 aircraft.

If H_4 is rejected then H_3 offers the best explanation — a fact confirmed by Jørgensen's findings.

Table 7.11 *Maximum likelihood estimates of μ and σ^2 and the log-likelihood contribution for each aircraft under Hypothesis 3 (for which $\gamma = 0$ and $p = 1$).*

Aircraft	n	$\hat{\mu}$	σ^2	log-likelihood contribution
1	6	36.03	28.14	-31.88
2	23	30.73	14.53	-126.18
3	29	42.45	43.87	-154.43
4	15	40.24	19.99	-85.42
5	14	61.09	53.48	-81.31
6	30	11.18	2.581	-151.14
7	27	14.50	3.376	-145.92
8	24	18.26	7.270	-123.39
9	9	8.59	0.3854	-53.96
10	6	17.68	3.519	-33.35
12	12	14.78	2.340	-67.21
13	16	42.89	47.05	-84.79
Total				-1138.97

Table 7.12 *Sample log-likelihoods for the nested hypotheses*

Hypothesis	Number of estimated parameters	Sample log-likelihood
1. Separate μ, σ^2, γ	36	-1134.55
2. Separate μ, σ^2 Common γ	25	-1138.69
3. Separate μ, σ^2 Common γ	24	-1138.97
4. Common $\mu, \sigma^2, \gamma = 0 (p=1)$	2	-1159.43

Remarks What did we accomplish by fitting the $M\text{-}IG(\mu, \lambda, p)$ model? Since the RIG model is a subset of the M-IG model we are testing the right hypothesis, when γ is set at zero.

When $\mathcal{L}(X) = M\text{-}IG(\mu, \lambda, p)$ we have seen that $\mathcal{L}(X) = \mathcal{L}(X_1 + V)$ where X_1 and V are independently distributed with $\mathcal{L}(X_1) = IG(\mu, \lambda)$ and V a compound Bernoulli variable. The variable V takes the value zero with probability $(1-p)$, that is to say a defect in the equipment occurs instantaneously with probability $(1-p)$. On the other hand if there is a random delay in the defect occurring, the probability is p and the randomness proceeds according to a χ_1^2 law. The random variable X_1 is the time — the first time that the defect develops into a failure of the equipment (ie, from the start of the defect to actual failure) — and is well represented by the first passage time of a Wiener process to reach a critical level. The parameter p describes the general state of conservation of the equipment. For perfect conservation $p = 1$ and the RIG law describes this state. Thus the M-IG model is well qualified to describe the physical process of the system we are examining.

7.8 Exponential-inverse Gaussian mixtures

Bhattacharya and Kumar (1986) proposed a model for a life time distribution by compounding the exponential distribution with the IG law, to obtain the $E\text{-}IG(\mu, \lambda)$ law. The density is

$$f(x) = \sqrt{\frac{\lambda}{2\pi}} \exp\left(\frac{\lambda}{\mu}\right) \int_0^\infty \theta^{-\frac{3}{2}-1} \exp\left\{-\frac{(2x+\lambda)}{2\theta} - \frac{\lambda\theta}{2\mu^2}\right\} d\theta$$

$$= \sqrt{\frac{2}{\pi}} \exp\left\{\frac{\lambda}{\mu}\left(1 - \sqrt{1 + \frac{2x}{\lambda}}\right)\right\} \mathbb{E}\left(\frac{1}{\theta}\right) 1_{R^+}(x)$$

where $\mathcal{L}(\frac{1}{\theta}) = IG\left(\mu\sqrt{\frac{\lambda+2x}{\lambda}}, \lambda + 2x\right)$ for $\lambda, \mu > 0$. Thus the density is

$$f(x) = \sqrt{\frac{2}{\pi}} \left(\frac{\sqrt{\lambda}}{\mu\sqrt{\lambda+2x}} + \frac{1}{\lambda+2x}\right) \exp\left\{\frac{\lambda}{\mu}\left(1 - \sqrt{1 + \frac{2x}{\lambda}}\right)\right\} 1_{\mathbb{R}^+}(x). \tag{7.24}$$

The moments of X are obtained by using conditional expectation. The mean and variance are

$$\mathbb{E}(X) = \mu, \quad Var(X) = \mu^2 \left(1 + \frac{2\mu}{\lambda}\right).$$

One can now routinely compute the hazard rate and reliability. Thus

$$R(x; \mu, \lambda) = P(X > x) = \sqrt{\frac{2}{\mu\pi}} \frac{\lambda^{\frac{3}{4}} \exp\left(\frac{\lambda}{\mu}\right)}{(\lambda+2x)^{\frac{1}{4}}} K_{\frac{1}{2}}\left(\sqrt{\frac{\lambda(\lambda+2x)}{\mu^2}}\right)$$

and the hazard rate is

$$h(x; \mu, \lambda) = \frac{1}{\lambda+2x} + \frac{1}{\mu}\sqrt{\frac{\lambda}{\lambda+2x}}.$$

Since $h'(x; \mu, \lambda) < 0$ for all $x \in (0, \infty)$ and $\lambda, \mu > 0$, the hazard rate is monotone decreasing in x. The asymptotic variances of the moment estimators of μ and λ are given by

$$Var(\tilde{\mu}) = \frac{\mu^2}{n}\left(1 + \frac{2\mu}{\lambda}\right)$$
$$Var(\tilde{\lambda}) = \frac{\lambda^4}{n\mu^2}\left\{1 + \frac{10\mu}{\lambda} + \frac{38\mu^2}{\lambda^2} + \frac{54\mu^3}{\lambda^3}\right\}$$
$$Cov(\tilde{\mu}, \tilde{\lambda}) = \frac{\lambda\mu}{n}\left(1 + \frac{3\mu}{\lambda}\right).$$

These authors have also suggested a p-variate E-IG model with density

$$f(\boldsymbol{x}) = \sqrt{\frac{2\lambda}{\pi}} \exp\left(\frac{\lambda}{\mu}\right) \left(\sqrt{\frac{\lambda\mu^{-2}}{\lambda + 2\sum_{i=1}^{p} x_i}}\right)^{p+\frac{1}{2}}$$

$$\times K_{p+\frac{1}{2}}\left(\frac{\lambda}{\mu}\sqrt{1 + \frac{2\sum^p x_i}{\lambda}}\right) 1_{\mathbb{R}_1^p}(\boldsymbol{x})$$

which yields as marginal laws density proportional to

$$\frac{1}{(\lambda+2x_i)^{\frac{3}{4}}} \; K_{\frac{3}{2}}\left(\frac{\lambda}{\mu}\sqrt{1+\frac{2x_i}{\lambda}}\right) 1_{\mathbf{R}^+}(x_i).$$

Pursuing this lead Whitmore and Lee (1991) define a multivariate survival model as an inverse Gaussian mixture of exponential laws.

Definition 7.1 Let $\mathcal{L}(T) = IG(\mu, \lambda)$, and $\mathbf{X} = (X_1, \ldots, X)$ a random vector such that $\mathcal{L}(\mathbf{X}|T=t) = \prod_{i=1}^{p}\mathcal{L}(X_i|T=t) = t^p \exp\{-tS_p\}1_{\mathbf{R}_+^p}(\mathbf{x})$ where $S_p = \sum_{i=1}^{p} x_i$. Then the unconditional law of the vector $\mathbf{X}$ is said to have a multivariate exponential-inverse Gaussian mixture distribution. It follows easily from the definition that

$$\begin{aligned}\mathcal{L}(\mathbf{X}) &= \int_0^\infty e^{\frac{\lambda}{\mu}} \sqrt{\frac{\lambda}{2\pi}} \, t^{-p\frac{3}{2}} \exp\left\{-\frac{\lambda}{2t} - \left(\frac{\lambda}{2\mu^2}+S_p\right)t\right\} dt \\ &= \exp\left\{\frac{\lambda}{\mu}\left(1+\frac{2\mu^2}{\lambda}S_p\right)^{\frac{1}{2}}\right\} \mathbb{E}(T_1^p)\end{aligned} \quad (7.25)$$

where

$$\mathcal{L}(T_1) = IG\left(\mu\sqrt{\frac{\lambda}{\lambda+2\mu^2 S_p}}, \lambda\right), \text{ and } S_p = \sum_{i=1}^{p} x_i.$$

In survival analysis T is considered as a hazard rate of a process, and each of the variables $X_1, \ldots, X_p$ has the same hazard rate T which is randomly distributed as the first passage time distribution of a Brownian motion with positive drift. The multivariate survival function is

$$R(\mathbf{X}) = P(X_1 > x_1, \ldots X_p > x_p) = \mathbb{E}(\exp{-TS_p}) \quad (7.26)$$

where the expectation is taken with respect to $IG(\mu, \lambda)$ and all $x_i's \geq 0$. Thus the survival function is proportional to the Laplace transform of $IG(\mu, \lambda)$ and depends on $(X_1, \ldots, X_p)$ only through the sum S_p. This construction introduces a dependency on the random variables X_i which were independent to begin with.

Whitmore and Lee establish a few useful properties associated with (7.25). Among the salient ones are

(A) The distribution (7.25) is dependent by total positivity of order 2 in each pair of variables (i.e., TP_2 in pairs) (i,j) with the remaining variables held fixed

(B) The distribution (7.25) is positively dependent being a mixture of independent p-variate distributions

(C)
$$\mathbb{E}(X_1^{k_1} \ldots X_p^{k_p}) = \prod_{i=1}^{p} k_i! \, \mathbb{E}(T^{-p})$$
$$= \prod_{i=1}^{p} k_i! \, \mathbb{E}(T^{p+1}) \left(\frac{1}{\mu^{2p+1}}\right)$$

(D) $0 < \rho_{ij} = \dfrac{2 + \frac{\lambda}{\mu}}{5 + \frac{4\lambda}{\mu} + \left(\frac{\lambda}{\mu}\right)^2} < \dfrac{2}{5}, \, 1 \leq i \neq j \leq p$

(E) If $(i_1, \ldots, i_k)$ is any subset of $(1, \ldots, p)$

$$\frac{R(x_{i_1}+h, \ldots x_{i_k}+h)}{R(x_{i_1}, \ldots \ldots, x_{i_p})} = \frac{\exp\left[\frac{\lambda}{\mu}\left\{1 + \frac{2\mu^2}{\lambda}S_{i_k}\right\}^{\frac{1}{2}}\right]}{\exp\left[\frac{\lambda}{\mu}\left\{1 + \frac{2kh\mu^2}{\lambda} + \frac{2\mu^2}{\lambda}S_{i_k}\right\}^{\frac{1}{2}}\right]}$$

is increasing in $(x_{i_1}, \ldots, x_{i_k})$ where x_i's > 0

(F) The conditional hazard rate function defined by

$$\frac{f(x_i \mid x_1, \ldots, x_{i-1})}{R(x_i \mid x_1, \ldots, x_{i-1})} = h(x_i \mid X_1 = x_1, \ldots, X_{i-1} = x_{i-1})$$

is decreasing in $(x_1, \ldots, x_{i-1})$ for every x_i

(G) When the hazard rate function is conditioned by the event $\{X_1 > x_1, \ldots, X_{k-1} > x_{k-1}\}$, the resulting hazard rate function is decreasing in $(x_1, \ldots x_{i-1})$ for any x_i.

Relation to the $P - IG(\mu, \lambda)$

Stein et al., (1987) have discussed a multivariate extension of the P-IG distribution, called the multivariate Sichel distribution (see also Holla 1966). Its construction is as follows.

Let $X_i (i = 1, \ldots, p)$ be p independent Poisson random variables with mean λt_i. That is

$$P(X_i = x_i \mid \lambda) = \frac{(\lambda t_i)^{x_i} \exp(-\lambda t_i)}{x_i!}.$$

Suppose now that the parameter λ has the $IG(1, w)$ law, then, using the parametrization of Stein et al.,

$$P(x_1, \ldots, x_p \mid \alpha, t_1, \ldots t_p)$$

$$= \frac{K_{\sum_{i=1}^{p} x_i - \frac{1}{2}}(\sqrt{w}(w + 2\sum_{i=1}^{p} t_i)^{\frac{1}{2}})}{K_{-\frac{1}{2}}(w)}$$

$$\times \left(\frac{w}{w + 2\sum_{i=1}^{p} t_i}\right)^{\frac{1}{2}(\sum_{i=1}^{p} x_i - \frac{1}{2})} \prod_{i=1}^{p} \frac{(t_i)^{x_i}}{x_i!}$$

for $x_i = 0, 1, \ldots$ and $\alpha, t_1, \ldots, t_p > 0$.

The model of Whitmore and Lee is based on the random time intervals $Y_1, \ldots Y_p$ of the consecutive events in a Poisson process where the parameter follows an IG law.

7.9 Birnbaum-Saunders distribution

Based on considerations of a fatigue process Birnbaum and Saunders (1969) introduced a two-parameter family of distributions. We have seen briefly that this is a subfamily of the M-$IG(\mu, \lambda, p)$ family for $p = \frac{1}{2}$. The fatigue process involves some key assumptions that are summarized below.

(a) fatigue failure is due to a continued cyclic stress on a material
(b) the cyclic stress induces a dominant crack in the material which grows until it attains a critical dimension w beyond which fatigue failure is certain
(c) the crack extensions $L_1, L_2, \ldots$ are independent
(d) the total crack extension at the nth stage $z_n = \sum_{i=1}^{n} L_i$ has an approximate normal law by an application of the Central Limit Theorem.

The failure time distribution is given by

$$f(t; \mu, \lambda) = \sqrt{\frac{\lambda}{2\pi}} \frac{1}{2} \left(\frac{t^{-\frac{1}{2}}}{\mu} + t^{-\frac{3}{2}}\right) \exp\left\{-\frac{\lambda(t-\mu)^2}{2\mu^2 t}\right\} 1_{\mathbb{R}^+}(t) \quad (7.27)$$

As noted in Section 7.7, Desmond (1986) pointed out a simple relationship between the Birnbaum-Saunders law and the IG law, thus predating a similar observation by Jørgensen et al. (1991).

Specifically if $\mathcal{L}(X_1) = IG(\mu, \lambda)$ and $\mathcal{L}(X_2^{-1}) = IG\left(\frac{1}{\mu}, \frac{\mu^2}{\lambda}\right)$ then, if $P(U = 0) = P(U = 1) = \frac{1}{2}$, the random variable $T = UX_1 + (1-U)X_2$ has the Birnbaum-Saunders law — with parameters μ and λ. Recall (from Seshadri (1993)) that Fletcher (1911) had first considered this law. Schrödinger had accused Konstantinowsky (1914) (who had cited Fletcher's work) of making a false claim that (7.27) represented the first passage time of Brownian motion with positive drift. (Schrödinger, later offered a retraction realizing that the claim was due to Fletcher.) In the case of $IG(\mu, \lambda)$ it is to be noted that

(A) the cumulative fatigue in time period $[0, t]$ is a Wiener process with positive drift

(B) the fatigue failure time is to be construed as the first passage time of the Wiener process to the critical level w.

Bhattacharyya and Fries (1982) pointed out that the assumption of normality of Z_n, the total crack extension, presupposes that the possibility of Z_n assuming negative values with non-zero probability is ignored. Therefore they argued that the Birnbaum-Saunders law should only be regarded as an approximate solution. Since the IG law was an exact solution they suggested that it was more appropriate in fatigue failure models.

Desmond (1986), however, argues that, since crack size is strictly a positive random variable hence crack increments tend to be positive. Thus probabilities of negative values should be excluded from consideration. Moreover, Desmond has shown that the assumption of normality is not a key issue in the derivation of the Birnbaum-Saunders law and that there exist many laws for crack size which lead to (7.27). Furthermore it is possible to allow the crack increment to depend on the crack size at the start of operations and still obtain a law of the Birnbaum-Saunders type. This dependency does not work in the IG model due to the independent increments implicit in a Wiener process. From the point of view of statistical analysis, however, the IG model comes out the winner, since censored data can be handled without problems due to the exponential family structure of the IG law. We present below an outline of Desmond's derivation of the Birnbaum-Saunders law. Random stress environments leading to failure are modelled as stationary continuous time stochastic processes, as for example, varying response like voltage or temperature. Desmond uses a biological model considered by Cramèr (1946) to develop (7.27). In this model

$$Y_{i+1} = Y_i + \pi_{i+1} g(Y_i) \quad i = 0, 1, \ldots \quad (7.28)$$

where π_i denotes a random variable describing the magnitude of the ith impulse, Y_i the crack size after the application of the ith impulse, $g(y)$ a continuous function of y, and Y_0 is the initial crack size. The crack growth at stage $(i+1), \Delta Y_i = Y_{i+1} - Y_i$ is considered sufficiently small. Hence

$$\pi_{i+1} = \frac{\Delta Y_i}{g(Y_i)}$$

and $\sum_{i=1}^{n} \pi_i$ can then be approximated by

$$\sum_{i=1}^{n} \pi_i = \int_{Y_0}^{Y_n} \frac{dy}{g(y)}.$$

By applying the Central Limit Theorem (assuming its applicability and that the π_i have a common mean μ and variance σ^2), the law of $\sum_{i=1}^{n} \pi_i$ is approximately normal. Hence if $Y(t)$, the crack size at time t is regarded as a continuous time stochastic process

$$G(Y(t)) = \int_0^{Y(t)} \frac{dy}{g(y)} \approx \frac{1}{\sqrt{2\pi t}\,\sigma} \exp\left\{-\frac{(Y(t)-\mu t)^2}{2t\sigma^2}\right\}.$$

Let $Y_c > Y_0$ be the critical crack size at which failure occurs. Then the time to fatigue failure T is

$$T = \inf\{t \mid Y(t) > Y_c\}.$$

Let $F_t(y) = P(Y(t) \le y)$ be the distribution function of $Y(t)$ at time t

$$\{T \le t\} = \{Y(t) \ge Y_c\}.$$

The distribution function of T is then equal to

$$\begin{aligned}P(Y(t) \ge Y_c) &= P[G(Y(t) > G(Y_c)] \\ &= 1 - P[G(Y(t)) \le G(Y_c)] \\ &= \Phi\left(\frac{t\mu - G(Y_c)}{\sigma\sqrt{t}}\right).\end{aligned}$$

Thus regardless of the form of $g(y)$ the model (7.28) leads to the failure laws in the Birnbaum-Saunders family. Desmond points out that the choice of $g(y)$ determines to what extent the rate of crack growth depends on the previous crack size. From empirical evidence he suggests using $g(y) = y^\delta$ ($\delta > 0$) (δ being a parameter relevant to the material under stress). Then the distribution function of T is

$$F_T(t;\delta) = \begin{cases} \Phi\left(\frac{Y_c^{1-\delta}-Y_0^{1-\delta}+(\delta-1)t\mu}{\sqrt{t}\sigma(\delta-1)}\right) & \delta > 1 \\ \Phi\left(\frac{Y_0^{1-\delta}-Y_c^{1-\delta}+(1-\delta)t\mu}{\sqrt{t}\sigma(1-\delta)}\right) & \delta < 1 \end{cases}$$

For $\delta = 0$ one obtains the Birnbaum-Saunders law.

Inference

Chang and Tang (1993) describe a graphical method for estimation of μ and λ as well as checking for goodness-of-fit. Since maximum likelihood estimation and confidence interval procedures are quite cumbersome (Birnbaum and Saunders, Engelhardt et al., (1969, 1981)), the graphical approach is very similar to probability plotting and seems quite practical. The distribution function of the Birnbaum-Saunders law is

$$F(t;\mu,\lambda) = \Phi\left(\sqrt{\lambda}\left(\frac{\sqrt{t}}{\mu} - \frac{1}{\sqrt{t}}\right)\right).$$

Let us first reparametrize by $\alpha = \sqrt{\frac{\mu}{\lambda}}, \beta = \mu$ and then solve the above equation for t. Thus we have

$$t = \beta + \alpha\sqrt{\beta}\sqrt{t}\Phi^{-1}(F^*(t;\beta,\alpha))$$

where $F^*(t;\beta,\alpha) = F(t;\mu,\lambda)$. We now write

$$p = \sqrt{t}\Phi^{-1}(F^*)$$

to obtain the linear equation

$$t = \beta + \alpha\sqrt{\beta}p = a + bp.$$

The idea is to plot the failure times $t_1, \ldots, t_n$ against the p_i values where

$$\hat{p}_i = \sqrt{t_i}\Phi^{-1}(\hat{F}^*(t_i)).$$

If the failure times follow the Birnbaum-Saunders law the plot should indicate an approximate linear relationship. The question now is, what is $\hat{F}^*(t_i)$? Following Johnson (1951) we use the median rank

$$\hat{F}^*(t_i) = \frac{i - 0.3}{n + 0.4}.$$

Thus a visual analysis provides a quick check of goodness-of-fit. One may also use the R^2 statistic proposed by Shapiro and Wilk (1965) to examine the fit. As for estimation we can use the method of least squares to obtain

$$\hat{\beta} = \hat{a}$$

and

$$\hat{\alpha} = \frac{\hat{b}}{\sqrt{\hat{a}}}. \tag{7.29}$$

Recalling that for the Birnbaum-Saunders law β is the median, we note that the intercept a gives us the estimate. Furthermore α increases with b and decreases with a. When a is fixed (7.29) gives us a unique value α in terms of b and conversely. Therefore

$$A = \{ \underline{a} \leq a \leq \bar{a},\ \underline{b} \leq \bar{b}\} \Rightarrow \{\underline{\alpha} \leq \alpha \leq \bar{\alpha}\} = B,$$

and we have since $A \subseteq B$,

$$P(B) \geq P(A)$$
$$\geq 1 - P[\{\underline{a} \leq a \leq \bar{a}\}]^c P[\{\underline{b} \leq b \leq \bar{b}\}^c]$$

by the Bonferroni inequality. Thus

$$P(\underline{\alpha} \leq \alpha \leq \bar{\alpha}) \geq 1 - \gamma_1 - \gamma_2$$

where $[\underline{a}, \bar{a}]$ and $[\underline{b}, \bar{b}]$ are the approximate $100(1-\gamma_1)\%$ and $100(1-\gamma_2)\%$ confidence intervals respectively for a and b. Finally we can write $[\underline{\alpha}, \bar{\alpha}]$ as $[\underline{b}/\sqrt{\bar{a}}, \bar{b}/\sqrt{\underline{a}}]$. Chang and Tang have analyzed the example originally studied by Birnbaum and Saunders (see Table 7.13). We consider this example to illustrate their graphical technique.

Example 7.7 This is an example given in Birnbaum and Saunders on the fatigue life (in cycle) of 6061-t6 aluminum coupons cut parallel to the direction of rolling. The data consist of $n = 101$ observations under a maximum stress of 31,000 psi with 18 cycle/s oscillation. The Shapiro-Wilk statistics $R^2 = 0.977$ suggests that the failure data are indeed conforming to the Birnbaum-Saunders distribution. From the slope and intercept, the least square estimates for the parameters together with their respective 90% confidence intervals are summarised in Table 7.14.

Besides the results obtained from complete data, a similar set of results with right censoring are also given in Table 7.14 where last 21 failure data are ignored. Both the R^2 value and the estimates compare favourably with that of the complete data.

Table 7.13 *Birnbaum-Saunders data*
70, 90, 96, 97, 99, 100, 103, 104, 104, 105, 107, 108, 108, 108, 109, 109, 112, 112, 113, 114, 114, 114, 114, 116, 119, 120, 120, 120, 121, 121, 123, 124, 124, 124, 124, 124, 128, 128, 129, 130, 130, 130, 130, 131, 131, 131, 131, 131, 132, 132, 132, 133, 134, 134, 134, 134, 134, 136, 136, 137, 138, 138, 138, 139, 139, 141, 141, 142, 142, 142, 142, 142, 142, 144, 144, 145, 146, 148, 148, 151, 151, 152, 152, 155, 156, 157, 157, 157, 157, 158, 159, 162, 163, 163, 164, 166, 166, 168, 170, 174, 196, 212.

Table 7.14 *Least square estimates and their 90% confidence intervals for complete and type II censored data*

	$\hat{\alpha}$	$\hat{\beta}$	$[\underline{\alpha}, \bar{\alpha}]$	$[\underline{\beta}, \bar{\beta}]$	R^2
complete	0.1686	131.9	[0.163, 0.174]	[131.3, 132.5]	0.977
TypeII censoring	0.1750	132.5	[0.168, 0.182]	[132.0, 133.0]	0.965
MLE	0.1704	131.8	[0.153, 0.193]	[128.3, 135. 5]	

In Table 7.14 we also list the MLE of β which is given by the positive root of
$$\beta^2 - B\beta + C = 0,$$
where
$$B = \left[\frac{2}{\bar{t}_-} + \frac{n}{\sum_{i=1}^{n}(\beta+t_i)^{-1}}\right], \quad C = \frac{1}{\bar{t}_-}\left[\frac{n}{\sum_{i=1}^{n}(\beta+t_i)^{-1}} + \bar{t}\right].$$

and MLE of α given by
$$\hat{\alpha} = \left[\frac{\bar{t}}{\hat{\beta}} + \hat{\beta}\bar{t}_- - 2\right]^{1/2}.$$

7.10 Linear models and the P-IG law

Since the $P\text{-}IG(\mu, \lambda)$ law is a useful model for fitting overdispersed data, Stein and Juritz (1988) employ it in modelling response which comes in the form of counts by introducing a linear model with a P-IG error distribution. The model proposed by them is
$$\mathcal{L}(Y_i) = IG(\mu_i, \alpha) \quad i = 1, 2, \ldots, n$$
where $g(\mu_i) = X_i\boldsymbol{\beta}$, $\boldsymbol{\beta}$ being a $(p \times 1)$ vector of unknown constants, X_i a set of explanatory variables and α a shape parameter. We describe their approach and discuss the assessment of the fit together with an example on fish species data examined by them. In what follows we stick to the parametrization of Stein et al., (see 4 of Table 7.1). Recall that the probability function is
$$p(y; \mu_i, \alpha) = \sqrt{\frac{2\alpha}{\pi}} e^{\sqrt{\mu_i^2+\alpha^2}-\mu_i}\left(\frac{\mu_i(\sqrt{\mu_i^2+\alpha^2}-\mu_i)}{\alpha}\right)^y \frac{1}{y!} K_{y-\frac{1}{2}}(\alpha).$$

The log-likelihood $\ell(\theta) = \ell(\boldsymbol{\beta}, \alpha)$ where $\theta = (\mu_1, \ldots, \mu_n, \alpha) = (\beta_0, \ldots, \beta_p, \alpha)$, is proportional to
$$\ell(\boldsymbol{\beta}, \alpha) \propto \frac{n}{2}\log\alpha + \sum_{i=1}^{n}\left(\sqrt{\mu_i^2+\alpha^2}-\mu_i\right) + \sum_{i=1}^{n} y_i \log\left(\frac{\mu_i}{\alpha}\right)$$
$$+ \sum_{i=1}^{n} y_i \log\left(\sqrt{\mu_i^2+\alpha^2}-\mu_i\right) + \sum_{i=1}^{n} \log K_{y_i-\frac{1}{2}}(\alpha)$$

The μ_i are related to $\boldsymbol{\beta}$ by
$$g(\mu_i) = \eta_i = \beta_0 + \sum_{j=1}^{p}\beta_j x_{ij}.$$

Now
$$\frac{\partial \ell}{\partial \beta_j} = \frac{\partial \ell}{\partial \mu_i} \frac{\partial \mu_i}{\partial \eta_i} \frac{\partial \eta_i}{\partial \beta_j}, \quad i=1,\ldots,n, \ j=0,1,\ldots,p.$$

But
$$\frac{\partial \ell}{\partial \mu_i} = \sum_{i=1}^{n}\left(\frac{\mu_i}{\sqrt{\mu_i^2+\alpha^2}}-1\right) + \sum_{i=1}^{n} y_i\left(\frac{1}{\mu_i}-\frac{1}{\sqrt{\mu_i^2+\alpha^2}}\right)$$
$$= \sum_{i=1}^{n}(y_i-\mu_i)\left[\frac{1}{\mu_i}-\frac{1}{\sqrt{\mu_i^2+\alpha^2}}\right]$$

and
$$\frac{\partial \eta_i}{\partial \beta_j} = x_{ij}.$$

Hence
$$\frac{\partial \ell}{\partial \beta_j} = \sum (y_i-\mu_i)\left[\frac{1}{\mu_i}-\frac{1}{\sqrt{\mu_i^2+\alpha^2}}\right]x_{ij}\frac{\partial \mu_i}{\partial \nu_i}, \quad j=0,1,\ldots,p$$

$$\frac{\partial \ell}{\partial \alpha} = \frac{n}{2\alpha} + \sum_{i=1}^{n}\frac{\alpha}{\sqrt{\mu_i^2+\alpha^2}} - \sum_{i=1}^{n}\frac{y_i}{\alpha} + \sum_{i=1}^{n}\frac{y_i}{(\sqrt{\mu_i^2+\alpha^2}-\mu_i)}\frac{\alpha}{\sqrt{\mu_i^2+\alpha^2}}$$
$$+ \frac{\sum_{i=1}^{n} K'_{y_i-\frac{1}{2}}(\alpha)}{K_{y_i-\frac{1}{2}})\alpha)}.$$

Since
$$\sum_{i=1}^{n}\frac{K'_{y_i-\frac{1}{2}}(\alpha)}{K_{y_i-\frac{1}{2}}(\alpha)} = -\sum_{i=1}^{n} R_{y_i-\frac{1}{2}}(\alpha) + \sum_{i=1}^{n}\frac{y_i}{\alpha} - \frac{n}{2\alpha}$$

(see (iv) of section 7.2), we have,

$$\frac{\partial \ell}{\partial \alpha} = \sum_{i=1}^{n}\frac{\alpha}{\sqrt{\mu_i^2+\alpha^2}} + \alpha\sum_{i=1}^{n}\left(\frac{y_i}{\mu_i^2+\alpha^2-\mu_i\sqrt{\mu_i^2+\alpha^2}}\right) - \sum_{i=1}^{n} R_{y_i-\frac{1}{2}}(\alpha)$$

The expected information matrix for fixed α is (using Fisher's scoring method) $X^t W X$, where $W = \mathrm{diag}(w_1,\ldots,w_n)$ and

$$w_i = \left(\frac{\partial \mu_i}{\partial \eta_i}\right)^2 \frac{1}{\mathrm{var}(Y_i)}, \quad \eta_i = g(\mu_i), \quad i=1,\ldots,n,$$

due to parameter orthogonality. Recall that

$$\mathrm{Var}(Y_i) = \frac{\mu_i^2}{\sqrt{\mu_i^2+\alpha^2}-\mu_i} + \mu_i$$
$$= \mu_i\left[\frac{\sqrt{\mu_i^2+\alpha^2}}{\sqrt{\mu_i^2+\alpha^2}-\mu_i}\right]$$

so that
$$\frac{1}{\operatorname{Var} Y_i} = \frac{1}{\mu_i} - \frac{1}{\sqrt{\mu_i^2 + \alpha^2}},$$
and finally
$$w_i = \left(\frac{\delta \mu_i}{\delta \eta_i}\right)^2 \left(\frac{1}{\mu_i} - \frac{1}{\sqrt{\mu_i^2 + \alpha^2}}\right).$$

The following steps are used by Stein and Juritz to estimate α and β.
(1) Use Newton-Raphson iteration to obtain $\hat{\alpha}$, the maximum likelihood estimate of α (assuming $\beta_0 = g(\mu_i)$)
(2) Use this value of $\hat{\alpha}$ and an iterated nonweighted least squares method to obtain the maximum likelihood estimate $\hat{\beta}$
(3) Now use $g(\mu_i) = X_i \hat{\beta}$ and reestimate α
(4) Repeat steps (2) and (3) until convergence is attained. The convergence will be rapid due to the asymptotic uncorrelated structure between $\hat{\beta}$ and $\hat{\alpha}$.(The fitting procedure uses GLIM.)

In order to assess the adequacy of the fit it is necessary to see if the extra parameter introduced, α, has contributed in improving the fit. If the P-IG fit is assessed appropriate it is then essential to verify the linearity of the model. These can be done as follows. First note that $P\text{-}IG(\mu, \alpha) \to$ the Poisson in the limit as $\alpha \to \infty$. Therefore the first assessment is checked by a test of $H_0 : \alpha = \infty$ (Poisson) against $H_A : (\alpha < \infty)$ (P-IG). When $\hat{\beta}$ is known this poses no problem since a likelihood ratio test solves the issue. Stein and Juritz suggest using the statistic
$$2(\ell(\hat{\beta}_{H_A}, \hat{\alpha}) - \ell(\hat{\beta}_{H_0}))$$
which is approximately χ_1^2. Here $\ell(\hat{\beta}_{H_A}, \hat{\alpha})$ and $\ell(\hat{\beta}_{H_0})$ denote the maximized log-likelihood under the P-IG model and the Poisson model respectively. Testing for linearity implies testing for $(\beta_0, \ldots, \beta_p)$. Two cases now arise depending on whether α is known or not. If α is known the adequacy of the linear model can be tested by the deviance
$$2\left[\ell_{\text{P-IG}}\left(\hat{\beta}_\Omega, \alpha\right) - \ell_{\text{P-IG}}\left(\hat{\beta}, \alpha\right)\right]$$
which has an asymptotic χ^2 law with $n-(p+1)$ degrees of freedom. When α is unknown the maximum likelihood estimate of α cannot be obtained under the saturated hypothesis (more parameters than observations). In this case one obtains an alternative estimate $\hat{\alpha}$ from the null model, namely
$$H_0 : \boldsymbol{\beta} = (\beta_0, \ldots, \beta_p)^t.$$

Now test using the deviance

$$2\left[\ell_{P-IG}(\hat{\beta}_\Omega, \hat{\alpha}_{H_0}) - \ell_{P-IG}(\hat{\beta}_{H_0}, \hat{\alpha}_{H_0})\right]$$

$$= 2\sum_{i=1}^{n}\left[\sqrt{y_i^2 + \hat{\alpha}^2} - y_i + y_i \log\left\{y_i\sqrt{y_i^2 + \hat{\alpha}^2} - y_i\right\}\right]$$

$$- 2\sum_{i=1}^{n}\left[\sqrt{\hat{\mu}_i^2 + \hat{\alpha}^2} - \hat{\mu}_i + \hat{\mu}_i \log\left\{\hat{\mu}_i\left(\sqrt{\hat{\mu}_i^2 + \hat{\alpha}^2} - \hat{\mu}_i\right)\right\}\right]$$

which has an approximate χ^2 law with $n - p - 2$ degrees of freedom. A sequence of nested hypothesis is tested in similar fashion. Stein and Juritz also advocate a graphical display plot using $\hat{p}_i = P(Y \geq y_i \mid \hat{\mu}_i, \hat{\alpha})$ $i = 1, \ldots, n$, the estimated exceedance probabilities to examine any departure from uniformity. (Compare with expected uniform order statistics). Barbour and Brown (1974) have analyzed the fish species diversity of 70 lakes of the world. They postulated a power function model for the data set. Stein and Juritz have examined the data set and obtained the following numerical results based on a log link with the log-lake area as the covariate.

Table 7.15 *Analysis of deviance*

Error law	Model	Deviance	degrees of freedom
Poisson	β_0	2646	69
	$\beta_0 + \beta_1 x_i$	1538	68
$P - IG$	β_0	104	68
	$\beta_0 + \beta_1 x_i$	64	67

Parameter estimates and their asymptotic standard errors are
$\hat{\alpha} = 11.62$ s.d. $= 10.12$
$\hat{\beta}_0 = 2.52$ s.d. $= 0.17$
$\hat{\beta}_1 = 0.15$ s.d. $= 0.03$

When testing for linearity the deviance value of 64 based on 67 degrees of freedom shows a good fit of the *P-IG* model. A test of $H_0 : \beta = \beta_0$ against $H_A : \beta = (\beta_0, \beta_1)^t$ yields a deviance of $104 - 64 = 40$. Based on one degree of freedom this indicates the need for the log lake area as an explanatory variable. Finally they produce a graphical display of the estimated exceedance probabilities plotted against the expected uniform order statistics (Figure 7.1) which shows concordance of uniformity of the computed statistics.

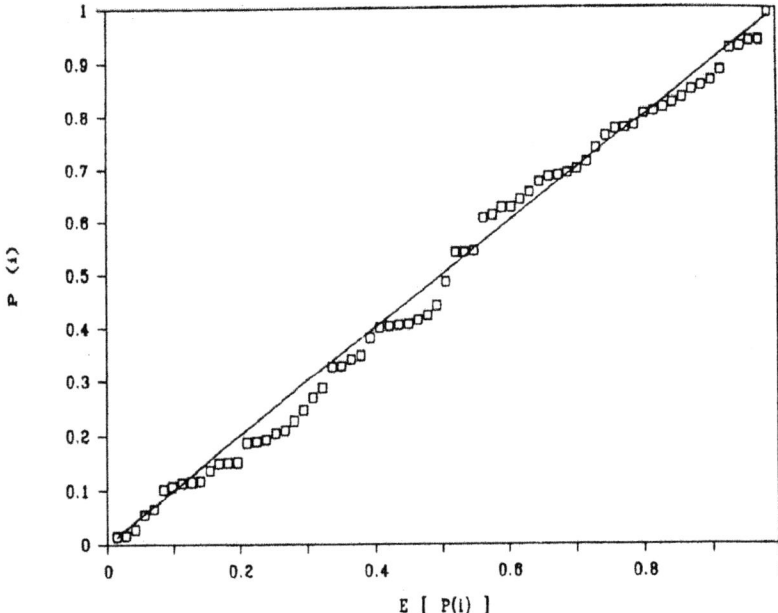

Figure 7.1 *Estimated exceedance probabilities vs expected uniform order statistics*

7.11 P-IG regression model

Dean et al. (1989) consider another model called a Poisson-inverse Gaussian regression model. It is a multiplicative Poisson-random effects model which takes into account a random effect to the response variable with a Poisson distribution.

To make the above more precise assume that with the fixed explanatory variables $X = (X_1, \ldots, X_p)^t$ (called covariates) there is associated a positive valued function $\mu(X)$ depending on a vector of unknown parameters $\beta = (\beta_1, \ldots, \beta_p)^t$. For fixed X, the random response Y is assumed to be Poisson distributed with mean $(\nu\mu(X))$ where the random variable ν has the IG$(1, \frac{1}{\tau})$ distribution for $\tau > 0$. Thus

$$P[Y = y \mid X] = \int_0^\infty \frac{e^{-\nu\mu(X)}}{y!} (\nu\mu(X))^y \frac{1}{\sqrt{2\pi\tau\nu^3}} \exp\left(-\frac{(\nu-1)^2}{2\nu\tau}\right) d\nu.$$

It follows from section 7.1 that the probability function is

$$p_y = \sqrt{\frac{2}{\pi}} \, e^{\frac{1}{\tau}} \left(\frac{1+2\tau\mu(X)}{\tau^2}\right)^{\frac{1}{4}} \left(\sqrt{\frac{\mu^2(X)}{1+2\tau\mu(X)}}\right)^y \frac{1}{y!}$$

$$\times K_{y-\frac{1}{2}}\left(\sqrt{\frac{1}{\tau^2}(1+2\tau\mu(X))}\right)$$

while the probability generating function is

$$\exp\left[\frac{1}{\tau}\left\{1 - (1+2\tau\mu(X)(1-s))^{\frac{1}{2}}\right\}\right].$$

The mean and variance are easily found to be $\mu(X)$ and $\tau\mu^2(X)$ respectively. The recurrence formulae for the probabilities is (see (f) section 7.1) for $y = 1, 2, \ldots$

$$p_{y+1} = \frac{\tau\mu(X)}{1+2\tau\mu(X)}\left(\frac{2y-1}{y+1}\right) p_y + \frac{\mu^2(X)}{1+2\tau\mu(X)} \frac{1}{y(y+1)} p_{y-1}$$

with

$$p_0 = \exp\left[\frac{1}{\tau}\left\{1 - (1+2\tau\mu(X))^{\frac{1}{2}}\right\}\right]$$

$$p_1 = \mu(X)\left\{1+2\tau\mu(X)\right\}^{-\frac{1}{2}} p_0.$$

Estimation

Consider the data $(Y_i, X_i), i = 1, \ldots, n$ where each Y_i takes the value $0, 1, \ldots$. We first write

$$p_{y_i} = \sqrt{\frac{2\alpha_i}{\pi}} \, e^{\frac{1}{\tau}} \left(\frac{\mu_i}{\alpha_i \tau}\right)^{y_1} K_{y_i-\frac{1}{2}}(\alpha_i)$$

where

$$\alpha_i = \tau^{-1}(1+2\tau\mu_i)^{\frac{1}{2}}, \quad \mu_i = \mu(X_i, \boldsymbol{\beta}).$$

Now

$$\frac{\partial \alpha_i}{\partial \beta_\gamma} = \frac{\partial \alpha_i}{\partial \mu_i} \frac{\partial \mu_i}{\partial \beta_\gamma}$$

$$= (1+2\tau\mu_i)^{-\frac{1}{2}} \frac{\partial \mu_i}{\partial \beta_\gamma}$$

$$\frac{\partial \alpha_i}{\partial \tau} = -\tau^{-2}(1+2\tau\mu_i)^{\frac{1}{2}} + \tau^{-1}(1+2\tau\mu_i)^{-\frac{1}{2}} \mu_i$$

$$= -\frac{1}{\tau^2}(1+\tau\mu_i)(1+2\tau\mu_i)^{-\frac{1}{2}}.$$

The log-likelihood of p_{y_i} is proportional to

$$\ell_i(\beta,\tau) \propto \frac{1}{2}\log \alpha_i + \frac{1}{\tau} + y_i(\log\mu_i - \log\tau - \log\alpha_i) + \log K_{y_i-\frac{1}{2}}(\alpha_i)$$

so that

$$\frac{\partial \ell_i}{\partial \beta_\gamma} = \left(\frac{1}{2\alpha_i} - \frac{y_i}{\alpha_i} + \frac{K'_{y_i-\frac{1}{2}}(\alpha_i)}{K_{y_i-\frac{1}{2}}(\alpha_i)}\right)\frac{\partial \alpha_i}{\partial \beta_\gamma} + \frac{y_i}{\mu_i}\frac{\partial \mu_i}{\partial \beta_\gamma}.$$

Recalling that

$$\left(\frac{K'_{y_i-\frac{1}{2}}(\alpha_i)}{K_{y_i-\frac{1}{2}}(\alpha_i)}\right) = -R_{y_i-\frac{1}{2}}(\alpha_i) + \frac{2y_i-1}{2\alpha_i} - \frac{1}{2\alpha_i},$$

we have for $r = 1,\ldots,p$

$$\frac{\partial \ell_i}{\partial \beta_r} = \left(\frac{y_i}{\mu_i} - \frac{R_{y_i-\frac{1}{2}}(\alpha_i)}{(1+2\tau\mu_i)^{\frac{1}{2}}}\right)\frac{\partial \mu_i}{\partial \beta_r} = U_r(\beta,\tau), \quad \text{(say)}.$$

Likewise we obtain

$$\frac{\partial \ell_i}{\partial \tau} = \frac{-(1+\tau y_i)}{\tau^2} + \frac{(1+\tau\mu_i)}{\tau^2(1+2\tau\mu_i)^{\frac{1}{2}}}R_{y_i-\frac{1}{2}}(\alpha_i) = U_{r+1}(\beta,\tau)$$

Gathering all these

$$\frac{\partial \ell}{\partial \beta_r} = \sum_{i=1}^n \left(\frac{y_i}{\mu_i} - \frac{R_{y_i-\frac{1}{2}}(\alpha_i)}{(1+2\tau\mu_i)^{\frac{1}{2}}}\right)\frac{\partial \mu_i}{\partial \beta_r}$$

and

$$\frac{\partial \ell}{\partial \tau} = \frac{1}{\tau^2}\sum_{i=1}^n \left\{\frac{1+\tau\mu_i}{(1+2\tau\mu_i)^{\frac{1}{2}}}R_{y_i-\frac{1}{2}}(\alpha_i) - (1+\tau y_i)\right\}$$

Dean et al., recommend that $U_r(\beta,\tau) = 0$ be solved first to obtain $\hat{\beta}(\tau)$, and then the profile likelihood $\ell(\hat{\beta}(\tau),\tau)$ maximized with respect to τ to yield $\hat{\tau}$ and $\hat{\beta}(\hat{\tau})$. The estimates $\hat{\beta}(\tau)$ are found using the Newton-Raphson iteration or the scoring algorithm. The information matrix $I(\hat{\beta},\hat{\tau})$ can be computed from the following.

$$\frac{\partial^2 \ell_i}{\partial \tau^2} = \frac{(2+\tau y_i)}{\tau^3} - \left\{\frac{1+(1+\tau\mu_i)^2}{\tau^3(1+2\tau\mu_i)^{\frac{1}{2}}}\right\}R_{y_i-\frac{1}{2}}(\alpha_i)$$
$$- \frac{(1+\tau\mu_i)^2}{\tau^4(1+2\tau\mu_i)}R'_{y_i-\frac{1}{2}}(\alpha_i)$$

$$\frac{\partial^2 \ell}{\partial \beta_r \partial \beta_s} = \sum_{i=1}^n \left(\frac{y_i}{\mu_i} - \frac{R_{y_i-\frac{1}{2}}(\alpha_i)}{(1+2\tau\mu_i)^{\frac{1}{2}}}\right)\frac{\partial^2 \mu_i}{\partial \beta_r \partial \beta_s}$$

$$+ \sum_{i=1}^{n} \left(-\frac{y_i}{\mu_i^2} \frac{\partial \mu_i}{\partial \beta_s} - \frac{R'_{y_i-\frac{1}{2}}(\alpha_i)}{(1+2\tau\mu_i)^{\frac{1}{2}}} \frac{\partial \alpha_i}{\partial \beta_s} - \frac{\tau R_{y_i-\frac{1}{2}}(\alpha_i)}{(1+2\tau\mu_i)^{\frac{3}{2}}} \frac{\partial \mu_i}{\partial \beta_s} \right) \frac{\partial \mu_i}{\partial \beta_r}$$

$$= \sum_{i=1}^{n} \left(\frac{y_i}{\mu_i} - \frac{R_{y_i-\frac{1}{2}}(\alpha_i)}{(1+2\tau\mu_i)^{\frac{1}{2}}} \right) \frac{\partial^2 \mu_i}{\partial \beta_r \partial \beta_s}$$

$$- \sum_{i=1}^{n} \frac{R'_{y_i-\frac{1}{2}}(\alpha_i)}{(1+2\tau\mu_i)} \left(\frac{\partial \mu_i}{\partial \beta_r} \right) \left(\frac{\partial \mu_i}{\partial \beta_s} \right)$$

$$- \sum_{i=1}^{n} \left(\frac{y_i}{\mu_i^2} + \frac{\tau R_{y_i-\frac{1}{2}}(\alpha_i)}{(1+2\tau\mu_i)^{\frac{3}{2}}} \right) \left(\frac{\partial \mu_i}{\partial \beta_r} \right) \left(\frac{\partial \mu_i}{\partial \beta_s} \right)$$

$$\frac{\partial^2 l}{\partial \beta_r \partial \tau} = \frac{1}{\tau^2} \sum_{i=1}^{n} \frac{(1+\tau\mu_i)}{(1+2\tau\mu_i)} R'_{y_i-\frac{1}{2}}(\alpha_i) \frac{\partial \mu_i}{\partial \beta_r}$$

$$+ \sum_{i=1}^{n} \frac{\mu_i R'_{y_i-\frac{1}{2}}(\alpha_i)}{(1+2\tau\mu_i)^{\frac{3}{2}}} \left(\frac{\partial \mu_i}{\partial \beta_r} \right)$$

Tests and confidence intervals on β_i's and τ are obtained by using asymptotics, in particular, $(\hat{\beta} - \beta, \hat{\tau} - \tau) \sim N_{p=1}(\underline{0}, I^{-1})$. As in the previous section a test of $\tau = 0$ corresponds to a test of a Poisson model and may be based on the statistic

$$\Lambda = 2 \left[l(\tilde{\beta}, \tilde{\tau}) - l(\hat{\beta}(0), 0) \right].$$

When $\tau = 0$ $P(\Lambda = 0) = \frac{1}{2}$ while $\mathcal{L}(\Lambda)$ follows a $\frac{1}{2}\chi_1^2$ law for $\Lambda > 0$ (a result due to Chernoff (1954)).

PART II APPLICATIONS
A. Actuarial Science
Claim Cost analysis

In the field of insurance mathematics it is important to specify the probability distribution for the cost of a single claim. From this specification one calculates the probability distribution for the total number of claims. According to Seal (1969) by far the greatest number of graduations of observed individual claim amounts have been based on the logmormal distribution, where it is assumed that the logarithm of the amount claimed follows the Gaussian law. The specification of a particular law for the cost of a single claim may not always be justifiable on axiomatic grounds, but nevertheless the tractability of the distribution is often a motivation behind its choice. Seal (1978) and Berg (1980) have advocated the use of the inverse Gaussian law in modelling claim cost distributions. In the following discussion we consider the approach taken by Berg to analyze loglinear claim cost analysis.

Let Y denote a claim cost and assume that $\mathcal{L}(Y) = IG(\mu, \lambda)$, parametrized by μ and $\phi = \frac{\lambda}{\mu}$ has density

$$f(y \mid \mu, \phi) = \sqrt{\frac{\mu\phi}{2\pi y^3}} \exp\left\{-\frac{\phi\mu}{2y} - \frac{\phi y}{2\mu} + \phi\right\} 1_{\mathbb{R}^+}(y).$$

Denote by Y_r the total claim costs based on n_r claims where $r = 1, \ldots, R$. Then the density function of Y_r, $f(y_r \mid n_r\mu_r, n_r\phi_r)$ is

$$= n_r \sqrt{\frac{\mu_r\phi_r}{2\pi y_r^3}} \exp\left\{-\frac{n_r\phi_r}{2}\left(\frac{n_r\mu_r}{y_r} + \frac{y_r}{n_r\mu_r} - 2\right)\right\} 1_{\mathbb{R}^+}(y_r).$$

Consider the following log-linear parametrization, in the spirit of the generalized linear models á la Nelder and Wedderburn (1972):

$$\left.\begin{array}{l}\log\mu_r = X_r'\beta \\ \log\phi_r = Z_r'\gamma\end{array}\right\} \quad r = 1, \ldots, R,$$

where the $(K \times 1)$ vector X_r and $(L \times 1)$ vector Z_r are explanatory variables characterizing a risk group, an insurance line or a time period. The vectors $((K \times 1)$ and $(L \times 1)$ respectively) β and γ are parameters. (In many instances X_r and Z_r could take the values 0 or 1.)

Furthermore we use the following matrix notation.

$$N = \text{diag}(n_1, \ldots, n_R),$$
$$\Phi = \text{diag}(\phi_1, \ldots, \phi_R),$$
$$X = [X_1, \ldots, X_R]^t \text{ an } R \times K \text{ matrix}$$
$$Z = [Z_1, \ldots, Z_R]^t \text{ an } R \times L \text{ matrix}$$

Here the first columns of X and Z are taken as the unit vector $e = (1,\ldots,1)^t$. The $(K \times K)$ matrix $X^t N X$ and the $L \times L$ matrix $Z^t N Z$ are both assumed to be non-singular. Although it is not essential for the analysis that follows, it is further assumed that as the n_r increase

$$\lim \frac{N}{\operatorname{tr} N} = \bar{N}$$

exists making $X^t \bar{N} X$ and $Z^t \bar{Z}$ both singular. Under these assumptions it is easily verified that $e^t \bar{N} e = \operatorname{tr} \bar{N} = 1$.

Using the parametrization introduced above the log-likelihood function $\ell(\beta, \gamma)$ is proportional to

$$\ell(\beta, \gamma) \propto \frac{1}{2} e^t X\beta + \frac{1}{2} e^t Z\gamma - \frac{1}{2} \sum_{r=1}^{R} y_r \exp\left(z_r^t \gamma - x_r^t \beta\right)$$

$$- \frac{1}{2} \sum_{r=1}^{R} n_r^2 \, y_r^{-1} \exp\left(x_r^t \beta + z_r^t \gamma\right) + \sum_{r=1}^{R} n_r \exp(z_r^t \gamma).$$

The likelihood equations are now obtainable by differentiation with respect to β and γ and we obtain

$$\frac{\partial \ell}{\partial \beta} = \frac{1}{2} X^t e + \frac{1}{2} \sum_{r=1}^{R} y_r \exp(z_r^t \gamma - x_r^t \beta) x_r$$

$$- \frac{1}{2} \sum_{r=1}^{R} n_r^2 \, y_r^{-1} \exp(x_r^t \beta + z_r^t \gamma) x_r$$

$$\frac{\partial \ell}{\partial \gamma} = \frac{1}{2} Z^t e - \frac{1}{2} \sum_{r=1}^{R} y_r \exp(z_r^t \gamma - x_r^t \beta) z_r$$

$$- \frac{1}{2} \sum_{r=1}^{R} n_r^2 y_r^{-1} \exp(x_r^t \beta + z_r^t \gamma) z_r + \sum_{r=1}^{R} n_r \exp(z_r^t \gamma) z_r.$$

To obtain the Hessian H of the log-likelihood we differentiate again and obtain $H = \begin{pmatrix} h_{\beta\beta} & h_{\beta\gamma} \\ h_{\beta\gamma}^t & h_{\gamma\gamma} \end{pmatrix}$ where

$$\frac{\partial^2 \ell}{\partial \beta \partial \beta^t} = -\frac{1}{2} \sum_{r=1}^{R} \phi_r \left(\frac{y_r}{\mu_r} + \frac{n_r^2 \mu_r}{y_r} \right) x_r \, x_r^t = h_{\beta\beta}$$

$$\frac{\partial^2 \ell}{\partial \beta \partial \gamma^t} = \frac{1}{2} \sum_{r=1}^{R} \phi_r \left(\frac{y_r}{\mu_r} - \frac{n_r^2 \mu_r}{y_r} \right) x_r \, z_r^t = h_{\beta\gamma}$$

$$\frac{\partial^2 \ell}{\partial \gamma \partial \gamma^t} = -\frac{1}{2} \sum_{r=1}^{R} \phi_r \left(\frac{y_r}{\mu_r} + \frac{n_r^2 \mu_r}{y_r} - 2 n_r \right) z_r \, z_r^t = h_{\gamma\gamma}.$$

Note that $h_{\beta\beta}$ and $h_{\gamma\gamma}$ are negative definite matrices and H is not negative definite for all β and γ implying that $\ell(\beta,\gamma)$ is not concave in the whole parameter space. But for fixed γ, $\ell(\beta,\gamma)$ is concave and vice-versa. Therefore by a lemma due to Oberhofer and Kmenta (1974) $\ell(\beta,\gamma)$ can be maximized iteratively by first using an initial value $\gamma^{(0)}$ of γ and maximizing $\ell(\beta,\gamma^{(0)})$ with respect to β. If $\hat{\beta}(\gamma^{(0)})$ is the value maximizing $\ell(\beta,\gamma^{(0)})$ one proceeds iteratively using this zig-zag procedure. The lemma guarantees convergence yielding the values $(\hat{\beta},\hat{\gamma})$. Equivalently one can use the E-M algorithm.

The information matrix $I(\beta,\gamma)$ is obtained from the Hessian matrix. Indeed

$$-\mathbb{E}(h_{\beta\beta}) = \frac{1}{2}\sum_{r=1}^{R}\frac{\phi_r}{\mu_r}(n_r\mu_r)x_r x_r^t +$$

$$\frac{1}{2}\sum_{r=1}^{R}\phi_r\mu_r n_r^2\left(\frac{1}{n_r\mu_r}+\frac{1}{\mu_r\phi_r n_r^2}\right)x_r x_r^t$$

$$= \frac{1}{2}\sum_{r=1}^{R}n_r\phi_r x_r x_r^t + \frac{1}{2}\sum_{r=1}^{R}(\phi_r n_r+1)x_r x_r^t$$

$$= \sum_{r=1}^{R}n_r\phi_r x_r x_r^t + \frac{1}{2}\sum_{r=1}^{R}x_r x_r^t$$

$$= X^t N\Phi X + \frac{1}{2}X^t X,$$

$$-\mathbb{E}(h_{\beta\gamma}) = -\frac{1}{2}\sum_{r=1}^{R}n_r\phi_r x_r z_r^t + \frac{1}{2}\sum_{r=1}^{R}n_r\phi_r x_r z_r^t + \frac{1}{2}\sum_{r=1}^{R}x_r z_r^t$$

$$= \frac{1}{2}X^t Z$$

and

$$-\mathbb{E}(h_{\gamma\gamma}) = Z^t N\Phi Z + \frac{1}{2}Z^t Z - Z^t N\Phi Z$$

$$= \frac{1}{2}Z^t Z.$$

Thus

$$I(\beta,\gamma) = \begin{pmatrix} X^t N\Phi X + \frac{1}{2}X^t X & \frac{1}{2}X^t Z \\ \frac{1}{2}Z^t X & \frac{1}{2}Z^t Z \end{pmatrix}.$$

This matrix can be inverted to yield the covariance matrix of $\begin{pmatrix}\hat{\beta}\\\hat{\gamma}\end{pmatrix}$.

$$\mathrm{Var}(\hat{\beta}) = \left[X^t[N\phi + \frac{1}{2}(I - Z(Z^tZ)^{-1}Z^t)]X\right]^{-1} = V_1(\hat{\beta}), \text{ say},$$

$$\mathrm{Var}(\hat{\gamma}) = 2(Z^tZ)^{-1} + (Z^tZ)^{-1}\left[Z^tX\mathrm{Var}(\hat{\beta})X^tZ\right](Z^tZ)^{-1},$$

and

$$\mathrm{Cov}(\hat{\beta}, \hat{\gamma}) = -\mathrm{Var}(\hat{\beta})X^tZ(Z^tZ)^{-1}.$$

It may be of interest to know if there is any loss of information in the estimation of β, when γ is fixed, due to considering total claims rather than individual claims. In this situation we form the likelihood function for individual claims. Thus

$$\ell(\beta, \gamma) \propto \sum_{j=1}^{R} n_j x_j^t \beta + \sum_{j=1}^{R} n_j z_j^t \gamma + \sum n_j \phi_j$$

$$- \frac{1}{2}\sum_{j=1}^{R} \phi_j \mu_j \left(\sum_{k=1}^{n_j} \frac{1}{y_{jk}}\right) - \frac{1}{2}\sum_{j=1}^{R} \frac{\phi_j}{\mu_j} y_j.$$

This shows that when we form the Hessian matrix, and then take expectations we have some modifications to contend with arising from the term $\sum_{j=1}^{R} \phi_j \mu_j \left(\sum_{k=1}^{n_j} \frac{1}{y_{jk}}\right)$. Then

$$-\mathbb{E}\left(\frac{\partial^2 \ell}{\partial \beta \partial \beta^t}\right) = \frac{1}{2}X^t N\Phi X + \frac{1}{2}\sum_{j=1}^{R} \mu_j \phi_j \left(\sum_{k=1}^{n_j}\left(\frac{1}{\mu_j} + \frac{1}{\mu_j \phi_j}\right)\right)x_j x_j^t$$

$$= \frac{1}{2}X^t N\Phi X + \frac{1}{2}X^t N\Phi X + \frac{1}{2}X^t N X$$

$$-\mathbb{E}\left(\frac{\partial^2 \ell}{\partial \gamma \partial \gamma^t}\right) = \frac{1}{2}Z^t N\Phi Z + \frac{1}{2}\sum_{j=1}^{R} \phi_j \mu_j \left(\sum_{k=1}^{n_j}\left(\frac{1}{\mu_j} + \frac{1}{\mu_j \phi_j}\right)\right)Z_j Z_j^t$$

$$- \sum_{j=1}^{R} n_j \phi_j z_j z_j^t$$

$$= \frac{1}{2}Z^t N Z,$$

and

$$-\mathbb{E}\left(\frac{\partial^2 \ell}{\partial \beta \partial \gamma^t}\right) = \frac{1}{2}\sum_{j=1}^{R} n_j \phi_j x_j z_j^t + \frac{1}{2}\sum_{j=1}^{R} n_j x_j z_j^t - \frac{1}{2}\sum n_j \phi_j x_j z_j^t$$

$$= \frac{1}{2}X^t N Z.$$

Claim cost analysis

In this instance the covariance matrix for $\hat{\beta}$ is

$$V_2(\hat{\beta}) = \left[X^t \left[N\Phi + \frac{1}{2}(N - NZ(Z^tNZ)^{-1}Z^tN) \right] X \right]^{-1}.$$

When $X = ZC$ (i.e. the columns of X are linear combinations of the columns of Z) both the expressions for $V(\hat{\beta})$ reduce to

$$V_1(\hat{\beta}) - V_2(\hat{\beta}) = (X^t N \Phi X)^{-1}.$$

This means that there is no loss of efficiency. On the other hand by considering the asymptotic case and examining $\lim \operatorname{tr} NV_1(\hat{\beta}) - \lim \operatorname{tr} NV_2(\hat{\beta})$ it is possible to obtain a positive semidefinite matrix, namely

$$\frac{1}{2} X^t \left[\bar{N} - \bar{N}Z(Z^t\bar{N}Z)^{-1}Z^t\bar{N} \right] X.$$

This clearly shows that there is a loss of efficiency, due to aggregation. Finally Berg considers a logarithmic transformation of the aggregate y_r in the hope that for large n_r a more symmetric density close to the normal will emerge. For this log transformation one has asymptotically

$$\begin{aligned}
w_r &= \log(y_r/n_r) \\
&= \log(\bar{y}_r) \\
&= \log \mu_r + \log \left[1 + \frac{\bar{y}_r - \mu_r}{\mu_r} \right] \\
&\approx \log \mu_r + \frac{\bar{y}_r - \mu_r}{\mu_r} - \frac{1}{2}\left(\frac{\bar{y}_r - \mu_r}{\mu_r}\right)^2. \\
\mathbb{E}(\bar{w}_r) &= \log \mu_r - \frac{\mu_r^3}{2n_r \lambda_r \mu_r^2} \\
&= \log \mu_r - \frac{1}{2n_r \phi_r} \\
&= x_r^t \beta - (2n_r \phi_r)^{-1}. \\
\operatorname{Var}(\bar{w}_r) &= (n_r \phi_r)^{-1}.
\end{aligned}$$

This model, when the term $(2n_r\phi_r)^{-1}$ is neglected has been analyzed by Harvey (1976). Berg considers a model of the form

$$\mathbb{E}(w_r) = x_r^t \beta - d(2n_r\phi_r)^{-1}$$

where d takes two values 0 and 1, and shows that regardless of whether $d = 0$ or 1 the asymptotic variance of $\hat{\beta}$ has the value $(X^t \bar{N} \Phi X)^{-1}$.

B. Analysis of reciprocals

Tweedie (1956) introduced a method for the analysis of residuals from an inverse Gaussian population paralleling the analysis of variance in normal theory. He called it the "analysis of reciprocals". The analysis of variance is invariant under linear transformations whereas the analysis of reciprocals is invariant under scale changes only. Furthermore the analysis of reciprocals is restricted to nested classifications thereby limiting its application. We first consider the one and two way layouts and then discuss an analysis of two factor experiments examined by Fries and Bhattacharyya (1983).

One way classification

In this model we assume that there are n_i items from the ith population each of which is distributed as $IG(\mu_i, \lambda)$ where $i = 1, \ldots, I$. Thus we have independent observations X_{ij}, $j = 1, \ldots, n_i$, $i = 1, \ldots, I$ such that $\mathcal{L}(X_{ij}) = IG(\mu_i, \lambda)$. In all there are $n = \sum_{i=1}^{I} n_i$ observations. We also assume that λ is an unknown positive constant which is the same from sample to sample and that $\mu_i > 0$. The parameters μ_i and λ lie in the set

$$\Omega = \{(\mu_1, \ldots, \mu_I, \lambda) \mid 0 < \mu_i < \infty, 0 < \lambda < \infty\}.$$

Let us now consider the following problem, namely, of testing $H_0 : \mu_1 = \ldots = \mu_I$; λ unknown against $H_A : H_0$ is false. Then the likelihood function is proportional to

$$L(\mu, \lambda) \propto \lambda^{\frac{n}{2}} \exp\left\{-\frac{\lambda}{2} \sum_{i=1}^{I} \sum_{j=1}^{n_i} \frac{(x_{ij} - \mu_i)^2}{\mu_i^2 x_{ij}}\right\}$$

and differentiation with respect to μ_i and λ easily yields the estimates

$$\hat{\mu}_i = \left(\sum_{j=1}^{n_i} x_{ij}\right) \Big/ n_i = \bar{x}_{i\cdot},$$

$$n\hat{\lambda}_\Omega^{-1} = \sum_{i=1}^{I} \sum_{j=1}^{n_i} \left(x_{ij}^{-1} - \bar{x}_{i\cdot}^{-1}\right).$$

Under H_0 the maximum likelihood estimates are

$$\hat{\mu} = \left(\sum_{i=1}^{I} \sum_{j=1}^{n_i} x_{ij}\right) \Big/ n = \bar{x},$$

$$n\hat{\lambda}_\omega^{-1} = \sum_{i=1}^{I} \sum_{j=1}^{n_i} \left(x_{ij}^{-1} - \bar{x}^{-1}\right).$$

One way classification

Hence the likelihood ratio Λ satisfies

$$\Lambda^{\frac{2}{n}} = \frac{n\hat{\lambda}_\Omega^{-1}}{n\hat{\lambda}_\omega^{-1}} = \frac{Q_0}{Q} \text{ (say)}.$$

Now Q can be decomposed as

$$Q = Q_0 + \sum_{i=1}^{I} n_i \bar{x}_{i.}^{-1} - n\bar{x}^{-1} = Q_0 + Q_1 \text{ (say)}.$$

Moreover, it is clear that

$$\mathcal{L}(\lambda Q_0) = \chi^2_{n-I}, \; \mathcal{L}(\lambda Q_1) = \chi^2_{I-1} \text{ and } Q_0 \perp\!\!\!\perp Q_1.$$

Thus the likelihood ratio test calls for rejection of H_0 at level α if

$$\frac{(n-I)}{(I-1)} \frac{Q_1}{Q_0} = F_{I-1, n-I} > F_{1-\alpha}.$$

The following example from Tweedie (1956) illustrates this test.

Example B.1 Four IG(μ_i, λ) populations with $n_i = 5$ yield the following values

Population i	1	2	3	4
	8.7	8.5	8.4	8.1
	9.0	8.6	9.0	8.4
	8.4	8.4	8.9	8.5
	8.6	8.3	8.5	8.1
	8.4	8.8	8.8	8.4
$\bar{x}_{i.}$	8.62	8.52	8.72	8.30
$\bar{x} =$	8.54			

$Q = 0.002058, Q_0 = 0.001274, Q_1 = 0.000784$.

The following is the analysis of reciprocals table.

Table B.1 *Analysis of Reciprocals*

Sources of variation	degrees of freedom	Sum of differences of reciprocals	F
Between	3	$Q_1 = 0.000784$	3.28
Within	16	$Q_0 = 0.001274$	
Total	19	$Q = 0.002058$	

The tabulated $F_{0.95} = 3.24 <$ observed $F = 3.28$ leading to rejection of H_0 at level $\alpha = 0.05$.

Suppose that we wish to test for homogeneity of the λ's, i.e., $H_0 : \lambda_1 = \lambda_2 = \ldots = \lambda_I$; μ_i unknown against $H_A : H_0$ is false.

In this instance

$$\Omega = \{(\mu_1, \ldots, \mu_I, \lambda_1, \ldots, \lambda_I) \mid 0 < \mu_i < \infty, 0 < \lambda_i < \infty\}.$$

The likelihood function is proportional to

$$\prod_{i=1}^{I} \lambda_i^{\frac{1}{2}} \exp\left\{-\frac{1}{2} \sum_{i=1}^{I} \frac{\lambda_i}{\mu_i^2} \sum_{j=1}^{n_i} \frac{(x_{ij} - \mu_i)^2}{x_{ij}}\right\}$$

and routine computations give

$$\hat{\mu}_i = \bar{x}_{i\cdot}, \quad n_i \hat{\lambda}_i^{-1} = \sum_{j=1}^{n_i} (x_{ij}^{-1} - \bar{x}_i^{-1}) = Q_i$$

while under H_0

$$\hat{\mu}_i = \bar{x}_{i\cdot}, \quad n\hat{\lambda}^{-1} = \sum_{i=1}^{I} \sum_{j=1}^{n_i} (x_{ij}^{-1} - \bar{x}_{i\cdot}) = Q$$

Then the likelihood ratio is

$$\Lambda = \frac{\prod_{i=1}^{I} (Q_i/n_i)^{\frac{n_i}{2}}}{(Q/n)^{\frac{n}{2}}}.$$

The distribution of Λ is complicated so that we use the fact that under H_0, $-2 \ln \Lambda$ is approximately a χ^2 variable. Since under H_0 $\mathcal{L}(\lambda Q_1) = \chi^2_{n_i-1}$ an approximation due to Bartlett (1937) provides a modified test function, namely

$$\Lambda^* = \frac{(n-I) \ln (Q/n - I) - \sum_{i=1}^{I} (n_i - 1) \ln (Q_i/n_i - 1)}{g(I)}$$

where

$$g(I) = 1 + \frac{1}{3(I-1)} \left[\sum_{i=1}^{I} \left(\frac{1}{n_i - 1} - \frac{1}{n - I}\right)\right].$$

An application in environmental sciences

Finally one has $\mathcal{L}(\Lambda^*) \approx \chi^2_{I-1}$.

Example B.2 For the data in Example B.1 let us omit $x_{15}, x_{34}, x_{35}, x_{44}$ and x_{45}. Then we have

$$\bar{x}_{1.} = 8.5, \bar{x}_{2.} = 8.52, \bar{x}_{3.} = 8.6, \bar{x}_{4.} = 8.43, \bar{x} = 8.51,$$

$Q = 0.00082, Q_1 = 0.00006, Q_2 = 0.00025, Q_3 = 0.00022, Q_4 = 0.00001$.

Writing $Q_0 = \sum_{i=1}^{4} n_i (\bar{x}_{i.}^{-1} - \bar{x}^{-1})$, we obtain $Q_0 = 0.00028$, and $g(I) = 1.0975$, so that

$$\Lambda^* = \frac{13 \ln\left(\frac{Q}{13}\right) - 3\left\{\ln\left(\frac{Q_1}{3}\right) + \ln\left(\frac{Q_3}{3}\right) + \ln\left(\frac{Q_4}{3}\right)\right\} - 4 \ln\left(\frac{Q_2}{4}\right)}{1.0975}$$

$= 11.96$.

The $\chi^2_{.05}$ value for 3 degrees of freedom being 7.81, we reject H_0.

Suppose that we had tested the hypothesis that the means are all different when λ is the same for all four groups we would then obtain the following analysis of reciprocals table.

Table B.2 *Analysis of Reciprocals*

Sources of variation	degrees of freedom	Sum of differences of reciprocals	F
Between	3	$Q_0 = 0.00028$	$F_{3,16} = 1.901$
Within	11	$Q - Q_0 = 0.00054$	
Total	14	$Q = 0.00082$	

With the tabulated $F_{0.95}$ at 3.59, we would conclude that H_0 is not significant at the 5% level.

An application in environmental sciences

The movement of chemicals in the soil media has been modelled using the Brownian motion process with drift by Helge Gydesen (1984). Problems concerning ground water pollution arise in the movement of chemicals in the soil. The amount of chemical leaching through a hand-packed soil column of given length can be observed as a function of time or of the volume of leaching water.

Water is supplied at a certain flow rate at the top of a soil column from a reservoir using a peristaltic pump. A chemical is applied at the upper surface of the soil column and leaching amounts are collected at the bottom of the soil column. Fractions of leachate are then taken for chemical and physical analysis. In modelling this experiment the position of a particle at time t, namely $X(t)$, is not observed but rather the travelling time of a particle measured from the top of the soil column to the bottom. Thus the quantity observed is

$$T = \sup\{\tau \mid X(t) \leq d,\ 0 \leq t \leq \tau\}$$

where d is the length of the soil column. Although the diffusion process is not unrestricted in that the particle cannot move out of the column (due to the flow being quite large) normal distribution theory serves as a reasonable approximation. Indeed if $X(t)$ is the position of a particle at time t, assuming $X(0) = x_0$, and infinitesimal drift ν and variance σ^2 are constant over time and position, the density $p(x_0, x; t)$ for $X(t)$ satisfies the differential equation

$$\frac{\partial p}{\partial t} = -\nu \frac{\partial p}{\partial x} + \frac{\sigma^2}{2} \frac{\partial^2 p}{\partial x^2}$$

and the solution for p is the normal density $N(x_0 + \nu t, \sigma^2 t)$, while the distribution of T is given by

$$\frac{d}{\sigma \sqrt{2\pi t^3}} \exp\left[-\frac{(d-\nu t)^2}{2\sigma^2 t}\right].$$

The parametrization $\mu = \frac{d}{\nu}$, $\lambda = \frac{d^2}{\sigma^2}$ gives us at once the standard form of the $IG(\mu, \lambda)$ density.

Now suppose that n_i particles have gone through the soil columns at times $t_i (i = 1, \cdots, k)$ we can use the formulae obtained in Section 1.2 to determine the maximum likelihood estimates of μ and λ (see Equation 1.4 with n_i replacing w_i).

If m different soil columns are observed and it is desirable to know if there is a difference between the behaviour of the soil columns, then, we test first for the homogeneity of the λ values and follow it up with a test of the equality of the μ values.

Example B.3 In the experimental analysis reported by Gydesen, three chemicals – ethylene glycol, ethanol and sodium chloride – were tested on several soil columns. Ethylene glycol was tested on two soil columns (1,2) and ethanol on six soil columns (3-8). The sodium chloride compound was tested on four soil columns, two together with ethylene glycol (1,2) and two together with ethanol (3,4). Undisturbed moraine sand from one metre depth was the soil used in the experiment. Porosity varied in the range 30-50% and percolation was performed in nearly saturated

An application in environmental sciences

conditions. The water flow was vertically downward in columns (1-3) and upward in columns (5-8) respectively. The variation in waterflow in single experiments was accounted for by estimation of a mean flow rate.

Table B.3 *Results for 8 soil columns*

Soil Column	Chemical	Measuring method	β	λ	Convective flow $v(mS^{-1})$	Dispersion coefficient $D(m^2 C^{-1})$
1	1 EG	^{14}C	53.93	171.40	1.5×10^{-7}	4.7×10^{-9}
	2 EG	GC	54.60	225.72	1.5×10^{-7}	3.6×10^{-9}
	3 C		53.28	163.59	1.5×10^{-7}	6.0×10^{-9}
2	4 EG	^{14}C	54.41	227.90	3.8×10^{-8}	9.1×10^{-9}
	5 EG	GC	57.87	281.34	3.6×10^{-8}	7.4×10^{-9}
	6 C		54.68	226.11	4.1×10^{-8}	1.1×10^{-9}
3	7 E	^{14}C	45.73	139.24	1.5×10^{-7}	5.0×10^{-9}
	8 E	GC	44.86	140.74	1.5×10^{-7}	4.9×10^{-9}
	9 C		51.20	182.21	1.3×10^{-7}	3.1×10^{-9}
4	10 E	^{14}C	55.29	98.81	1.1×10^{-7}	6.6×10^{-9}
	11 E	GC	54.14	92.78	1.2×10^{-7}	7.0×10^{-9}
	12 C		57.56	97.69	1.1×10^{-7}	6.6×10^{-9}
5	13 E	^{14}C	42.89	94.07	4.3×10^{-8}	2.0×10^{-9}
	14 E	GC	41.12	96.63	4.5×10^{-8}	1.9×10^{-9}
6	15 E	^{14}C	39.42	48.42	6.5×10^{-8}	4.1×10^{-9}
	16 E	GC	37.94	46.33	6.7×10^{-8}	5.0×10^{-9}
7	17 E	^{14}C	61.20	178.65	1.1×10^{-7}	3.8×10^{-9}
	18 E	GC	58.11	182.85	1.2×10^{-7}	3.7×10^{-9}
8	19 E	^{14}C	55.65	215.75	1.2×10^{-7}	3.2×10^{-9}
	20 E	GC	54.90	212.75	1.3×10^{-7}	3.2×10^{-9}

EG=Ethylene glycol, C=Chloride, E=Ethanol.

Table B.3 gives the maximum likelihood estimates of μ and λ as well as ν and σ^2. Two methods were used in the estimation process. In the first method radioactive (^{14}C) labelled chemicals were detected and quantified by unspecific counting of the sum of unchanged and possible degradation products marked with ^{14}C while in the second method gas chromatography was used for measuring the amount of specific test chemical.

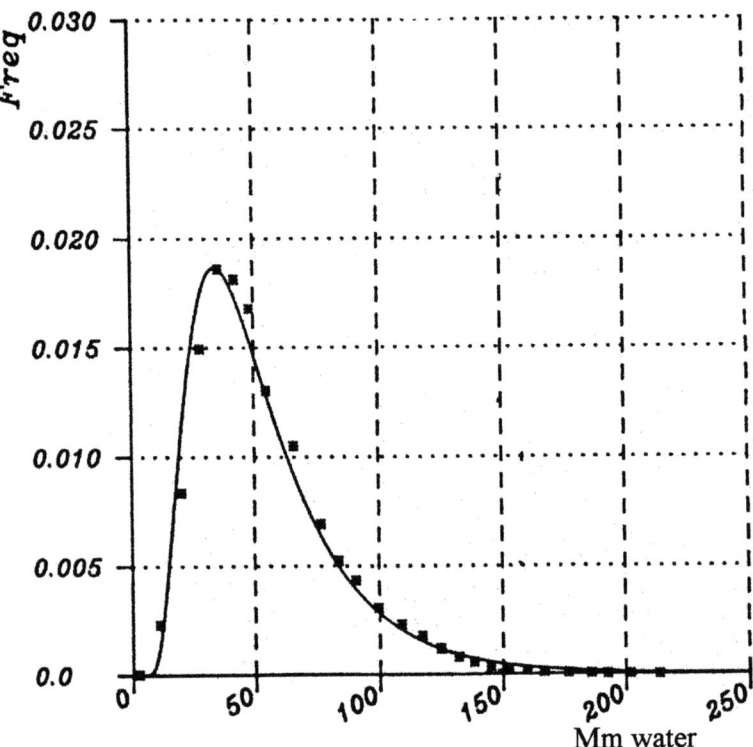

Figure B.1 *Estimated density and observed frequencies for EG measured by ^{14}C in soil column 1*

Gydesen notes that
(a) the effect of diffusion was small in comparison with convective dispersion. The flow rate was not always constant and the amount of water passing through a soil column was used as a time variable. The actual measurement was a quantity proportional to the number of particles.
(b) The exact leaching time was not measured and only the quantity of chemical which had been leaching during fixed time intervals was measured.
(c) The time of occurence was assumed to be the mid-point of the time interval.

An application in environmental sciences

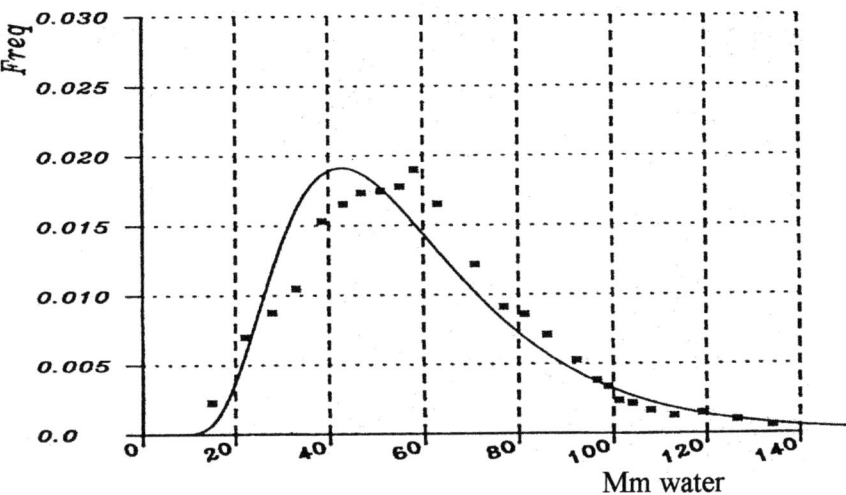

Figure B.2 *Estimated density and observed frequencies for Ethylene Glycol measured by GC in soil column 2*

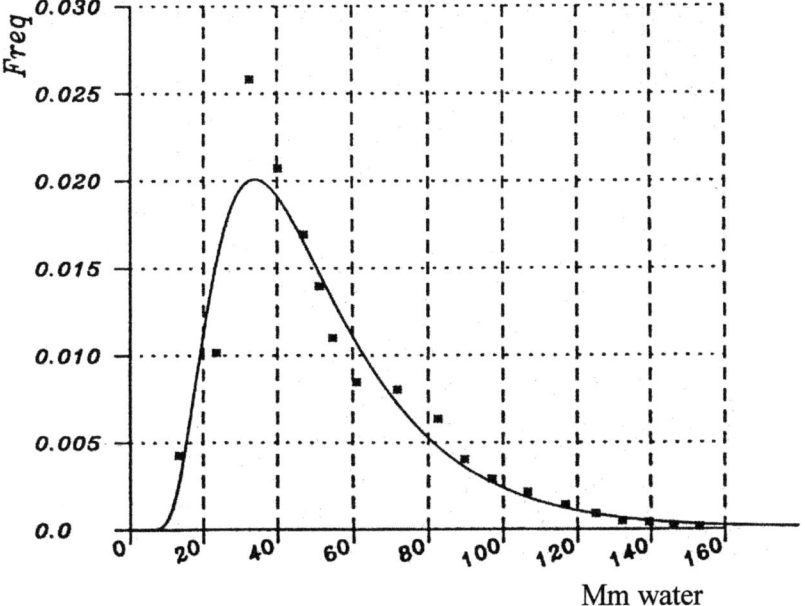

Figure B.3 *Estimated density and observed frequencies for chloride in soil column 3*

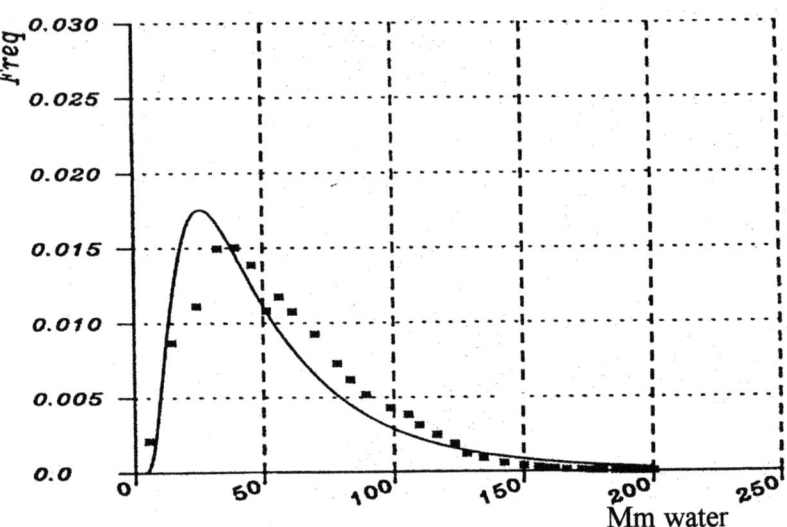

Figure B.4 *Estimated density and observed frequencies for Ethylene Glycol measured by ^{14}C in soil column 4*

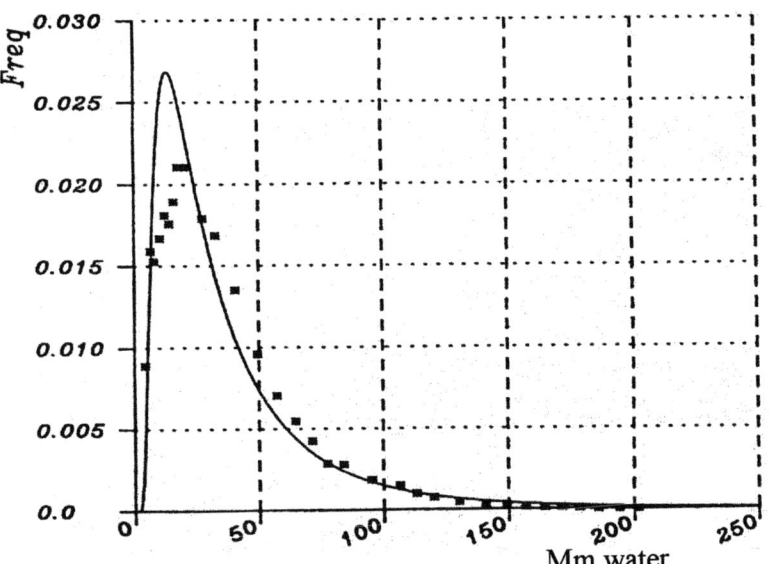

Figure B.5 *Estimated density and observed frequencies for Ethanol measured by GC in soil column 6*

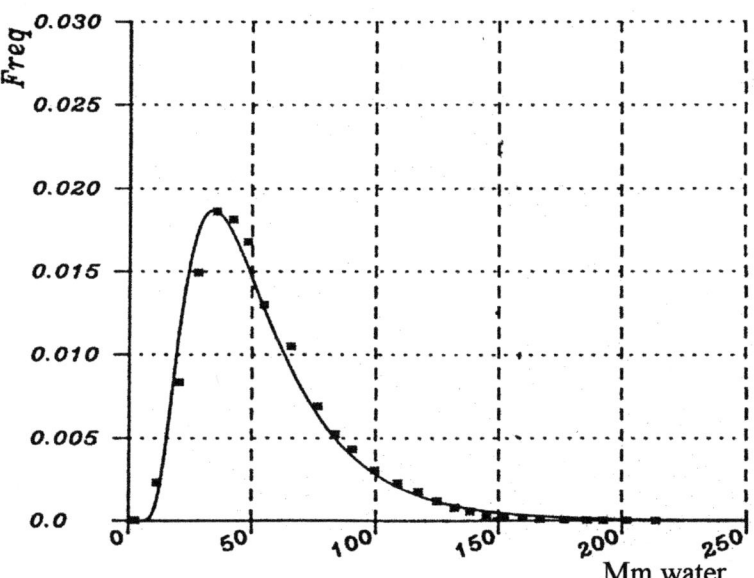

Figure B.6 *Estimated density and observed frequencies for Ethanol measured by ^{14}C in soil column 8*

Conclusions From the figures it is clear that the IG model is quite adequate for six of the 20 experiments performed by Gydesen. Since there is a tendency of the estimated density to peak a little ahead of the observed frequencies, it maybe that a GIG model is called for.

Using the tests of hypothesis outlined in the theoretical discussion, it turned out that the calculated test values were found highly significant. Gydesen surmises that this could have arisen from the fact that the only randomness in the model originates from the movement of each particle and concludes that the randomness arising from the experimental error could have been estimated, had there been replications in the experimental set-up. On the other hand the hypothesis of homogeneity of the parameters for the two methods of measurement seems acceptable.

Analysis of two-factor experiments

In this layout we assume that the independent random variables X_{ijk} have the distribution $\mathcal{L}(X_{ijk}) = IG(\mu + \alpha_i + \beta_j, \lambda)$ where $\mu > 0, \mu + \alpha_i + \beta_j > 0, \sum_{i=1}^{I} \alpha_i = 0$ and $\sum_{g=1}^{J} \beta_j = 0$. Further we assume that there are n_{ij} observations in cell (i,j). The row (i) and column (j) effects are denoted by α_i and β_j respectively. As Tweedie was quick to point out, the algebraic identity for the main effects and interactions in the analysis of reciprocals, does not give independent components. Indeed we have

$$\sum_{i=1}^{I}\sum_{j=1}^{J}(x_{ij}^{-1}-\bar{x}^{-1}) = \sum_{i=1}^{I}\sum_{j=1}^{J}(\bar{x}_{i.}^{-1}-\bar{x}^{-1}) + \sum_{i=1}^{I}\sum_{j=1}^{J}(\bar{x}_{.j}^{-1}-\bar{x}^{-1})$$
$$+ \sum_{i=1}^{I}\sum_{j=1}^{J}(x_{ij}^{-1}-\bar{x}_{i.}^{-1}-\bar{x}_{.j}^{-1}+\bar{x}^{-1}).$$

Unfortunately the third sum does not have a chi-square law since it has a finite probability of taking negative values, the reason why crossed classifications cannot be handled.

Miura (1978) has studied the problem of testing for row (column) effects for this model, i.e., testing
$H_0 : \alpha_1 = \cdots = \alpha_I$
against
$H_A : H_0$ is false.
Routine computations of the likelihood ratio statistic yield $\Lambda = \frac{Q_{ij}}{Q}$ where

$$Q_{ij} = \sum_{i=1}^{I}\sum_{j=1}^{J}\sum_{k=1}^{n_{ij}}(x_{ijk}^{-1}-\bar{x}_{ij.}^{-1}) \Big/ \sum_{i=1}^{I}\sum_{j=1}^{J}(n_{ij}-1),$$

$$Q = \sum_{i=1}^{I}\sum_{j=1}^{J}\sum_{k=1}^{n_{ij}}(x_{ijk}^{-1}-\bar{x}_{.j}^{-1}) \Big/ \sum_{i=1}^{I}\sum_{j=1}^{J}(n_{ij}-1),$$

$$\bar{x}_{ij.} = \frac{\sum_{k=1}^{n_{ij}} x_{ijk}}{n_{ij}}, \quad \bar{x}_{.j} = \frac{\sum_{i=1}^{I}\sum_{k=1}^{n_{ij}} x_{ijk}}{\sum_{i=1}^{I} n_{ij}}.$$

The denominator Q can be further decomposed as

$$\sum_{i=1}^{I}\sum_{j=1}^{J}\sum_{k=1}^{n_{ij}}(x_{ijk}^{-1}-\bar{x}_{ij.}^{-1}) + \sum_{i=1}^{I}\sum_{j=1}^{J}\sum_{k=1}^{n_{ij}}(\bar{x}_{ij.}^{-1}-\bar{x}_{.j}^{-1})$$

so that

$$\Lambda = \frac{Q_{ij}}{Q_{ij}+Q_j}\frac{(n-J)}{(n-IJ)}.$$

Observe that Q_{ij} is an estimate of λ^{-1} by cells and Q is an estimate of λ^{-1} by columns and that Q_{ij} is independent of $\bar{x}_{ij.}$ for every (i,j).

The distribution of Λ, apart from a constant is a Beta$((n-IJ)/2, (IJ-J)/2)$ random variable, which can be transformed to an F random variable with $J(I-1)$ and $n-IJ$ degrees of freedom. Note that unlike Q_{ij}, Q is unbiased for λ only under H_0. It can be shown that $\mathbb{E}(\Lambda) = 1$ and H_0 is then rejected if Λ is close to unity.

Shuster and Miura (1972) have considered a modified analysis of reciprocals when $\mathbb{E}(X)/\text{Var}(X)$ remains constant from sample to sample. Thus if we have I populations and X_{ij}, $j = 1, \ldots, n$, $i = 1, \ldots, I$ are all independent such that $\mathcal{L}(X_{ij}) = IG(\mu_i, \lambda)$ where $\xi = \frac{\lambda}{\mu_i^2}, i = 1, \ldots, n$, then it follows that a one-way analysis of reciprocals can be carried out with the test function

$$\Lambda = \frac{\sum^I (\bar{x}_{i.}^{-1} - \bar{x}^{-1})}{\sum_{i=1}^I \sum_{j=1}^n (x_{ij}^{-1} - \bar{x}_{i.}^{-1})}.$$

The details are omitted and the verification left as an exercise. A similar study for a two-way layout considered by these authors assumes $\mathcal{L}(X_{ijk}) = IG(\mu_i + \beta_j + \alpha_{ij}, \lambda), i = 1, \ldots, I, j = 1, \ldots, J$ and $k = 1, \ldots, K$. Further it is assumed that $\min(I, J, K) > 1$ and $\sum_{i=1}^I \alpha_{ij} = \sum_{j=1}^J \alpha_{ij} = 0$. A test for main effects can be reduced to a test of $H_0 : \beta_1 = \ldots = \beta_J$ against $H_A : H_0$ is false.

This test again reduces to a one-way analysis of reciprocals by considering

$$\mathcal{L}(X_{ijk}) = \mathcal{L}\left(\sum_{i=1}^I X_{ijk}\right) = IG\left(\sum_{i=1}^I \mu_i + I\beta_j, \lambda\right).$$

A test for interaction is blemished by the fact that it requires an equal number of replicates per cell.

Fries and Bhattacharyya (1983) have investigated a reciprocal linear model for the factor effects in a factorial experiment. The inference procedures for a balanced two factor experiment are outlined in the ensuing discussion.

Suppose that there are two factors, A at I levels and B at J levels and that there are n observations per cell (i,j). Then we can consider the independent random variables Y_{ijk} ($i = 1, \ldots, I$, $j = 1, \ldots, J$, $k = 1, \ldots, n$) as distributed according to the law $\mathcal{L}(Y_{ijk}) = IG(\delta_{ij}, \sigma)$ where $\sigma > 0$ $\delta_{ij} > 0$ and

$$\delta_{ij}^{-1} = \mu + \alpha_i + \beta_j.$$

In standard terminology μ represents a general (main) effect, the α_i the row effect and the β_j the column effect. The parameter space Ω is of dimension $I+J$ and

$$\Omega = \{\theta^* = (\mu, \alpha^t, \beta^t, \sigma) \mid e^t\alpha = e^t\beta = 0, \mu + \alpha_i + \beta_j > 0, \sigma > 0\}$$

where $e^t = (1,\ldots,1), \alpha = (\alpha_1, \cdots, \alpha_I)$ and $\beta = (\beta_1, \cdots, \beta_J)$.

Let us define

$$s_{ij} = \sum_{k=1}^{n} Y_{ijk} = n\bar{Y}_{ij.}, \quad s_{i.} = \sum_{j=1}^{J}\sum_{k=1}^{n} Y_{ijk} = nJ\bar{Y}_{i.}$$

$$s_{.j} = \sum_{i=1}^{I}\sum_{k=1}^{n} Y_{ijk} = nI\bar{Y}_{.j.}, \quad s_{..} = \sum_{i=1}^{I}\sum_{j=1}^{J}\sum_{k=1}^{n} Y_{ijk} = nIJ\bar{Y}_{...}$$

$$nIJ\bar{Y} = \sum_{i=1}^{I}\sum_{j=1}^{J}\sum_{k=1}^{n} Y_{ijk}^{-1},$$

and

D=diag $(\bar{Y}_{11}, \ldots, \bar{Y}_{IJ})$ an $(IJ \times IJ)$ matrix.

From this it follows that

$$\mathbb{E}(\bar{Y}_{ij}^{-1}) = \delta_{ij}^{-1} + \frac{\sigma}{n}, \quad \text{Var}(\bar{Y}_{ij}^{-1}) = \delta_{ij}^{-1}\frac{\sigma}{n} + 2\left(\frac{\sigma}{n}\right)^2.$$

A method of obtaining estimates of the unknown parameters consists of first using the constraints to eliminate two parameters α_I and β_J and then redefining

(1) $\psi = (\theta^t, \sigma)^t$ where $\theta = (\mu, \alpha_1, \ldots, \alpha_{I-1}, \beta_1, \ldots, \beta_{J-1})^t$
(2) x_{ij} is an $(I+J-1) \times 1$ vector of $-1, 0$ and 1 such that $\theta^t X_{ij} = \mu + \alpha_i + \beta_j$
(3) the $(I+J-1) \times IJ$ matrix $X = (x_{11}, \ldots, x_{IJ}^t)^t$
(4) M the symmetric positive definite random matrix $(I+J-1) \times (I+J-1)$ is defined as

$$M = X^t D X$$

(5) $X^t e = IJe_1^t = \delta$ (say) where $e_1^t = (1, 0, \ldots, 0)$. With this notation the log likelihood is proportional to

$$\ell(\theta) \propto -\frac{n}{2}IJ\log\sigma - (2\sigma)^{-1}\left[nIJ\bar{Y}_- - 2n\theta^t\delta + n\theta^t M\theta\right].$$

The estimates of θ and σ are obtained from

$$\frac{\partial \ell}{\partial \theta} = \frac{n}{\sigma}[\delta - M\theta] = 0$$
$$\frac{\partial \ell}{\partial \sigma} = -\frac{nIJ}{2\sigma} + \frac{1}{2\sigma^2}\left[nIJ\bar{Y} - 2n\theta^t\delta + n\theta^t M\theta\right] = 0$$

These equations have a unique solution

$$\hat{\theta} = M^{-1}\delta$$
$$\hat{\sigma} = (IJ)^{-1}\left[IJ\bar{Y} - \delta^t M^{-1}\delta\right]. \qquad (B.1)$$

Moreover the matrix

$$-\frac{\partial^2 \ell}{\partial \psi \partial \psi^t} = n \begin{bmatrix} \hat{\sigma}^{-1}M & 0 \\ 0^t & \frac{IJ\hat{\sigma}^{-2}}{2} \end{bmatrix}$$

is positive definite, so that (B.1) indeed maximizes the likelihood in terms of ψ. For $I = J = 2$, $\hat{\psi}$ is the maximum likelihood estimator inside Ω as well, as shown by Fries (1982). For $I, J > 2$ the problem is still unresolved.

Nevertheless $\hat{\psi}$ is unique and as $n \to \infty$, $\hat{\psi}$ tends to the maximum likelihood estimator with probability one.

The following theorem due to Fries and Bhattacharyya concerns the consistency and asymptotic distribution of $\hat{\theta}$ and $\hat{\sigma}$.

Theorem B.1 *The estimators* $\hat{\theta} = M^{-1}\delta$ *and* $\hat{\sigma} = (IJ)^{-1}[IJ\bar{Y} - \delta^t M^{-1}\delta]$ *are strongly consistent and furthermore*
(a) $\sqrt{n}(\hat{\theta} - \theta)$ *has a limit law*
$N_{I+J-1}(0, \sigma\Gamma^{-1})$
(b) $\sqrt{n}(\hat{\sigma} - \sigma)$ *has a limit law* $N(0, 2\sigma^2(IJ)^{-1})$ *where* $\Gamma = X^t \nabla X$, *and* $\nabla = \text{diag}(\delta_{11}, \ldots, \delta_{IJ})$.

Tests for model adequacy

Consider a sequence of nested hypothesis

$$\Omega^n = H_0 \supseteq H_1 \ldots \supseteq H_4$$

where

$H_4 : \delta_{ij}^{-1} = \mu + \alpha_i + \beta_j$, $H_3 : \delta_{ij}^{-1} = \mu + \alpha_i + \beta_j$; $e^t\alpha = e^t\beta = 0$ (additive model), $H_2 : \delta_{ij}^{-1} = \mu + \alpha_i$; $e^t\alpha = 0$ (no B effects), $H_1 : \delta_{ij}^{-1} = \mu + \beta_j$; $e^t\beta = 0$ (no A effects), $H_0 : \delta_{ij}^{-1} = \mu$ (no factor effects).

In each of these models σ is an unknown nuisance parameter. H_0 is the general model and corresponds to the so-called saturated hypothesis or unrestricted hypothesis.

Let $\ell(\hat{\Omega}_{H_i})$ represent the maximized log-likelihood under $H_i (i = 0, \ldots, 4)$ and $D(H_i)$ the deviance associated with H_i, namely

$$D(H_i) = -2\left[\ell(\hat{\Omega}_{H_i}) - \ell(\hat{\Omega}_{H_0})\right].$$

Then the statistic $D(H_i) - D(H_{i-1}) = 2\left[\ell(\hat{\Omega}_{H_{i-1}}) - \ell(\hat{\Omega}_{H_i})\right]$ and in terms of the likelihood ratio statistic, in order to test H_i under H_{i-1} we use

$$\Lambda_{i-1,i} = 2\left[\ell(\hat{\Omega}_{H_i}) - \ell(\hat{\Omega}_{H_{i-1}})\right].$$

Since under H_i $\ell(\hat{\Omega}_{H_i}) = \left(-\frac{nIJ}{2}\right)(\log(\hat{\sigma}_i) + 1)$ for $j > i$ we have

$$\Lambda = nIJ \log(\hat{\sigma}_i / \hat{\sigma}_j).$$

The corresponding $\hat{\sigma}_i$, being the MLE under H_i, are

$$n\, IJ\hat{\sigma}_4 = n\, I\, J\, \bar{Y}_{-} - n \sum_{i=1}^{I} \sum_{j=1}^{J} \bar{Y}_{ij}^{-1}.$$

$$n\, IJ\hat{\sigma}_3 = n\, I\, J\, \bar{Y}_{-} - n \sum_{i=1}^{I} \sum_{j=1}^{J} \hat{\delta}_{ij}^{-1}.$$

$$n\, IJ\hat{\sigma}_2 = n\, I\, J\, \bar{Y}_{-} - n\, J \sum_{j=1}^{J} \bar{Y}_{i..}^{-1}.$$

$$n\, IJ\hat{\sigma}_1 = n\, I\, J\, \bar{Y}_{-} - n\, I \sum_{i=1}^{I} \bar{Y}_{.i.}^{-1}.$$

$$n\, IJ\hat{\sigma}_0 = n\, I\, J\, \bar{Y} - n\, I\, J\, \bar{Y}_{...}^{-1}.$$

For large n we use the fact the Λ_{ij} ($i < j$) has an approximate χ^2 law with degrees of freedom $= \dim(H_j) - \dim(H_i)$.

In order to use the analysis of reciprocals we define

$$T_{ij} = n\, I\, J \left(\frac{\hat{\sigma}_i - \hat{\sigma}_j}{\hat{\sigma}_j}\right)$$

The analysis of reciprocals

and observing that

$$\Lambda_{ij} = n \, I \, J \, \log\left[1 + \frac{\hat{\sigma}_i - \hat{\sigma}_j}{\hat{\sigma}_j}\right]$$

$$= n \, I \, J \, \log\left[1 + \frac{T_{ij}}{nIJ}\right]$$

we can use the test based on T_{ij} instead.

The analysis of reciprocals

The analysis of reciprocals for the respective effects are defined in terms of the sums of the differences of reciprocals as follows.

The variation due to the A effects denoted by SSA is estimated by

$$SSA = n \, I \, J \, (\hat{\sigma}_1 - \hat{\sigma}_3) = n \sum_{i=1}^{I} \sum_{j=1}^{J} (\hat{\delta}_{ij}^{-1} - \bar{Y}_{.j.}^{-1}),$$

that due to the B effects is

$$SSB = n \, I \, J \, (\hat{\sigma}_2 - \hat{\sigma}_3) = n \sum_{i=1}^{I} \sum_{j=1}^{J} (\hat{\delta}_{ij}^{-1} - \bar{Y}_{i..}^{-1}),$$

the variation due to the interaction $SSAB$ is

$$SSAB = n \, I \, J \, (\hat{\sigma}_3 - \hat{\sigma}_4) = n \sum_{i=1}^{I} \sum_{j=1}^{J} (\bar{Y}_{ij.}^{-1} - \hat{\delta}_{ij}^{-1}),$$

and the variation due to random error is

$$SSE = n \, I \, J \, \hat{\sigma}_4 = n \, I \, J \, \bar{Y}_{-}^{-1} - n \sum_{i=1}^{I} \sum_{j=1}^{J} \bar{Y}_{ij.}^{-1}.$$

A justification for these formulae can be advanced by noting the easily verified identity

$$Y_{ijk}^{-1} = \bar{Y}_{-} + (\hat{\delta}_{ij}^{-1} - \bar{Y}_{.j.}^{-1}) + (\hat{\delta}_{ij}^{-1} - \bar{Y}_{i..}^{-1}) + (\bar{Y}_{ij.}^{-1} - \hat{\delta}_{ij}^{-1})$$

$$+ (Y_{ijk}^{-1} - \bar{Y}_{ij.}^{-1}) + (\bar{Y}_{i..}^{-1} - \bar{Y}_{.j.}^{-1} - \bar{Y}_{-} - \hat{\delta}_{ij}^{-1}).$$

Note that $\bar{Y}_{ij.}^{-1}$ estimates δ_{ij}^{-1} in the absence of a constraint and $\hat{\delta}_{ij}^{-1}$ estimates δ_{ij}^{-1}. Based on a reciprocal scale $(Y_{ijk}^{-1} - \bar{Y}_{ij.}^{-1})$ can be considered as a residual. The very last term can be regarded as a non-orthogonality component. Summing the identity over i, j and k yields SSA etc. The analysis of reciprocals is

Table B.4 *Analysis of Reciprocals*

Source	d.f.	Sum of differences of reciprocals	Approximate F
Factor A	$I - 1$	SSA	$\dfrac{(n-1)IJ\ SSA}{(I-1)\ SSE}$
Factor B	$J - 1$	SSB	$\dfrac{(n-1)IJ\ SSB}{J-1\ SSE}$
$A \times B$	$(I-1)(J-1)$	$SSAB$	$\dfrac{(n-1)IJ\ SSAB}{(I-1)(J-1)}$
Residual	$(n-1)IJ$	$SSE.$	

We remark that the null distribution of $\frac{SSE}{\sigma}$ is an exact χ^2 law with $(n-1)IJ$ degrees of freedom and $SSE \perp\!\!\!\perp SSAB$. The laws of SSA, SSB and $SSAB$ are only in an asymptotic sense. We omit the proof.

A least squares approach based on a preliminary reduction of the data through sufficiency is also possible. For details we refer the reader to Fries and Bhattacharyya. We conclude with the example considered by them. Shuster and Miura had analyzed it by making the assumption of a constant λ/μ^2 across samples.

Example B.4 From each of 5 lots of insulating material, 10 lengthwise specimens and 10 crosswise specimens are cut. Table B.5 gives the impact strength in foot-pounds from tests on the specimens. We shall analyze and interpret the data.

The analysis of reciprocals

Table B.5 *Impact strength in foot-pounds*

Type of Cut		Lot	Number		
	I	II	III	IV	V
	1.15	1.16	0.79	0.96	0.49
	0.84	0.85	0.68	0.82	0.61
	0.88	1.00	0.64	0.98	0.59
	0.91	1.08	0.72	0.93	0.51
Lengthwise	0.86	0.80	0.63	0.81	0.53
specimens	0.88	1.01	0.59	0.79	0.72
	0.92	1.14	0.81	0.79	0.67
	0.87	0.87	0.65	0.86	0.47
	0.93	0.97	0.64	0.84	0.44
	0.95	1.09	0.75	0.92	0.48
	0.89	0.86	0.52	0.86	0.52
	0.69	1.17	0.52	1.06	0.53
	0.46	1.18	0.80	0.81	0.47
	0.85	1.32	0.64	0.97	0.47
Crosswise	0.73	1.03	0.63	0.90	0.57
specimens	0.67	0.84	0.58	0.93	0.54
	0.78	0.89	0.65	0.87	0.56
	0.77	0.84	0.60	0.88	0.55
	0.80	1.03	0.71	0.89	0.45
	0.79	1.06	0.59	0.82	0.60

The analysis of reciprocals test is presented in Table B.6.

Table B.6 *Analysis of Reciprocals*

Source	Sum of reciprocals	Degrees of freedom	MR	F Ratio (approx p value)
Cut	.11985	1	.11985	5.39 (.024)
Lot	6.56334	4	1.64084	73.78 ($\ll$ .001)
Interaction	.25922	4	.06481	2.91 (.025)
Error	2.00150	90	.02224	

Table B.7 *MLE and LSE of impact strengths (10 reps. per cell)*

Type of Cut		I	II	III	IV	V
			Lot	Number		
Lengthwise	$\bar{y}$	0.919	0.997	0.690	0.870	0.511
	MLE	0.855	1.051	0.673	0.916	0.550
	LSE	0.854	1.058	0.676	0.922	0.552
Crosswise	$\bar{y}$	0.743	1.022	0.624	0.899	0.526
	MLE	0.803	0.972	0.640	0.856	0.527
	LSE	0.795	0.969	0.638	0.853	0.526

$\hat{\sigma} = .02261$, $\hat{\phi} = (1.342, -.039, -.134, -.352, .182, -.212)^t$, $\tilde{\sigma} = .02224$, $\tilde{\phi} = (1.342, -.044, -.128, -.354, .181, -.214)^t$.

The corresponding estimates for the cell means are calculated by using the relation $X\phi = \Theta^{-1}e$. These estimated cell means and the sample cell means are given in Table B.7.

C. Demography

In the theory of population projection (Keyfitz 1968) an integral equation of fundamental importance is

$$B(t) = G(t) + \int_0^t B(t-x)\, p(x)\, m(x)\, dx. \tag{C.1}$$

Here $G(t)$ represents the children born at time t for all parents surviving from time zero, $B(t)$ is the density of births and $p(x)\, m(x)$ is called the net maternity function. In demographic language $p(x)$ is the life table number-living on radix unity and $m(x)$, the age-specific birth rate. The net maternity function, written $\phi(x)(=p(x)m(x))$ is then fitted by a curve involving three constants, the moments of order zero, one and two. Thus the constants describing $\phi(x)$ are

$$R_i = \int_\alpha^\beta x^i\, \phi(x)\, dx \quad, i = 0, 1, 2$$

where R_0 is called the net reproduction rate, $R_1/R_0 = \mu$ is the mean age of child bearing in the stationary population, $\frac{R_2}{R_0} - \left(\frac{R_1}{R_0}\right)^2 = \sigma^2$, the variance of age at childbearing in the stationary population.

It is customary to substitute a "graduated form" of $\phi(x)$ in the integral equation (C.1) to obtain an approximate solution to $B(t)$.

Three graduations are popular, due respectively to Lotka (1939), Wicksell (1931) and Hadwiger (1940). Lotka used the Gaussian law for $\phi(x)$ while Wicksell found the gamma a superior fit by virtue of the fact that the net maternity function is based on the inverse Gaussian law and is due to Hadwiger. He introduced a reproduction function that fitted successive generations from purely mathematical considerations. We give below a brief account of his derivation.

Let $\phi_1(x)$ denote the probability of a girl child born from a girl J_0 at age x, the age being measured not from her own birth date but from the birth date of the ancestor J_0. Hadwiger calls this the reproduction function. The child born of J_0 will be denoted by J_1, that born to J_1 by J_2 and so on. Then the set of all J_n forms a generation of a population descending from J_0. If $\phi(x)$ represents the probability of a girl child J_n produced at time x, assuming some stability in the population. Hadwiger arrives at the equation

$$\phi_{n+1}(x) = \int_0^x \phi_n(x-t)\, \phi_1(t)\, dt$$

and hence by iteration

$$\phi_{n+m}(x) = \int_0^x \phi_n(x-t)\, \phi_m(t)\, dt. \tag{C.2}$$

Suppose we introduce a parameter $a > 0$ and let

$$\phi_1(x) = \psi(x, a) \;(\Rightarrow \phi_n(x) = \psi(x, na))$$

it follows from (C.2) that

$$\psi(x, na + ma) = \int_0^x \psi(x - t, na)\, \psi(t, ma)\, dt.$$

Given $p > 0$, $q > 0$ and an arbitrary $\epsilon > 0$ one can find a number a and two integers n and m such that

$$|p - na| < \epsilon \text{ and } |q - ma| < \epsilon.$$

Hence

$$\psi(x, p + q) = \int_0^x \psi(x - t, p)\, \psi(t, q)\, dt. \tag{C.3}$$

Hadwiger now claims that a solution to (C.3) is

$$\psi(x, a) = \frac{a}{\sqrt{\pi x^3}} \exp\left\{ ca - \left(\frac{a^2}{x} + Ax\right)\right\}$$

for some constants $A, c > 0$.

Since the reproduction law must be the same for the individual in the nth generation G_n, one gets

$$\varphi_n(x) = \frac{na}{\sqrt{\pi x^3}} \exp\left\{ n\, ca - \left(\frac{n^2 a^2}{x} + Ax\right)\right\}.$$

Now let $B(x)$ be the children born at time x for all parents starting from J_0. Then

$$B(x) = \sum_{n=1}^{\infty} \phi_n(x)$$
$$= \frac{a e^{-Ax}}{\sqrt{\pi x^3}} \sum_{n=1}^{\infty} n \exp\left\{ n\, ca - \frac{n^2 a^2}{x}\right\} \tag{C.4}$$

represents a convergent series in each finite interval of time.

In terms of the reproductions from J_0 at time x and the births during time $x - t$ one obtains the integral equation similar to (C.1) namely

$$B(x) = \phi_1(x) + \int_0^x B(x - t)\, \phi_1(t)\, dt. \tag{C.5}$$

Demography

One can again verify that (C.4) is a solution to (C.5). In demographical studies one is interested in equating the Laplace transform of $\phi(x), L(\theta)$ to unity and solve for θ in terms of the moments of ϕ. This is equivalent to solving
$$\exp\left\{a(c - 2\sqrt{A+\theta})\right\} = 1.$$

Thus $a(2\sqrt{A+\theta} - c) = 2n\pi i$, and

$$\theta = \left(\frac{c}{2} \pm \frac{\pi n i}{a}\right)^2 - A, \quad n = 1, 2, \ldots.$$

Letting $\theta = u + iv$, we get for $n = 0, 1, 2, \ldots$

$$u = \frac{c^2}{4} - \frac{\pi^2 n^2}{a^2} - b$$
$$v = \pm \frac{\pi n c}{a}.$$

These quantities are used in the Hadwiger graduation scheme. We shall refrain from going into the details and refer the reader to Keyfitz (1968).

D. Histomorphometry

Villman et al., (1990) have studied the histomorphometrical analysis of the influence of soft diet on masticatory muscle development in the muscular dystrophic mouse. The muscle fibre size distributions were fitted by an inverse Gaussian law. We discuss the nature of the problem and the investigation as developed by these authors.

Muscles of mastication and postcranial muscles are differentially affected in the muscular dystrophic mouse. It turns out that the masseter muscle, regardless of the age, is more seriously affected than the digrastic muscle. One of the reasons is that slow fibres being more resistant to dystrophy than fast fibres, most muscles containing slow fibres suffer less damage. Masticatory muscles in rodents contain a large amount of fast fibres. A second explanation is that muscles of survival (respiratory) contain protective mechanism. The third reason is that the severity of the disease in a given muscle is dependent on the amount of exertion and the muscle that does more work becomes more affected. The main muscle that is activated during chewing is the masseter and it shares the most work load in contrast to the digrastic muscle. The experiment conducted by Villman et al., examines this third premise. This experiment is justified by the observation that substitution of the normal laboratory diet of hard food pellets by a soft diet induces a marked change in the masticatory pattern in rodents.

The experiment involved 40 mice, 20 of which were dystrophic and the rest normal. When they were 3 weeks old, 10 dystrophic and 10 normal mice were fed on a soft diet and the rest were given food comprising hard pellets. The soft diet group was caged in an environment that had no hard parts. The mice were weighed at weekly intervals until they were seven weeks of age. At the end of seven weeks the mice were put to death and the masseter and the anterior belly of the digrastic muscles were dissected free. The muscles were frozen in isopentene cooled to -150°C with liquid nitrogen and later cut in $6\mu m$ sections. The stained sections were studied by light microscopy, photographed and analyzed.

Fibre size within a given area was measured as the minimum diameter of the cross-section of each fibre in the area studied. Quantitative differences between the muscles were calculated in terms of the variances in fibre size. The results can be summarized as follows.

	Normal Mice	Dystrophic Mice
Weight at 3 weeks	10.4g $\sigma = 1.1$	7.8g $\sigma = 1.4$

	Ordinary diet	Soft diet	Ordinary diet	Soft diet
Weight at 7 weeks	18.7g $\sigma = 1.7$	18.2g $\sigma = 1.6$	9.9g $\sigma = 1.8$	12.6g $\sigma = 1.3$

Measurements were made of muscle fibre from 8 groups of mice, namely normal-masseter-soft, normal-masseter-ordinary, normal-digrastic-soft, normal-digrastic-ordinary and a similar classification for the dystrophic mice.

Table D.1 *Estimated means and standard errors for varying sample sizen*

M/F	Di	n	Normal	Group		
			μ	s.d.	ϕ	s.d.
Ma	soft	499	85.09	(0.77)	24.64	(1.57)
Ma	ord	499	106.41	(0.80)	35.21	(2.24)
D	soft	495	78.68	(0.72)	24.22	(1.56)
D	ord	500	98.35	(0.78)	31.78	(2.03)

M/F	Di	N	Dystrophic	Group		
			μ	s.d.	ϕ	s.d.
Ma	soft	500	93.31	(1.93)	4.70	(0.31)
Ma	ord	483	101.27	(2.35)	3.84	(0.26)
D	soft	498	87.99	(1.15)	11.85	(0.77)
D	ord	500	98.63	(1.24)	12.58	(0.81)

M=Muscle, F=Fibre, Ma=Masseter, D=Digrastic, Di=Diet.

Their estimated mean values, the ϕ values and their corresponding standard errors based on different sample sizes n are given in Table D.1, based on the assumption that the fibre diameters all come from $IG(\mu, \lambda)$.

Due to the large sample sizes, hypotheses about the mean and ϕ values were based on the Wald statistic for the normal law. Table D.2 gives a summary of a survey of differences between the measured muscles and the associated p-values.

Vilmann et al., conclude that the experimental findings confirm their suspicion that the greater weight gain of dystrophic mice on a soft diet relative to those on an ordinary diet may be because they cannot chew enough hard food and are hence underfed.

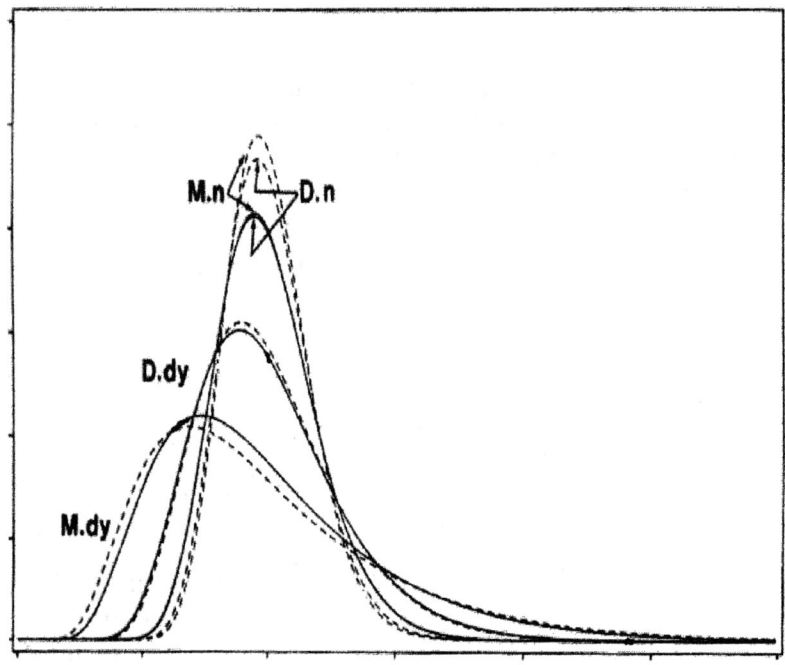

Figure D.1 *Estimated inverse Gaussian density functions for muscle fibre-size distributions. The dotted curves refer to the soft-diet groups and the solid curves refer to the ordinary-diet groups. M.n-masseter normal M.dy-masseter dystrophic, D.n- digrastric normal, D-dy-digrastic dystrophic.*

Table D.2 *Survey of differences between the measured muscles*

Muscles groups	A. Differences between means p	
s.no./o.no.	0.000	$o > s$
s.dy/o.dy	0.000	$o > s$
m/m s.no./s.dy	0.000	$dy > no$
o.no./o.dy	0.04	$no > dy$
s.no./o.no.	0.000	$o > s$
s.dy/o.dy	0.000	$o > s$
d/d s.no./s.dy	0.000	$dy > no$
o.no./o.dy	0.85	
s.no./s.no.	0.000	$m > d$
o.no./o.no.	0.000	$m > d$
m/d s.dy/s.dy	0.02	$m > d$
o.dy/o.dy	0.32	

Muscles groups	B. Differences between dispersions p	
s.no./o.no.	0.000	$s > o$
s.dy/o.dy	0.03	$o > s$
m/m s.no./s.dy	0.000	$dy > no$
o.no./o.dy	0.000	$dy > no$
s.no./o.no.	0.003	$s > 0$
s.dy/o.dy	0.51	
d/d s.no./s.dy	0.000	dy/no
o.no./o.dy	0.000	dy/no
s.no./s.no.	0.85	
o.no./o.no.	0.19	
m/d s.dy/s.dy	0.000	$m > d$
o.dy/o.dy	0.000	$m > d$

m= masseter, d= digastricus, s=soft diet, o=ordinary diet, no=normal, dy=dystrophic.

E. Electrical networks

Barndorff-Nielsen (1994) considers a finite tree whose edges are endowed with random resistances, and shows that, subject to suitable restrictions on the parameters, if the resistances are either inverse Gaussian or reciprocal inverse Gaussian random variables, then the overall resistance of the tree follows a reciprocal inverse Gaussian law. Barndorff-Nielsen and Koudou (1996) have generalized the idea to infinite trees.

Random electrical networks which possess the structure of rooted trees can be endowed with independent stochastic edge resistances. The overall resistance R_T of such networks is then determined from the edge resistances by the laws of Ohm and Kirchoff. The general problem of determining the distribution of R_T is indeed difficult (Grimmet 1991), but is computable when the edge resistances follow the inverse Gaussian or reciprocal inverse Gaussian laws.

In order to understand the theory we begin with some terminology and review some properties of the inverse Gaussian law.

A tree $T = (V, E)$ is an undirected and connected graph where V denotes the set of vertices v and E the set of edges e (E is contained in the set of subsets of V with two elements) without loops. When the number of vertices is finite, T is called a *finite* tree. If we select one vertex s in V, then (V, E, s) becomes a *rooted* tree. The vertex s is called the initial vertex or the root. This induces a natural map $\gamma : V\backslash\{s\} \to V$ such that $\{\gamma(v), v\} \in E$.

There is a natural distance d on V defined by $d(v, v')$, namely the length of the shortest path from v to v' and $d(\gamma(v), s) = d(v, s) - 1$. In addition there is a partial order on V : $v < v'$ if either $v = v'$ or there exists $n > 0$ such that $\gamma^n(v') = v$.

We also denote by V_v the subset of V consisting of all terminal vertices that can be reached by a path starting from v. We write R_e for the resistance (positive) of an edge and R_T for the overall resistance.

If the edges are numbered and R_1 and R_2 are two resistances connected in series, the overall resistance $R_T = (R_1 + R_2)$ while if R_2 and R_3 are two resistances connected in parallel the overall resistance is given by $(R_2^{-1} + R_3^{-1})^{-1}$. Thus the total resistance for the network is $R_1 + (R_2^{-1} + R_3^{-1})^{-1}$.

In order to simplify calculations we adopt the notation used by Barndorff-Nielsen rather than the conventional one for the inverse Gaussian and reciprocal inverse Gaussian laws. Let t and τ be two positive real numbers.Then we write

$$IG(t,\tau)(dx) = \frac{t}{\sqrt{2\pi x^3}} e^{t\tau} \exp\left\{-\frac{t^2}{2x} - \frac{\tau^2}{2}x\right\} 1_{(0,\infty)}(x)dx$$

for the inverse Gaussian law with parameters $(t^2\tau^2)$ and

$$RIG(t,\tau)(dx) = \frac{\tau}{\sqrt{2\pi x}} e^{t\tau} \exp\left\{-\frac{t^2}{2x} - \frac{\tau^2}{2}x\right\} 1_{(0,\infty)}(x) dx$$

for the reciprocal inverse Gaussian law with parameters (t^2, τ^2). The Gamma law with parameter τ^2 is defined as

$$Ga(\tau) dx = \frac{\tau}{\sqrt{2\pi x}} \exp\left\{-\frac{\tau^2}{2}x\right\} 1_{(0,\infty)}(x) dx.$$

The following convolution results will be needed in the sequel.

$$IG(t_1, \tau) * IG(t_2, \tau) = IG(t_1 + t_2, \tau)$$
$$RIG(t_1, \tau) * IG(t_2, \tau) = RIG(t_1 + t_2, \tau)$$
$$IG(t, \tau) * G_a(\tau) = RIG(t, \tau)$$
$$\mathcal{L}(X) = IG(t, \tau) \Leftrightarrow \mathcal{L}(X^{-1}) = RIG(\tau, t).$$

Some compatibility criteria are needed for deriving convolution laws relating to the inverse Gaussian and reciprocal inverse Gaussian laws.

When working with three resistances R_1, R_2, R_3 suppose that
(1) $t_1 + t_2 = t_1 + t_3$, (2) $\tau_1 = \tau_2 + \tau_3$
then, if

$$\mathcal{L}(R_1) = IG(t_1, \tau_1), \ \mathcal{L}(R_2) = RIG(t_2, \tau_2), \ \mathcal{L}(R_3) = IG(t_3, \tau_3)$$

we have

$$\mathcal{L}(R_2^{-1}) = IG(\tau_2, t_2), \ \mathcal{L}(R_3^{-1}) = IG(\tau_3, t_3)$$

and

$$\mathcal{L}(R_2^{-1} + R_3^{-1}) = IG(\tau_2 + \tau_3, t_2) \quad \text{(since } t_2 = t_3\text{)}$$
$$= IG(\tau_1, t_2)$$

so that

$$\mathcal{L}((R_2^{-1} + R_3^{-1})^{-1}) = RIG(t_2, \tau_1)$$

and finally

$$\mathcal{L}(R_1 + (R_2^{-1} + R_3^{-1})^{-1}) = RIG(t_1 + t_2, \tau_1).$$

When a tree T has only two vertices, then R_T is the resistance of the single edge. Suppose now that R_T is defined whenever the number of edges, say $|E|$, is $\leq m-1$.

Consider a tree T with initial vertex s and such that $|E| = m$. Further let $\gamma^{-1}(s) = \{s_1, \cdots, s_k\}$ be the set of vertices linked with s. Denote by T_i the subtree of T with initial vertex s_i, $1 \leq i \leq k$. For $k = 4$ Figure E.1 illustrates the situation when $|E| = 23 = m$.

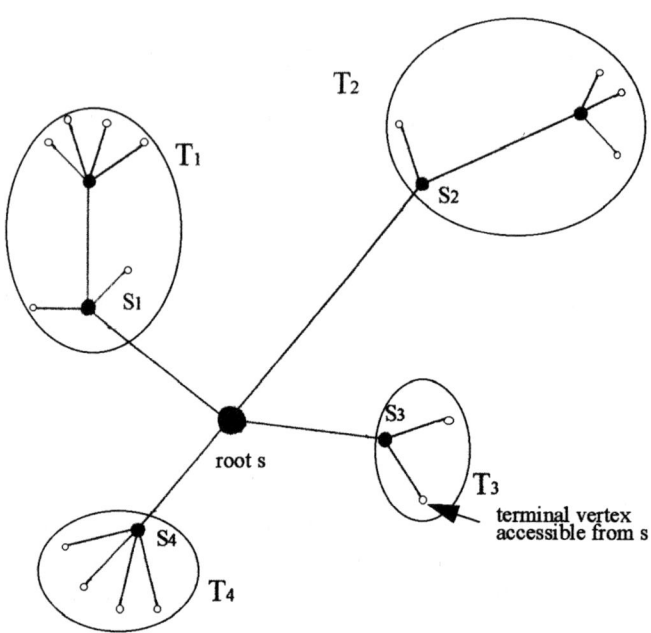

Figure E.1 *A finite tree with four vertices*

Define R_T as

$$R_T = \left[\sum_{i=1}^{k}\left(R_{T_i} + R_{\{s,s_i\}}\right)^{-1}\right]^{-1}. \tag{E.1}$$

With each vertex $v \in V$, we associate two real numbers $t_v > 0$ and $\tau_v > 0$ such that

(1) $$\tau_v = \sum_{v' \in V_v} \tau_{v'}$$

where V_v is the set of all terminal vertices that are accessible from v

(2) $$t = \sum_{v \in \pi \setminus \{s\}} t_v$$

and the sum t is the same along any path π starting from s and ending at a terminal vertex.

These two conditions will be known as the compatibility conditions. Barndorff-Nielsen's result on the distribution of the total resistance R_T of the tree T can now be stated as

Theorem E.1 *Let $X_v, v \in V \backslash \{s\}$ be a collection of independent random variables such that X_v is the random resistance of the edge $\{\gamma(v), v\}$ and suppose that*

$$\mathcal{L}(X_v) = RIG(t_v, \tau_v) \quad \text{if } v \text{ is a terminal vertex}$$
$$= IG(t_v, \tau_v) \quad \text{otherwise.}$$

Further let t_v and τ_v satisfy the conditions
a) $t = \sum_{v \in \pi \backslash \{s\}} t_v$ is the same along all paths π starting from s and ending at a terminal vertex
b) $\tau_v = \sum_{v' \in V_v} \tau_{v'}$
V_v being the set of all terminal vertices accessible from v. Then the total resistance R_T of T follows the $RIG(t, \tau)$ law where

$$\tau = \sum_{v \in V_s} \tau_v.$$

Proof The theorem is proved by induction on the number of edges of T. The conclusion of the theorem is obvious if T has only one edge. Suppose that the theorem is true when the number of edges $\mid E \mid \leq m-1$. Let T be a tree such that $\mid E \mid = m$, s be the initial vertex and

$$\gamma^{-1}(s) = \{s_1, \cdots, s_k\}$$

denote the set of vertices linked with s.

For $1 \leq i \leq k$ let $T_i = (U_i, E_i)$ be the subtree with root s_i. Thus for all i the number of edges of T_i namely $\mid E \mid \leq m - 1$. Finally let

$$K_v = \sum_{v' \in \pi_v \backslash \{v\}} t_{v'}$$

for $v \in V$ and any path π_v starting from v and ending at a terminal vertex. (As a consequence of (2) the sum is independent of the path.)(Now $v \to K_v$ is a decreasing function on the vertices of T.)

Clearly the restriction of the map $v \to (t_v, \tau_v)$ to the vertices of T_i satisfies (1) and (2). Thus the theorem is true for the tree T_i. This implies that the distribution of the total resistance of T_i, namely $\mathcal{L}(R_{T_i}) = RIG(K_{s_i}, \tau_{s_i})$. Denoting by R_T the total resistance we have from (E.1)

$$R_T = \left[\sum_{i=1}^{k} (X_{s_i} + R_{t_i})^{-1} \right]^{-1}.$$

Now
$$\mathcal{L}(R_{T_i}) = RIG(K_{s_i}, \tau_{s_i})$$
$$\mathcal{L}(X_{S_i}) = IG(t_{s_i}, \tau_{s_i})$$

so that
$$\mathcal{L}(X_{S_i} + R_{T_i}) = IG(\tau_{s_i}, K_{s_i} + t_{s_i})$$
$$= IG(\tau_{s_i}, K_s).$$

Then by the laws relating to convolution

$$\mathcal{L}\left(\sum (X_{s_i} + R_{t_i})^{-1}\right) = IG\left(\sum_{i=1}^{k} \tau_{s_i}, K_s\right)$$

$$\mathcal{L}\left[\sum_{i=1}^{k}(X_{s_i} + R_{t_i})^{-1}\right]^{-1} = RIG\left(K_s, \sum_{i=1}^{k} \tau_{s_i}\right)$$
$$= RIG(K_s, \tau_s).$$

Since K_s and τ_s are respectively equal to t and τ, the theorem is proved.

We remark in conclusion that the number of free τ parameters equals the number of terminal vertices and it can be proved that d_T, the number of free t parameters is

$$d_T = \mid T \mid - p$$

where $\mid T \mid =$ the number of vertices and p is the number of terminal vertices. ♣

F. Hydrology

The analysis of floods, or the monthly flow in rivers by statistical methods began as early as 1913. Although the normal law was used by Horton (1913) to describe annual floods, it soon became evident that annual flood series form positively skewed data and that other models were needed. Hazen (1914) used the log-normal law with some success. Foster (1924) gave detailed methods for modelling floods by the Pearson's Type I and III. Later on Gumbel's theory of extreme values (1941, 1958) became the vehicle for the analysis of flood data. In the early studies it was generally accepted that flood distribution be endowed with a minimum of three parameters. This was what motivated Halphen to develop the generalised inverse Gaussian law in 1941. Halphen fitted several series of mean monthly flow of water in the rivers with the harmonic law. Not satisfied with the adequacy of the fit Halphen introduced the Type B distribution which assumes a variety of forms leading to unimodal laws with positive or negative skewness with various kinds of algebraic decrease near the origin. Towards 1956 one of Morlat's co-workers, Larcher introduced a third distribution known as Type B^{-1}, obtained by considering the reciprocal of a Type B law. These laws came to be known as Halphen's system of distributions.

Denoting by $\mu'_{-1}, \mu'_0, \mu'_1$ the moments of order -1, quasi-zero and 1 respectively (Bobèe and Ashkar, 1988, Morlat 1956) the moment ratios $\delta_1 = ln(\mu'_1/\mu'_0)$ and $\delta_2 = ln(\mu'_0/\mu'_1)$ provide a useful display of the symmetry of the Halphen system of distributions.

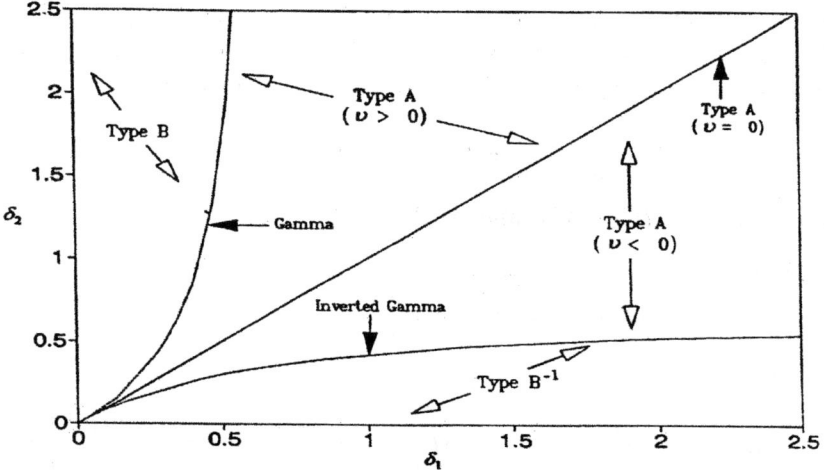

Figure F.1 *Halphen's laws in the (δ_1, δ_2) scale*

Emptiness of a dam

Gani and Prabhu (1963) presented a general theory of storage for dams subject to a steady release. At time $t \geq 0$ consider a dam with infinite capacity having a random content $Z(t) \geq 0$. Let $X(t) \geq 0$ be the input during time t (one often assumes that the law of $X(t)$ is infinitely divisible), while the dam is subject to a steady release at a constant unit rate except when it is empty in which case the release stops. The transient behaviour of $Z(t)$ can then be evaluated. Hasofer (1964) introduced an inverse Gaussian type input for $X(t)$ which permits the calculation of the probability of emptiness of the dam and the mean content in a nice closed form. Hasofer's analysis uses the inversion of Laplace transforms and the equation

$$\eta(s) = s + \xi(\eta(s)) \tag{F.1}$$

where $\exp(-t\xi(s))$ is the Laplace transform of the input law and $\exp(-z_0\eta(s))$ is the Laplace transform of the time of the first emptiness $T(z_0)$, z_0 being the initial content of the dam.

Kendall (1957) had conjectured a relation between the two densities whose cumulant transforms are related by (F.1). Jain and Khan (1979) used this relation (to be made precise) to generate exponential families (see also Hassairi 1995).

Assume that the volume X of a river discharge which flows into a semi-infinite reservoir during time t is a homogeneous process with non-negative and independent increments with probability density $p(x;t)$. Let a certain substance which is uniformly distributed in the fluid (x units of the fluid contain rx units of the substance, say) be analyzed at a unit rate per unit of time. Suppose the analysis starts when the reservoir holds z_0 units. Then the probability $g(x; z_0)dx$ that a quantity X of the fluid flows into the reservoir before it gets emptied for the first time after a quanity $z_0 + rx$ has been analyzed is given by

$$g(x; z_0) = \frac{z_0}{z_0 + rx} \, p(x; z_0 + rx) \text{ for } x > 0, r > 0 \; \nu r < 1$$

where $\mathcal{L}(X) = IG(\nu t, t^2)$ and $\mathbb{E}(X) = \nu t$.

Now let

$$p(x;t) = t\sqrt{\frac{1}{2\pi x^3}} \exp\left\{-\frac{t^2}{2x} - \frac{x}{2\nu^2} + \frac{t}{\nu}\right\}. \tag{F.2}$$

Then

$$g(x; z_0) = \frac{z_0}{z_0 + rx}(z_0 + rx)\sqrt{\frac{1}{2\pi x^3}} \exp\left\{-\frac{(x - \nu(z_0 + rx))^2}{2\nu^2 x}\right\}$$

$$= z_0\sqrt{\frac{1}{2\pi x^3}} \exp\left\{-\frac{(x - \nu(z_0 + rx))^2}{2\nu^2 x}\right\}.$$

Thus

$$g(x; z_0) = z_0 \sqrt{\frac{1}{2\pi x^3}} \exp\left\{-\frac{z_0^2}{2x} - \frac{(1-r\nu)^2 x}{2\nu^2} + \frac{z_0(1-r\nu)}{\nu}\right\}. \quad (F.3)$$

This shows that we still have an inverse Gaussian law $IG\left(\frac{\nu z_0}{(1-r\nu)}, z_0^2\right)$.

Note that if $T = T(z_0)$ denotes the time of first emptiness, then taking $r = 1$ and writing $x = T - z_0$, the density of $T(z_0)$ is for $T \geq z_0 > 0$

$$\frac{z_0}{\sqrt{2\pi(T-z_0)^3}} \exp\left\{-\frac{z_0^2}{2(T-z_0)} - \frac{(1-\nu)^2(T-z_0)}{2\nu^2} + \frac{z_0(1-\nu)}{\nu}\right\}.$$

It is clear that if we had assumed for $p(x,t)$ the stable law with exponent $\frac{1}{2}$ namely,

$$p(x;t) = \frac{t}{\sqrt{2\pi x^3}} \exp\left(-\frac{t^2}{2x}\right) 1_{\mathbf{R}^+}(x)$$

we would have obtained for $T \geq z_0$ the density

$$g(T(z_0); z_0) = \frac{z_0 \ \exp(-z_0)}{\sqrt{2\pi(T-z)^3}} \exp\left\{-\frac{z_0^2}{2(T-z_0)} - \frac{(T-z_0)}{2}\right\}$$

(a result given by Wasan, 1968).

G. Life tests

Shelf life failures

The definition of time to failure in sensory testing is purely subjective. The mean time to failure based on simple averages is biased by the inclusion of unfailed data. One way to overcome this is by defining failure time as the time required for a sample to reach a median or an average panel score of 3.5 on a 7-point rating scale.

"Storage life", a term associated with failure time, is defined as the time required for various deteriorative changes in a freshly packed food product to accumulate for a panel to judge it at the limit of its acceptability.

The term "just noticeable difference" is defined as the first time when a difference between a product and a control could be detected by a panel of experts.

The distribution of shelf life of a product by sensory evaluation is a key to the improvement of the method of estimation of shelf life.

A sample of n items of a product is evaluated for failures at pre-designated periods and the age to failure is recorded. When the alloted experimental time elapses, the age of the samples that did not fail are also recorded. The total experimental time is arbitrarily chosen by the examiner. Thus life testing produces two sets of data, the time to failure of the flawed items and the running time of the unflawed items.

The basis of recording failure was based on the average panel score of 3.5 (on a 7 point off-flavour rating scale - the contribution from the presence of yeasts and high bacterial count rendering the product unsuitable for taste was also a factor). Either the lack of samples or the expiration of the total time from the date of production was a factor deciding running time, the samples being obtained from the manufacturer and sent to the laboratory for sensory testing.

Gacula and Kubala (1975) have analyzed shelf life of several products using the normal, log-normal, exponential, Weibull and extreme-value distributions. The adequacy of the fit was examined using both the graphical method of Nelson (1969, 1972) and the Kolmogorov-Smirnov statistic. The data set given below was studied by Chhikara and Folks and the IG law was found to be a good fit.

The times to failure (days) for product M reported by Gacula and Kubala are: 24, 24, 26, 26, 32, 32, 33, 33, 33, 35, 41, 42, 43, 47, 48, 48, 48, 50, 52, 54, 55, 57, 57, 57, 57, 61.

Accelerated life tests

The IG distribution is useful in modelling fatigue growth in a material, as for instance when the failure is caused by accumulated fatigue or depletion of its strength exceeding a critical limit $a > 0$. Let us suppose that fatigue develops over a period of time according to a Wiener process with drift $\nu > 0$ and diffusion constant $\sigma^2 > 0$, so that the first passage time to failure is $IG\left(\mu = \frac{a}{\nu}, \lambda = \frac{a^2}{\sigma^2}\right)$. Further let x denote the intensity of the stress which causes the fatigue. Then the mean fatigue growth can be expressed as a linear function of x. By considering a reciprocal linear regression structure á la Bhattacharyya and Fries (1982) we then have

$$\mu^{-1} = \alpha + \beta x, \quad \alpha \geq 0, \ \beta \geq 0, \ \alpha + \beta > 0, \ x > 0$$

and

$$\lambda = \text{ constant } > 0.$$

The constancy of λ is quite analogous to the homoscedasticity of the variance in the normal model. In consonance with the range of the stress x, one requires that $\alpha + \beta x > 0$ on a finite interval of x.

Observations (x_i, Y_i), $i = 1, \cdots, n$ are taken from n replicates of an accelerated life test, Y_i denoting independent failure times corresponding to the stress levels x_i. Further we assume that $\mathcal{L}(Y_i) = IG(\mu_i, \lambda)$ where:

$$\mu_i^{-1} = \alpha + \beta x_i.$$

Denote by Θ the parameter space where

$$\Theta = \{\alpha, \beta, \lambda) \mid \alpha \geq 0, \beta \geq 0, \alpha + \beta > 0, \lambda > 0\}.$$

From general exponential theory it can be seen from the log likelihood $\ell = \ell(\alpha, \beta, \lambda)$ where

$$\ell = \frac{n}{2}\log\left(\frac{2\pi}{\lambda}\right) - \left\{\frac{\lambda\alpha^2}{2}n\overline{y} + \lambda\alpha\beta\sum_{i=1}^{n}x_i y_i + \frac{\lambda\beta^2}{2}\sum_{i=1}^{n}x_i^2 y_i + \frac{\lambda}{2}n\overline{y}_{-}\right\}$$

$$- \frac{3}{2}\sum_{i=1}^{n}\log y_i$$

that $(\overline{Y}, \sum_{i=1}^{n} x_i Y_i, \sum_{i=1}^{n} x_i^2 Y_i, \overline{Y}_{-})$ constitutes a 4- dimensional sufficient statistic for the 3-dimensional natural parameter space Θ, thereby complicating statistical inference.

However, ignoring this set-up, maximization of ℓ with respect to $(\alpha, \beta, \lambda) = \psi$ yields

$$\frac{\partial \ell}{\partial \alpha} = \lambda \sum_{i=1}^{n}\{1 - (\alpha + \beta x_i)y_i\} = 0$$

$$\frac{\partial \ell}{\partial \beta} = \lambda \sum_{i=1}^{n}\{1 - (\alpha + \beta x_i)y_i\}x_i = 0$$

and
$$\frac{\partial \ell}{\partial \lambda} = \frac{n}{2\lambda} - \frac{1}{2}\sum_{i=1}^{n}\{1 - (\alpha + \beta x_i)y_i\}^2 y_i^{-1} = 0.$$

Letting $V_j = \frac{1}{n}\sum_{i=1}^{n} x_i^j y_i \quad j = 0, 1, 2$, we have

$$\alpha V_0 + \beta V_1 = 1$$
$$\alpha V_1 + \beta V_2 = \bar{x}$$

and the likelihood equations give the unique root $(\hat{\alpha}_L, \hat{\beta}_L, \hat{\lambda}_L) = \hat{\psi}_L$ where

$$\hat{\alpha}_L = \frac{V_2 - \bar{x}V_1}{V_0 V_2 - V_1^2} = \bar{y}^{-1}(1 - V_1\hat{\beta}_L)$$

$$\hat{\beta}_L = \frac{\bar{x}V_0 - V_1}{V_0 V_2 - V_1^2} = -\frac{1}{n}\frac{\sum_{i=1}^{n}(x_i - \bar{x})(y_i - \bar{y})}{V_0 V_2 - V_1^2}$$

and

$$\hat{\lambda}_L^{-1} = \frac{1}{n}\left[\sum_{i=1}^{n}(y_i^{-1} - \hat{\alpha}_L - \hat{\beta}_L x_i)\right]$$
$$= \bar{y}_- \hat{\alpha}_L - \hat{\beta}_L \bar{x}.$$

From the Cauchy-Schwarz inequality, $V_0 V_2 - V_1^2 \geq 0$ with equality holding if and only if $\sqrt{y_i} = c\sqrt{y_i}x_i$ for some constant c and $i = 1, \cdots, n$, it can be shown that $\hat{\alpha}_L$ and $\hat{\beta}_L$ are well defined and thus with probability 1, they are the unique solutions to the likelihood equations.

When $\bar{x} > 0$ we observe that not both $\hat{\alpha}_L$ and $\hat{\beta}_L$ are negative. Moreover we note that $\hat{\lambda}_L = 0$ with probability 1.

In order to show that $\hat{\alpha}_L, \hat{\beta}_L$ and $\hat{\lambda}_L$ maximize ℓ we consider the negative of the matrix of second partial derivatives of ℓ, namely

$$-\frac{\partial^2 \ell}{\partial \psi \partial \psi^t} = n\hat{\lambda}_L \begin{pmatrix} V_0 & V_1 & 0 \\ V_1 & V_2 & 0 \\ 0 & 0 & \frac{1}{2}\hat{\lambda}_L^{-3} \end{pmatrix}$$

and note that the right side is positive definite with probability 1 since $\hat{\lambda}_L > 0$ and $V_0 V_2 - V_1^2 > 0$. Thus we can conclude that the maximum likelihood root estimators $\hat{\alpha}_L, \hat{\beta}_L, \hat{\lambda}_L$ as they will be referred to, indeed maximize ℓ. We still have to examine to see if they lie in Θ.

Firstly $\hat{\lambda}_L > 0$ with probability and it is seen that at most one of $\hat{\alpha}_L$ and $\hat{\beta}_L$ can be negative. In the situation where the solutions lie outside Θ, we should consider the maximization of ℓ on the boundaries $\alpha = 0$, $\beta = 0$.

Now with $\alpha = 0$, the equations

$$\frac{\partial \ell(0, \beta, \lambda)}{\partial \beta} = 0, \quad \frac{\partial \ell(0, \beta, \lambda)}{\partial \lambda} = 0$$

yield the unique solutions

$$\hat{\beta} = \overline{x} V_2^{-1}, \quad \hat{\lambda}_\alpha^{-1} = \left(\overline{y}_- - \overline{x}^2 V_2^{-1} \right)$$

and the maximum value

$$\ell_\alpha = \ell(0, \hat{\beta}_\alpha, \hat{\lambda}_\alpha) = \text{constant} + \frac{n}{2} \log \hat{\lambda}_\alpha.$$

Similarly when $\beta = 0$ the corresponding solutions turn out to be

$$\hat{\alpha}_\beta = V_0^{-1}, \quad \hat{\lambda}_\beta^{-1} = \left(\overline{y}_- - V_0^{-1} \right)$$

with the maximum equal to

$$\ell_\beta = \text{constant} + \frac{n}{2} \log \hat{\lambda}_\beta.$$

Using the same arguments as before it can be shown that w.p.1 all these estimators are positive. From a comparison of ℓ_α and ℓ_β we note that $\ell_\alpha > \ell_\beta$ if and only if $\hat{\lambda}_\beta > \hat{\lambda}_\alpha$ or equivalently $\hat{\beta}_L \overline{x} - \hat{\alpha}_L > 0$. When $\hat{\alpha}_L < 0$ and $\hat{\beta}_L > 0$ we have $\hat{\beta}_L \overline{x} - \hat{\alpha}_L > 0$ giving us the maximum likelihood estimators as $\hat{\alpha} = 0$, $\hat{\beta} = \hat{\beta}_\alpha$. Thus if $\hat{\alpha}_L < 0$ the maximum likelihood estimate of α is pulled to the value zero and the estimate of β changes to $\hat{\beta}_\alpha$. Similar arguments hold when $\hat{\beta}_L < 0$. Collecting these results we state the following theorem.

Theorem G.1 *Based on a random sample of n observations from $IG\left((\alpha + \beta x_i)^{-1}\right)$ The maximum likelihood estimator $(\hat{\alpha}, \hat{\beta}, \hat{\lambda})$ of $(\alpha, \beta, \lambda) \in \Theta$ is*

$$\left(\hat{\alpha}, \hat{\beta} \right) = \begin{cases} (\hat{\alpha}_L, \hat{\beta}_L) & \text{if } V_1 < \min(\overline{x} V_0, \overline{x}^{-1} V_2) \\ (0, \hat{\beta}_L) & \text{if } \overline{x}^{-1} V_2 \leq V_1 < \overline{x} V_0 \\ (\hat{\alpha}_L, 0) & \text{if } \overline{x} V_0 \leq V_1 < \overline{x}^{-1} V_2 \end{cases}$$

while

$$\hat{\lambda}^{-1} = \left[\overline{y}_- - \left(\hat{\alpha} + \hat{\beta} \overline{x} \right) \right].$$

The behaviour of these estimators under scale changes in x or y is provided in the next theorem.

Theorem G.2 *Define $\hat{\psi}_L = (\hat{\alpha}_L, \hat{\beta}_L, \hat{\lambda}_L) = \hat{\psi}_L(x, y)$ and $\hat{\psi} = \hat{\psi}(x, y) = \hat{\psi}(\hat{\alpha}, \hat{\beta}, \hat{\lambda})$, $x^t = (x_1, \cdots, x_n)$, $y^t = (y_1, \cdots, y_n)$ and $e^t = (1, \cdots, 1)$ the*

$(1 \times n)$ unit vectors. Further let d and k be two constants such that $d > 0$ and $k > -\min(x_1, \cdots, x_n)$. Then
(a) $\hat{\psi}_L(x, dy) = (d^{-1}\hat{\alpha}_L(x,y), d^{-1}\hat{\beta}_L(x,y), d\hat{\lambda}_L(x,y))$,
(b) $\hat{\psi}_L(dx, y) = (\hat{\alpha}_L(x,y), d^{-1}\hat{\beta}_L(x,y), \hat{\lambda}_L(x,y))$,
(c) $\hat{\psi}_L(x + ke, y) = (\hat{\alpha}_L(x,y) - k\hat{\beta}_L(x,y), \hat{\beta}_L(x,y), \hat{\lambda}_L(x,y))$.
Moreover $\hat{\psi}$ satisfies only (a) and (b).

In concluding this discussion we remark that the above theory can be extended to a multiple regression model where $\mathcal{L}(Y_i) = IG((\beta^t X_i)^{-1}, \lambda)$, $\beta \in \mathbb{R}^p$ $X_i \in \mathbb{R}^p$ and have positive entries for all i, $X_{i_1} = 1$, for every i and the $p \times n$ matrix $(X_1, \cdots, X_n)$ has rank p.

Bhattacharyya and Fries conclude their discussion of the maximum likelihood estimator by proving the strong consistency of $\hat{\psi}_L$ and $\hat{\psi}$ as well as the limiting normal distribution of $\sqrt{n}(\hat{\psi}_L - \hat{\psi})$ and the almost sure convergence of the sample information matrix.

The approximate variances of $\hat{\alpha}, \hat{\beta}$ and $\hat{\lambda}$ are given by

$$\text{Var}\hat{\alpha} = (n\hat{\lambda}D)^{-1}V_2$$
$$\text{Var}\hat{\beta} = (n\hat{\lambda}D)^{-1}V_0$$
$$\text{Var}\hat{\lambda} = 2n^{-1}\hat{\lambda}^2$$

where $D = V_0 V_2 - V_1^2$.

These can be used in constructing large sample confidence intervals for (α, β, λ).

The next section is devoted to a least squares approach for replicated designs.

Least squares

In the least squares approach one assumes that k stress levels $x_i, \cdots, x_k$ are available and that $(y_{i1}, \cdots, y_{in_i})$ are n_i independent failure times observed at level x_i. The y_{ij} are all assumed to be independent for $j = 1, \cdots, n_i$, $i = 1, \cdots k$ and $\mathcal{L}(y_{ij}) = IG\left((\alpha + \beta x_i)^{-1}, \lambda\right)$.

Consider the quantities $(\alpha + \beta x_i) = \mu_i^{-1}$ as free parameters; it is immediately clear that $\overline{y}_i = \left(\sum_{j=1}^{n_i} y_{ij}\right)/n_i$ and $Q = \sum_{i=1}^{k} \sum_{j=1}^{n_i} \left(y_{ij}^{-1} - \overline{y}_i^{-1}\right)$ are jointly sufficient for (μ_i, λ). From the distribution theory seen in Chapter 1

(1) $\mathcal{L}(\overline{Y}_i) = IG(\mu, \lambda)$
(2) $\mathcal{L}(\lambda Q) = \chi^2_{N-k}$ where $N = \sum_{i=1}^{k} n_i$
(3) $\mathbb{E}(\overline{Y}_i^{-1}) = \mu_i^{-1} + (n_i \lambda)^{-1}$
(4) $\text{Var}(\overline{Y}_i^{-1}) = (n_i \mu_i \lambda)^{-1} + 2(n_i \lambda)^{-2}$

Least squares

(5) $\overline{Y}_i^{-1} - (n_i\tilde{\lambda})^{-1} = t_i \perp\!\!\!\perp \tilde{\lambda}^{-1} = \dfrac{Q}{N-k}$

(6) t_i and $\tilde{\lambda}^{-1}$ are the uniformly minimum variance unbiased estimators of μ_i^{-1} and λ.

Since t_i are the bias corrected reciprocal means one has the linear model $\mathbb{E}(t_i) = \alpha + \beta x_i$ with covariance structure given by

$$\mathrm{Var}(t_i) = (\alpha + \beta x_i)(n_i\lambda)^{-1} + 2(n_i\lambda)^{-2}\left[1 + (N-k)^{-1}\right],$$
$$\mathrm{Cov}(t_i, t_j) = 2\left[n_i n_j (N-k)^{-1}\right]\lambda^{-2},\ i \neq j.$$

One can employ the weighted least squares $\sum_{i=1}^{k} n_i[t_i - \mathbb{E}(t_i)]^2$ and minimizing this gives the following estimators of α and β

$$\tilde{\alpha} = \dfrac{1}{(m_2 - m_1^2)}\left[\dfrac{m_2}{N}\sum_{i=1}^{k} n_i t_i - \dfrac{m_1}{N}\sum_{i=1}^{k} n_i t_i x_i\right]$$

$$\tilde{\beta} = \dfrac{1}{m_2 - m_1^2}\left[\dfrac{1}{N}\sum_{i=1}^{k} n_i t_i x_i - \dfrac{m_1}{N}\sum_{i=1}^{k} n_i t_i\right],$$

where

$$m_j = \dfrac{1}{N}\sum_{i=1}^{k} n_i x_i^j,\ j = 1,2,3.$$

Simplifying we obtain

$$\tilde{\alpha} = \dfrac{1}{N}\sum_{i=1}^{k} n_i \overline{Y}_i^{-1} - \tilde{\beta} m_1 + \dfrac{\tilde{\lambda}^{-1}}{kN}$$

$$\tilde{\beta} = \dfrac{1}{N(m_2 - m_1^2)}\sum_{i=1}^{k}\left(n_i \overline{Y}_i^{-1} - \tilde{\lambda}^{-1}\right)(x_i - m_1).$$

When $n_1 = \cdots = n_k = n$, $\tilde{\beta} \perp\!\!\!\perp \tilde{\lambda}$.

Ignoring terms of order N^{-2} one can show that

A) $\mathrm{Var}\left(\tilde{\lambda}^{-1}\right) = \dfrac{2}{(N-k)\lambda^2}$, $\mathrm{Cov}\left(\tilde{\alpha},\tilde{\lambda}^{-1}\right) = \mathrm{Cov}\left(\tilde{\beta},\tilde{\lambda}^{-1}\right) = 0$

B) $\mathrm{Var}(\tilde{\alpha}) = \dfrac{\alpha m_2 d_2 + \beta m_1 d_3}{N\lambda d_2^2}$

C) $\mathrm{Var}(\tilde{\beta}) = \dfrac{\alpha d_2 + \beta m_1^{-1}(d_2^2 + d_3)}{N\lambda d_2^2}$

D) $\mathrm{Cov}(\tilde{\alpha},\tilde{\beta}) = -\dfrac{\alpha m_1 d_2 + \beta d_3}{N\lambda d_2^2}$

where

$$d_2 = m_2 - m_1^2,\ d_3 = m_1 m_3 - m_2^2$$

E) $\tilde{\psi} = (\tilde{\alpha}, \tilde{\beta}, \tilde{\lambda})^t$ is strongly consistent and $\sqrt{N}(\tilde{\psi} - \psi) \to N_3(0, \Gamma)$ the entries of Γ being the limits of N times the corresponding expressions given in (A) - (D).

Example G.1 The illustrative data set studied by Nelson (1971) and considered by Bhattacharyya and Fries is reproduced in Table G.1. The data represent times to first failure of insulation material in a motorette test at temperature levels of 190°, 222°, 240° and 260°. Ten units were put on test at each temperature level. Duplicate failure times indicate that the test was not monitored continuously. A recorded time is the time midway between the last time of inspection when the motorette was operative and the first time of inspection for which the motorette was inoperative.

The original purpose of the experiment was to determine if the mean time to failure at a temperature of 180°C exceeded a specified minimum requirement. It was later discovered that the data set for 260°C had been taken on a batch different from the batch used at other stress levels. It was therefore important to investigate whether or not the data set at 260°C was consistent with the remainder. Nelson's graphical and analytic techniques are based on the assumptions that the log failure times are normally distributed with constant variance and a mean depending on the absolute temperature T_i through the Arrhenius relationship, namely, mean $= a + bT_i^{-1}$.

Table G.1 *Times to failure (thousands of hours)in an accelerated life test of insulation material.*

190°	220°	240°	260°
7.228	1.764	1.175	0.600
7.228	2.436	1.175	0.744
7.228	2.436	1.521	0.744
8.448	2.436	1.569	0.744
9.167	2.436	1.617	0.912
9.167	2.436	1.665	1.128
9.167	3.108	1.665	1.320
9.167	3.108	1.713	1.464

| 10.511 | 3.108 | 1.761 | 1.608 |
| 10.511 | 3.108 | 1.953 | 1.896 |

Bhattacharyya and Fries assume that the distribution of failure times is $IG(\mu_i, \lambda)$ where

$$\mu_i^{-1} = \alpha + \beta x_i, \quad x_i = 10^{-8}(T_j^3 - 180^3)$$

T_i being measured in degrees centigrade. They implicitly assume that the wear of the insulation material increases until a critical amount has disintegrated. Since the insulating materials of the first three levels have a common source a common λ seems appropriate. The relationship of μ_i indicated above is based on the following considerations
(1) regressing Y_i^{-1} on T_i^3 resulted in a value of $R^2 = 99.9\%$
(2) (Sample variances)$^{\frac{1}{3}}$ were approximately linear in T_i^3
(3) the change $T_i \to T_i^3 - 180^3$ gave positive maximum likelihood estimates.

In Table G.2 the estimates based only on the first three levels are displayed. Calculations were done using the formulae developed for $\hat{\alpha}, \tilde{\alpha}$ etc.

Table G.2 *Parameter estimates and standard errors for the(TableG.1) data using the first three levels*

	α	β	λ^{-1}
MLE	.0371	7.3260	.0102
	(.0129)	(.3557)	(.0026)
LSE	.0320	7.4316	.0097
	(.0141)	(.3747)	(.0026)

The standard errors are based on large sample normal approximations for the MLE while for the LSE they are based on the expressions (A) - (D). Both methods of estimation seem comparable.

The next table, Table G.3 gives the estimated mean failure times. The standard errors are based on the approximation

$$\widehat{\text{Var}}\hat{\mu} = \hat{\mu}^2 \widehat{\text{Var}}\left(\hat{\mu}^{-1}\right)$$

where $\hat{\mu}$ is either the one based on the MLE or the LSE.

The MLEs obtained exclusively from the 260°C data set are then compared with the estimates using only the first three levels.

Table G.3 *Estimated mean failure times and standard errors (thousands of hours) based on* (TableG.2)

Temp(°C)	190	220	240	260
Sample Mean	8.782	2.638	1.581	1.116
MLE	8.902 (0.094)	2.565 (0.029)	1.606 (0.033)	1.114 (0.037)
LSE	8.863 (0.101)	2.565 (0.030)	1.598 (0.034)	1.105 (0.038)

$\hat{\mu}_4$ corresponding to 260°C is, for example,

$$\hat{\mu}_4 = \left[0.0371 + (7.3260)10^{-8}\left(260^3 - 180^3\right)\right]^{-1}$$
$$= [0.0371 + 0.897446]^{-1}$$
$$= 1.114.$$

The MLE of λ_4 for the data set corresponding to $T = 260°C$, namely $\hat{\lambda}_4^{-1} = 0.1311$ differs from $\hat{\lambda}^{-1}$ and $\tilde{\lambda}^{-1}$ found in Table G.2. An F test with 9 and 27 degrees of freedom was performed to assess this discrepancy, and resulted in a value of 15.01 with a p-value $<<0.001$. In agreement with Nelson, Bhattacharyya and Fries also reach the same conclusion that the two batches of insulating material are significantly different.

Variable stress accelerated tests

Accelerated life tests are intended to obtain quick information on the lifetimes of products. Experimental units are subjected to severe stress conditions than normal so that more failure data can be generated in a limited time. In addition the stress level is increased at specific times on the surviving units. This method is commonly known as step-stress ALT or VALT (for variable step-stress accelerated life test) and reduces losses due to censoring. When the initial stress level x_0 is higher than the normal stress we have a fully accelerated life test, as opposed to a partial accelerated test. Applications of such models are cited in Nelson (1980, 1990), De Groot and Goel (1979), Bhattacharyya and Soejetti (1989) and the references therein. Doksum and Hoýland (1992) consider such a step-stress model in which accumulated decay is governed by a Wiener process $W(y)$. We describe in the ensuing sections their model assumptions and analysis.

Model Assumptions

(1) n units operating independently are put on test at time $t = 0$, simultaneously at a stress level (normal level) x_0 and the failure times over a specified interval $[0, t]$ are recorded.

(2) Starting at time t, the surviving units are subjected to different stress level x_i, (accelerated) until they all fail.

(3) The accumulated decay $W_0(y)$ is assumed to be a Wiener process $\{W_0(y) \mid y \geq 0\}$ with positive drift ν and diffusion constant $\sigma^2 > 0$. Furthermore $W_0(0) = 0$ and $\mathbb{E}(W_0(y)) = \nu y$.

(4) Failure of a unit occurs when $W_0(y)$ crosses a threshold value (barrier) $a > 0$.

(5) When the stress is changed at time t from x_0 to x, on the units that have not failed, there begins a new process $W_1(y)$ with initial value $W_0(t)$ such that

$$W_1(y) = \begin{cases} W_0(y), & y < t \\ W_0(t + \alpha[y - t]), & y \geq t, \ \alpha > 0 \end{cases}$$

(observe that when $\alpha > 1, x_1 > x_0$)

(6) Define $\xi(y) = \mathbb{E}(W(y))$ for $y > 0$, and let $\xi'(y)$ exist at y. Then we will say that $\xi'(y)$ is the **decay rate** at y and $\xi(y)$ is the cumulative (integrated) decay rate
(In a two step-stress model the decay rate changes from ν to $\alpha\nu$ as y crosses the time point t at which time the stress is changed from x_0 to x_1)

(7) The failure time Y of a unit is the first time point at which $W(y) \geq a$

(8) The non-accelerated time (effective time) is defined as $\tau_\alpha(y) = Z$ (as opposed to the true (calendar) time Y) where

$$\tau_\alpha(y) = \begin{cases} y, & y \leq t \\ t + \alpha(y - t), & y > t \end{cases}$$

(9) $\mathcal{L}(Z) = IG\left(\mu = \frac{a}{\nu}, \ \lambda = \frac{a^2}{\sigma^2}\right)$.

(Thus the model says that a monotone map of the failure time Y has an inverse Gaussian law.)

Denote by $F_0(\cdot)$ the distribution function of the random variable $Z = \tau_\alpha(y)$ corresponding to the normal stress level x_0. As the stress changes from x_0 to cx_0 for known c the decay process $W(y)$ can be modelled as a Wiener process with drift ν changing to $c\nu$ in $[t, \infty)$ implying that $\alpha = c$.

Given a sample of failure times $Y_1, \ldots, Y_n$ and the number r of failures at or before time t, we note that $1 \le r \le n-1$. Let the likelihood be $L_\alpha(\mu, \lambda)$. Then

$$L_\alpha(\mu_1 \lambda) = \alpha^{n-r} \prod_{i=1}^{n} f_0(\tau_\alpha(y_i)).$$

Thus when α is assumed known, the parameter estimates of μ and λ are quite straightforward since all we do is transform from the Y_i to $\tau_\alpha(Y_i) = Z_i$ and invoke the method of Chapter 3.

When α is unknown we have another parameter to contend with and so the log-likelihood now becomes

$$l(\mu, \lambda, \alpha) \propto (n-r) \log \alpha + n \left(\frac{\lambda}{\mu} + \frac{1}{2} \log \lambda \right)$$

$$- \frac{3}{2} \sum_{i=1}^{n} \log \tau_\alpha(y_i) - \frac{\lambda}{2\mu^2} \sum_{i=1}^{n} \tau_\alpha(y_i) - \frac{\lambda}{2} \sum_{i=1}^{n} \frac{1}{\tau_\alpha(y_i)}.$$

Once again the maximum likelihood estimates can be obtained by using the methods of Chapter 3. The estimators $\hat{\alpha}, \hat{\mu}, \hat{\lambda}$ are jointly asymptotically normal and standard techniques are employed to obtain asymptotic confidence intervals. This is left as an exercise.

The extension to multiple stress levels will be considered next.

Corresponding to $(k+1)$ stress levels $x_0, x_1, \ldots, x_k$ over the intervals $[0, t_1), \ldots, [t_k, \infty)$, accumulated decay $W(y)$ is again modelled by the Gaussian process $W(y)$.

$$W(y) = \begin{cases} W_0(y), & 0 \le y < t, \\ W_i(y), & t_i \le y < t_{i+1}, \ i = 1, \ldots, k \ (t_{k+1} = \infty). \end{cases}$$

As in the two step-stress case $W_0(y)$ is a Wiener process with drift ν and diffusion constant $\sigma^2 > 0$. Furthermore for $i = 1, \ldots, k$

$$W_i(y) = W_{i-1}(t_i + \alpha(y - t_i)) \quad t_i \le y < t_{i+1}.$$

Finally we can write $W(y)$ compactly in terms of $W_0(\tau(y))$ through the function $\tau(y)$ as follows

Define $\beta_i = \prod_{g=0}^{i} \alpha_j$, $\alpha_0 = 1$, $\beta(y) = \beta_i$ for $y \in [t_i, t_{i+1})$, $i = 0, 1, \ldots, k$, $t_0 = 0$. Then

$$\tau(y) = \begin{cases} y, & 0 \le y < t_1 \\ t_1 + \beta_1(t_2 - t_1) + \ldots + \beta_{i-1}(t_i - t_{i-1}) \\ \quad + \beta_i(y - t_i), & t_i \le y < t_i. \end{cases}$$

The function $\beta(y)$ holds the key in describing $\tau(y)$. Indeed we have

$$\tau(y) = \int_0^y \beta(x)dx.$$

Hence $W(y) = W_0(\tau(y))$ $y \geq 0$.

Thus we have a continuous Gaussian process with decay rate increasing from ν to $\nu(\alpha_0 \alpha_1 \ldots \alpha_i) = \nu\beta_i$ in the time interval $[t_i, t_{i+1})$. Here α_i is descriptive of the change in stress level from x_{i-1} to x_i during $[t_i, t_{i+1})$ while β_i describes the change in stress level from x_0 to x_i as y crosses successively the time points $t_1, \ldots, t_i$. In general observe that α_i is a function of x_{i-1} and x_i and can be parametrized in several forms two of which are

(a) $\alpha_i = \exp[\theta(x_i - x_{i-1})] \Rightarrow \beta_i = \exp[\theta(x_i - x_0)]$

(b) $\alpha_i = \left(\frac{x_i}{x_{i-1}}\right)^\theta \Rightarrow \beta_i = \left(\frac{x_i}{x_0}\right)^\theta$

for some real unknown parameter θ.

These parametrizations have the interesting feature which models the decay rate $\nu\beta_i$ as a function of the initial stress x_0 and the current stress only in the interval $[t_i, t_{i+1})$.

The failure time Y is again modelled by the time transform $Z = \tau(y)$ assumed to follow an inverse Gaussian law $IG(\mu, \lambda)$. Recall that $\tau(y) = \int_0^y \beta(x)dx$ where $\beta(x) = \beta_i$ for $x \in [t_i, t_{i+1})$, $i = 0, 1, \ldots, k$.

The question of estimation depends again on whether the β_i are known or not. When the drift $\nu\beta_1 = \nu(\alpha_0, \ldots, \alpha_i)$ with α_i known and the stress levels are regarded as $\alpha_i x_{i-1}$ then we have a straightforward extension to the two step-stress level case. The case when the β_i are unknown requires care. When a model stipulates that the decay rate $n\beta_i$ is proportional to the stress level, as for example $\exp\{\theta(x_i - x_0)\}$ or $\exp\{\theta(\log x_i - \log x_0)\}$ for some unknown θ, then the likelihood becomes

$$L(\mu, \lambda, \beta) = \prod_{j=1}^n \beta(y_j) f_0(\tau(y_j)).$$

Suppose further that a random sample of n units are divided into c groups with respective counts $n_1, \ldots, n_c$ in each group. If the stress levels are indicated by x_{ij} at time t_{ij} ($i = 0, 1, \ldots, k$, $j = 1, \ldots, c$) ($t_{0j} = 0$). The initial stress levels $x_{01}, \ldots, x_{0c}$ are all different for each group,

$\nu\alpha_{01}, \ldots, \nu\alpha_{0c}$ being the corresponding initial decay rates. Then

$$\beta_{ij} = \prod_{r=0}^{i} \alpha_{rj}, \quad \beta_j(y) = \beta_{ij}, \; y \in [t_{ij}, t_{i+1\,j}),$$

$$i = 0, \ldots, k_j - 1$$

$$= \tau_j(y) = \alpha_{0j} y \quad 0 \le y < t_{ij}$$

$$= \sum_{r=0}^{i-1} \beta_{rj}(t_{r+1\,j} - t_{rj}) + \beta_{ij}(y - t_{ij}),$$

$$t_{ij} \le y < t_{i+1\,j}, \; i = 1, 2, \ldots, k_j$$

and of course we have in compact form

$$\tau_j(y) = \int_0^y \beta_j(x)\,dx.$$

Corresponding to the failure times $(Y_{j1}, \ldots, Y_{jn_j})$, $j = 1, \ldots, c$, one can write the likelihood as

$$\prod_{j=1}^{c} \prod_{i=1}^{n_j} \beta_j(y_{ij}) f_0(\tau_j(y_{ij}))$$

extending the notation defined for the two step-stress case. The stress change modeled by α_{ij} can be parametrized by $\left(\frac{x_{ij}}{x_{i-1\,j}}\right)^{\theta}$ for example, so that $\beta_{ij} = \exp\{\theta(\log x_{ij} - \log x_{0j})\}$.

In case there are censored observations with $Y_{ij} = \min(Y'_{ij}, C_{ij})$, $\delta_j = I\,(Y'_{ij} \le C_{ij})$ where Y'_j is the failure time such that $\tau(Y'_{ij}) \sim IG(\mu, \lambda)$, and $\{C_{ij}\}$ are i.i.d. censoring times independent of $\{Y'_{ij}\}$, the likelihood now takes the form

$$\prod_{j=1}^{c} \prod_{i=1}^{n_j} [f_j(y_{ij})]^{\delta_{ij}} [1 - F_j(y_{ij})]^{1-\delta_{ij}}$$

where $f_j(y_{ij}) = \beta_j(y_{ij}) f_0(\tau_j(y_{ij}))$.

Example G.2 The example considered by Doksum and Hóyland is taken from Nelson (1980) and relates to an analysis of step stress test of cable insulation involving 2 groups (Nelson used 6 groups). Time was measured in minutes, the initial stress level being 400 volts/millimeter. 5 units were tested in group I at 10 time intervals of length 240. The stress levels $x_{i1} = 896.6 + i(86.2)$ ($i = 0, 1, \ldots, 9$). In the 5th interval $[960, 1200]$ where the stress was 1,241.4 the observed failure times were

1,056.9 and 1,057.9. In the 6th time interval $[1200, 1440)$ at a stress level of 1,326.6 the observed data was 1,209, 1,293, 1,293* (censored). Nine items were tested in group II at 10 time intervals of length 960. The stress levels $x_{i_2} = 866.7 + i(83.3)$, $i = 0, 1, \ldots, 9$. The data consists of $Y'_{12} = 323.9^*, Y'_{22}\ 858.4^*$ in the first interval at a stress level of 866.7, $Y_{32} = 1,120.0$ at a stress level of 950 observed in the 2nd interval, $Y_{42} = 2,420.9$, $Y'_{52} = 2,420.9^*$, $Y_{62} = 2,660.4$, $Y_{72} = 1,922.9$ observed in the 3rd interval corresponding to a stress level of 1,033.3, $Y_{82} = 2,833.9$ observed in the 4th interval where the stress level was 1,116.6 and $Y_{92} = 4,102.1$ in the 5th interval at a stress level of 1,199.9.

It was assumed here that $\beta_j(y) = \exp\{\theta(\log x_{ij} - \log x_0)\}$ with $x_0 = 400$. Doksum and Hóyland report the following estimates based on the non-linear least squares maximum likelihood routine of the SAS. $\hat{\mu} = 786,024.5$, $\hat{\lambda} = 1,830,477.7$, $\hat{\theta} = 6.38$ with estimated standard errors of 358,170.0, 740,251.6 and 0.328.

The estimate of the point beyond which one expects 95% of the insulation material to last under normal stress, namely $\hat{F}_0^{-1}(0.5) = 786,024.4$. Finally they give a percentage-percentage plot of $\{(\hat{F}_i(y_{(j)i}), \hat{F}_0(\hat{\tau}_i(y_{(j)i})); j = 1, \ldots d_i\}$ where $y_{(1)i} < y_{(2)i} \ldots < y_{(d_i)i}$ are the distinct ordered failure times for group i. $\hat{F}_i$ is the Kaplan-Meier estimate of F_i. Based on this plot the fit appears reasonably good.

There is another aspect to the changing stress levels in that it is conceivable that stress is increased on a continuous scale. In this situation $\beta(y)$ can be expressed in terms of the changing stress $x(y)$.

More generally if $\tau(y)$ is a non-negative strictly increasing continuous function of y on $[0, \infty)$ with $\tau(0) = 0$, then for a Wiener process $\{W_0(y) \mid y > 0\}$ with drift ν and diffusion constant $\sigma^2 > 0$, the first passage time Y for $W_0(y)$ to cross the barrier it yields the result that $\mathcal{L}(\tau(y)) = IG\left(\mu = \frac{a}{\nu}, \lambda = \frac{a^2}{\sigma^2}\right)$.

As a final word of caution note that, in the estimation problem, when there are no failures under normal stress (i.e, $r = 0$) it is impossible to obtain estimates of α, μ and λ. On the other hand when all items fail under normal stress α cannot be estimated.

H. Management

Labour turnover

Whitmore (1979) uses the inverse Gaussian model to explain employee job attachment and length of service in particular. In his formulation $X(t)$ is the employee's level of job attachment as a function of length of service. With the standard assumptions of a Wiener process for $X(t)$, the first passage time T from the initial level of attachment denoted by $X(0) = c$ to the separation threshold arbitrarily set at $X(T) = 0$ has a defective inverse Gaussian law

$$f(t) = \left(\frac{c^2}{2\pi t^3 \sigma^2}\right)^{\frac{1}{2}} \exp\left\{-\frac{(c+\nu t)^2}{2\sigma^2 t}\right\} 1_{\mathbb{R}^+}(t)$$

ν being the drift and σ^2 the diffusion constant. When $\delta < 0$ there is a tendency to drift towards the separation threshold. For $\delta \leq 0$ separation occurs in finite time with probability one. For $\delta > 0$ separation theoretically never occurs. Thus we have a defective (improper) probability law with positive probability concentrated at $T = \infty$, which says that the probability that the length of service is infinite when $\delta > 0$ is $P(T = \infty) = 1 - \exp(-2c\nu/\sigma^2)$. The mean and variance of T given that the length of service is finite are

$$\text{mean} = \frac{c}{|\nu|}, \quad \text{variance} = \frac{c\sigma^2}{|\nu|^2}.$$

The model assumes that the level of job attachment is a stationary process, the level of the separation threshold is fixed and that length of service is determined purely by the person's level of personal job attachment and not by any external factors.

The empirical results of Whitmore relate to the validation of the defective inverse Gaussian law as applied to homogeneous employee cohort studies (no parametric variation within an employee cohort). Four completed length of service distributions compiled by Silcock (1954) have been studied by Whitmore. The data relate to male and female entrants of J. Bibby and Sons Ltd, for the years 1950 and 1951, for a single factory of a British firm engaged in manufacturing animal feeds and soap. Furthermore only employees who left the firm and had to be replaced are in the study while those declared redundant have been excluded. It is assumed that the cohorts are homogeneous.

Table H.1 *Distributions of employees of J. Bibby and Sons Ltd, by completed length of service*

(a) Entrants of 1950

Length of service (in months)	Males Observed	Males Expected	Females Observed	Females Expected
0-under 3	182	181.3	25	23.5
3-under 6	103	103.2	26	30.2
6-under 9	60	54.6	22	19.4
9-under 12	29	34.9	13	13.3
12-under 15	31	24.7	15	9.8
15-under 18	23	18.7	7	7.6
18-under 21	10	14.7	5	6.1
21-under 24	8	12.0	1	5.0
24-under 27	7	10.0	4	4.2
27 or more	176	174.9	51	49.8
Total	629	629.0	169	169.0
	$\chi^2 = 7.94$, d.f.=7		$\chi^2 = 7.18$, d.f.=7	

(b) Entrants of 1951

Length of service (in months)	Males Observed	Males Expected	Females Observed	Females Expected
0-under 3	147	144.2	38	38.6
3-under 6	54	68.2	29	26.6
6-under 9	47	34.5	15	14.6
9-under 12	21	21.4	9	9.5
12-under 15	12	14.8	5	6.8
15 or more	237	234.9	77	76.9
Total	518	518.0	173	173.0
	$\chi^2 = 8.14$, d.f.=3		$\chi^2 = 0.71$, d.f.=3	

Table H.2 *A comparison of minimum χ^2 and maximum likelihood fits for Bibby CLS data*

Data set	Sample size	d.f.	Minimum χ^2 δ	ν	χ^2
M 1950	629	7	0.0007	0.2964	7.94
M 1951	518	3	0.0624	0.3576	8.14
F 1950	169	7	-0.0243	0.1356	7.18
F 1951	173	3	0.0177	0.2417	0.71
Total		20			23.97

Data set	Sample size	d.f.	Maximum likelihood δ	ν	χ^2
M 1950	629	7	0.0024 (0.0047)	0.3015 (0.0885)	7.99
M 1951	518	3	0.0673 (0.0241)	0.3665 (0.0219)	8.22
F 1950	169	7	-0.0219 (0.0081)	0.1406 (0.0100)	7.27
F 1951	173	3	0.0187 (0.0120)	0.2432 (0.0317)	0.71
Total		20			24.19

M=Males, F=Females.

The results presented in Table H.1 show that the model (defective inverse Gaussian) is indeed good. A pooled chi-square based on the four cohorts amounts to 23.97 corresponding to 20 degrees of freedom yielding a p-value of about 0.75. Whitmore points out that the IG model provides a better fit than that obtained by Silcock except in the case of the male entrants of 1951. Whitmore also gives a comparison of the estimates of the parameters and their asymptotic standard errors. The estimates permit the prediction of cohort attrition and of long-service experience for different types of employees. From Table H.2 a negative estimate of the drift for female entrants of 1950 is indicative of the fact that all employees in the cohort will leave the company eventually. For the remaining entrants the drift is away from the separation threshold ($\delta > 0$) indicative of a steady increase in the expected level of job attachment with longer service. On the other hand for male entrants of 1950 the estimate of δ is close to zero and in all likelihood the employees might eventually leave.

Examination of the survival function reveals that the probability of eventual separation, $P(T < \infty) = \exp(-2\hat{c}\hat{\nu}/\hat{\sigma}^2) \approx 0.7055$ for the male entrants of 1951, implying that of the 237 male entrants who have been with the company for 15 months or more about $(0.2945)(518) = 153$ can be expected to stay for a long time. Other deductions can be made regarding employee satisfaction or discontent. For a fuller analysis and interpretation the reader is referred to Whitmore (1979).

Duration of strikes

A strike is a stoppage of work with a dispute over terms and conditions of employment. Two measures of strike duration can be used. The first, derived from the strikers, involves the average number of working days lost per striker, and is independent of the number of work stoppages and strike magnitude. The second measure is derived from the actual length of individual strikes. Lancaster (1972) constructs a model where the strike duration is a random variable with a probability distribution whose parameters "embody the systematic determinants of duration" and such that these parameters are approximately the same for some observed set of strikes. Suppose that a dispute emerges between a group of workers and management over a pay claim and let the workers demand $b while the management is willing to offer $a. As the strike progresses the demand and the offer change and at some point of time when this difference has fallen to zero the strike is over. Lancaster postulates the existence of a scalar measure of the difference between the parties and assumes that the duration of the strike is a function of this difference.

The model assumptions

Let t be the time, varying on a continuous scale measured from an origin, say, the start of the strike and the unit time period be a working day. For each t, $X(t)$ denotes the measure of difference between the parties at time t. (We can change the origin and start at time $t = 0$ and let $X(0) = 0$.) Suppose the strike stops when $X(t) = 1$. Then $X(t)$ is a random variable such that

(1) for any $t_1 < t_2$, $\mathbb{F}(X(t_2) - X(t_1)) = \nu(t_2 - t_1)$, $\nu > 0$, which simply says that the average progress towards settlement is proportional to the time involved. Since $X(0) = 0$ and $X(t) = 1$, ν represents the mean proportionate rate of drift to settlement per working day.

(2) for any pair of non-overlapping intervals the changes in $X(t)$ are independent. This implies that given our knowledge of the difference between the parties when the strike began, knowledge of the differences before the strike started is of no help in predicting their differences at a future time and gives no clue as to how long the strike will last.

(3) for any $t_1 < t_2$, $X(t_2) - X(t_1) \sim N(\nu(t_2 - t_1), \sigma^2(t_2 - t))$. This assumption implies that the change in the differences can be negative (drifting apart) and the probability of this event is smaller, the longer the time interval and the more rapid the drift towards agreement.

These assumptions make $X(t)$ a Wiener process, and the first time T when $X(t) = 1$ is the $IG(\mu = \frac{1}{\nu}, \lambda = \frac{1}{\sigma^2})$.

The data sets analyzed and fitted by Lancaster pertain to the list of strikes recorded by the Ministry of Labour beginning in 1956. The data was divided into 8 industries and a scatter diagram of duration and number of men (size) involved was inspected to see if there was any evidence of lack of independence of duration and size. (All strikes of duration less than a day and involving fewer than 10 workers were not recorded.) For seven industries the evidence was in favour of independence. Thus it appeared that the recorded proportion of strikes of different durations were not systematically different from the corresponding proportions of all strikes. Moreover the data reported represents truncated data (at $T = 1$). The observations arising in grouped form are reproduced from Lancaster in the following tables. Maximum likelihood estimates of μ and σ^2 as well as a comparison of the observed and fitted data together with the Pearson goodness-of-fit statistics are provided.

Newby and Winterton (1983) found that the duration of unofficial strikes is log normally distributed and that the duration of official strikes is exponential. Chhikara and Folks (1983) applied Bartlett's test for testing the homogeneity of $\lambda = \frac{1}{\sigma^2}$ of the data set for the eight industries and concluded that the idea of a common λ should be rejected. They point out that a careful scrutiny should be made about the distribution of the truncated data, and the estimates of λ and μ that Lancaster used in his analysis.

Table H.3 *Observed(O) and predicted(P) numbers of strikes by duration(D) in eight industry groups*

1. Metal Manufacture			2. Non-electrical Engineering		
D	O	P	D	O	P
2	43	47.1	2	41	43.3
3	37	30.1	3	28	24.9
4	21	21.0	4	18	16.1
5	19	15.3	5	8	11.3
6	11	11.6	6	9	8.3
7	8	9.2	7	3	6.4
8	8	7.4	8	7	5.1
9	9	6.1	9	5	4.1
10	3	5.1	10	3	3.4
11-15	16	16.7	11-15	11	10.6
16-20	4	9.1	16-20	4	5.4
21-25	4	5.6	21-25	4	3.1
26-30	3	3.7	> 25	8	6.0
31-40	3	4.3		$\overline{149}$	
41-50	5	2.3			
> 50	4	3.7			
	$\overline{198}$				

$\hat{\mu} = 0.137, \hat{\sigma} = 0.612$ $\qquad$ $\hat{\mu} = 0.197, \hat{\sigma} = 0.721$
$\hat{\sigma}/\hat{\mu} = 4.47, \chi^2_{13} = 12.4$ $\qquad$ $\hat{\sigma}/\hat{\mu} = 3.66, \chi^2_{10} = 5.8$

3. Distributive Trades			4. Vehicles		
2	14	16.7	2	34	34.0
3	13	9.6	3	19	18.2
4	4	6.2	4	10	11.3
5	6	4.3	5	8	7.7
> 5	17	17.3	6	6	5.6
	$\overline{54}$		7	5	4.2
			8	2	3.3
			9	3	2.6

Table H.3 (ctd.)

D	O	P	D	O	P
			10	2	2.1
			11-15	6	6.5
			16-20	4	3.0
			> 20	4	4.3
				$\overline{103}$	

$\hat{\mu} = 0.231, \hat{\sigma} = 0.697$ | $\hat{\mu} = 0.255, \hat{\sigma} = 0.811$
$\hat{\sigma}/\hat{\mu} = 3.02, \chi_2^2 = 3.1$ | $\hat{\sigma}/\hat{\mu} = 3.18, \chi_9^2 = 1.3$

5.Construction			6.Shipbuilding		
2	44	44.3	2	27	28.2
3	33	34.6	3	19	18.4
4	28	25.3	4	19	12.5
5	23	19.0	5	7	9.0
6	11	14.7	6	4	6.8
7	12	11.7	7	5	5.3
8	8	9.4	8	3	4.2
9	6	7.8	9	1	3.5
10	13	6.5	10	2	2.8
11-15	16	21.0	11-15	11	9.0
16-20	7	11.0	16-20	6	4.6
21-25	6	6.4	> 20	8	7.8
26-30	6	4.0		$\overline{112}$	
> 30	11	9.2			
	$\overline{225}$				

$\hat{\mu} = 0.134, \hat{\sigma} = 0.502$ | $\hat{\mu} = 0.165, \hat{\sigma} = 0.605$
$\hat{\sigma}/\hat{\mu} = 3.75, \chi_{11}^2 = 13.1$ | $\hat{\sigma}/\hat{\mu} = 3.67, \chi_9^2 = 8.3$

7.Transport			8.Electrical Machinery		
2	50	43.0	2	24	23.5
3	19	18.3	3	5	11.6
4	10	10.3	4	16	7.1
5	5	6.7	5	6	4.9
6	2	4.7	6-7	5	6.3

Duration of strikes

Table H.3 (ctd.)

	7. Transport			8. Electrical machinery	
D	O	P	D	O	P
7-20	13	16.5	8-10	6	5.5
> 20	3	2.6	> 10	10	13.1
	$\overline{102}$			$\overline{72}$	
$\hat{\mu} = 0.988, \hat{\sigma} = 2.895$			$\hat{\mu} = 0.233, \hat{\sigma} = 1.093$		
$\hat{\sigma}/\hat{\mu} = 2.93, \chi_4^2 = 4.0$			$\hat{\sigma}/\hat{\mu} = 4.69, \chi_4^2 = 16.2$		

Table H.4 *All industry groups apart from transport and electrical machinery*

Duration	Observed	Predicted
2	203	212
3	149	136
4	100	92
5	71	66
6	49	50
7	33	39
8	29	31
9	26	26
10	23	21
11	14	18
12	12	15
13	9	13
14	11	11
15	15	10
16	6	9
17	7	7.7
18	6	6.9
19	4	6.1
20	4	5.5
21-25	17	20.5
26-30	16	12.9
31-35	8	8.6
36-40	8	5.8

Table H.4(ctd.)

Duration	Observed	Predicted
41-50	12	7.0
> 50	8	8.8
	840	

$$\hat{\mu} = 0.160, \quad \hat{\sigma} = 0.617, \quad \hat{\sigma}/\hat{\mu} = 3.86$$
$$\chi^2_{22} = 17.3$$

Table H.5

Duration of stoppages in connection with disputes over claims for wage increases			Duration of stoppages in connection with disputes over the employment of particular persons or classes		
Industries 1-6 pooled			Industries 1-6 pooled		
Duration	Observed	Predicted	Duration	Observed	Predicted
2	88	92.3	2	53	56.0
3	65	60.5	3	42	37.0
4	52	51.5	4	22	25.4
5	26	29.7	5	29	18.5
6	21	22.4	6	13	14.1
7	16	17.5	7	10	11.1
8	15	14.0	8	6	9.0
9-10	16	21.0	9	9	7.4
11-15	31	30.4	10	4	6.3
16-20	13	15.7	11-15	16	20.5
21-30	14	14.9	16-20	9	11.2
> 30	16	13.3	21-30	11	11.4
	373		> 30	17	13.0
				241	

$\hat{\mu} = 0.159, \hat{\sigma} = 0.603$
$\hat{\sigma}/\hat{\mu} = 3.79, \chi^2_9 = 6.4$

$\hat{\mu} = 0.133, \hat{\sigma} = 0.604$
$\hat{\sigma}/\hat{\mu} = 4.54, \chi^2_{10} = 12.3$

Duration of strikes

Table H.6 *Duration of stoppages in connection with disputes over working arrangments, rules, discipline*

Industries 1-6 pooled

Duration	Observed	Predicted
2	43	43.9
3	24	21.6
4	12	13.2
5	8	9.0
6	9	6.6
7	3	5.1
8	5	4.0
9	4	3.3
10	4	2.7
11-15	6	8.6
16-20	4	4.6
21-30	5	4.5
> 30	5	4.9
	$\overline{132}$	

$\hat{\mu} = 0.249, \hat{\sigma} = 1.093$

$\hat{\sigma}/\hat{\mu} = 4.39, \chi^2_{10} = 4.2$

I. Meteorology

Frequency distributions of wind speed near the surface of the earth are of special interest to meteorologists because of the potential derivation of electrical energy from wind power. Most distributions are known to be skewed to the right with the mean far exceeding the median. Stewart and Essenwanger (1978) have fitted a three-parameter Weibull distribution and found that it adequately describes the frequency of occurrence of high speed winds. They show that the three parameter Weibull is preferable to the two parameter model particularly for predicting 90-99% thresholds in missile climatology.

In modelling windspeed frequency distributions it is customary to regard the wind vector as a function of wind speed and wind direction and use a bivariate distribution system. Essenwanger (1959) shows that a square or cube root transformation of wind speed results in approximate normality. Large masses of data are divided into classes with boundaries corresponding to the square or cube of the wind speed.

Stewart and Essenwanger have observed that the introduction of the third parameter (the threshold parameter) brings in difficulties in the estimation stage. A positive value of the location parameter leads to an unrealistic condition of zero probability of wind speeds less than the parameter value. Data arises from potential wind energy sites possessing low probabilities of low wind speeds when the location parameter is positive.

Bardsley (1980) recommends the use of the inverse Gaussian law $IG(\mu, \lambda)$ as an alternative to the three-parameter Weibull for describing wind speed with low frequencies of low speeds. Bardsley observes that when $\phi = \frac{\lambda}{\mu}$ is quite large the inverse Gaussian exhibits an abrupt increase in the probability density, corresponding to the 0.0001 quantile.

Let $\xi_{.0001}$ be such that $P(X < \xi_{.0001}) = 0.0001$. Then by plotting $\left(\mu_3/(\mu_2)^{3/2}, \ T = \frac{\xi_{.0001}}{\mu_1}\right)$, Bardsley is able to mimic the Weibull law by the inverse Gaussian law whose 0.0001 quantile is set equal to the Weibull location parameter (γ). Thus for the $IG(\mu, \lambda)$ we have

$$F_{IG}(x) = \Phi\left(\sqrt{\frac{\phi\mu}{x}}(x\mu^{-1} - 1)\right) + e^{2\phi}\phi\left(-\sqrt{\frac{\phi\mu}{x}}(x\mu^{-1} + 1)\right)$$

and for the Weibull (γ, θ, β) we have

$$G_W(\omega) = 1 - \exp\left\{-\left(\frac{\omega - \gamma}{\theta}\right)^\beta\right\} \qquad (\omega \geq \gamma).$$

The Weibull mean is

$$\mu_W = \gamma + \theta\, T(1 + \beta^{-1}).$$

A measure of similarity suggested by Bardsley is

$$M = \max |F_{IG}(x) - G_W(x)|$$

where $F_{IG}(\gamma) = 0.0001$, i.e., $\xi_{.0001} = \gamma$. The comparison between the two laws is made with $\gamma = \xi_{.0001}$ and the mean of the IG, namely μ set equal to $\mu_W = \gamma + \theta\, T(1 + \beta^{-1})$. Since M is independent of the units of measurement of X or W, μ is taken equal to unity. Therefore if

$$R = \frac{\gamma}{\gamma + \theta\, T(1 + \beta^{-1})} = \gamma$$

one can plot M for values of (R, β). To see this, note that $\gamma + \theta\, T(1 + \beta^{-1}) = 1 \Rightarrow R = \gamma \Rightarrow \theta = \frac{1-R}{T(1+\beta^{-1})}$ so that $G_W(\omega)$ is a function of R and β only.

Values of M near zero indicate the closeness of the two laws and contours of M will depend on R and β. Bardsley finds that an approximate correspondence between the two laws occurs when the variances are equal. The similarity contours can then be constructed for choices of (R, β). Finally points corresponding to positive values of γ from the Tables of Stewart and Essenwanger were superimposed on the contours and this revealed the fact that a sufficient number of points (R, β) lay within the $M = 0.1$ contour to justify the use of the IG law. Bardsley recommends the use of the $IG(\mu, \lambda)$ law as a substitute for the three-parameter Weibull when ϕ is large. On the other hand when ϕ is low the IG law is useful for describing frequency laws with high peaks near zero and long right tails. Although such wind speed distributions may be uncommon the IG law could be a very viable alternative to the two-parameter Weibull.

The wind energy flux $Z = \frac{1}{2\rho} X^3$ where X is the wind speed, ρ being the air density. The energy flux distribution denoted by $h_Z(z)$ can be shown to tend to zero for $Z \to 0$. Thus $P[Z < z] = H_Z(z)$ can be expressed in terms of $F(\cdot)$ and power duration curves can be constructed for wind speed.

The mean wind energy flux μ_Z is estimated by

$$\hat{\mu}_Z = \frac{1}{2\rho} \hat{\mu}^3 \left\{ 1 + \frac{3}{\hat{\phi}} + \frac{3}{\hat{\phi}^2} \right\}.$$

When data available at a site is given in terms of speed measurements averaged over a long time interval ΔT, this can be converted into speed values averaged over small intervals Δt where $\Delta t = \frac{\Delta T}{K}$, for an integer K.

J. Mental health

Whitmore and Neufeldt (1970) examine a model to describe the mental health of an individual. Mental health is a function of several variables and cannot be qualitatively assessed in terms of the factors such as socio-economical, physiological, psychological or temporal. Nevertheless these factors do play a key role to the proper understanding of the characteristics underlying the mental function.

Individuals are often grouped into two categories, non-patient and inpatient at time t and their state at time $t + dt$ can be modelled by a Markov process. The parameters defining the conditional probabilities of moving from one group into the other can be taken to determine the characteristics of the state of an individual's mental health. The parameters specify an individual's rate of admission or rate of discharge and establish the pattern of entry and departure. In the jargon of mental health, length of stay in hospitals(los), the expected time spent in the non-patient state after each discharge (called time off books), the expected observation period spent in the inpatient state (being on books) and the expected number of times the individual is admitted to a psychiatric unit are important statistics in assessing the mental health status of a representative patient. Quite often the parameters of the model are functions of the socio-economic characteristics, the diagnostic classification and the method of treatment, the age and sex of the individual. In the absence of knowledge of the functional relation the analysis becomes quite complex and statistical tools are needed for its estimation.

Whitmore and Neufeldt also consider a more sophisticated model which regards the patient's state of mental health X as measured on a continuous scale. Since X changes with time the modelling can be achieved in terms of a diffusion process depending on endogenous as well as exogenous factors. To be more specific, let the state of mental health $X(t)$ be a Wiener process. At the start of the observation period $t = 0, X(0)$ represents the state of mental health of an individual. Suppose further that a and b ($a < b$) are two limits such that when $X(t)$ drops to the value a, the individual is admitted to a psychiatric unit for observation and treatment while when $X(t)$ reaches level b the patient indicates recovery and is discharged from the hospital. Indicating the drift of the Wiener process by ν (the propensity of the individual to move towards the barriers) and the diffusion constant σ^2 (which measures the extent to which an individual's mental health varies from an average mental health state), the probability for the time t to reach the state $c = b - a$ is (assuming $X(0) = 0$)

$$f(t) = \frac{c}{\sqrt{2\pi\sigma^2 t^3}} \exp\left\{-\frac{(c-\nu t)^2}{2\sigma^2 t}\right\}.$$

Mental health

Table J.1 *Cumulative distribution for observed data (Obs) and four LOS models. Cumulative proportion released by end of interval*

Time	Obs	Wei	Γ	LN	IG
10	.049	.094	.097	.049	.015
20	.130	.167	.168	.131	.096
30	.212	.230	.230	.209	.189
40	.283	.286	.286	.277	.270
50	.337	.337	.336	.337	.338
60	.392	.383	.382	.389	.394
70	.445	.425	.424	.435	.442
80	.481	.463	.462	.475	.484
90	.520	.499	.498	.511	.519
100	.549	.532	.531	.543	.550
110	.582	.562	.562	.572	.578
120	.604	.590	.590	.598	.602
130	.627	.616	.616	.621	.624
140	.643	.640	.641	.642	.644
150	.662	.662	.664	.662	.662
160	.681	.683	.685	.680	.679
170	.691	.702	.705	.696	.694
180	.709	.720	.724	.711	.707
190	.723	.737	.741	.725	.720
200	.732	.753	.757	.738	.732
300	.802	.865	.872	.828	.815
400	.848	.924	.932	.878	.863
500	.882	.957	.963	.909	.894
600	.903	.975	.980	.920	.916
700	.915	.985	.989	.945	.932
800	.925	.991	.994	.955	.944
900	.931	.995	.997	.963	.953
1,100	.940	.998	.999	.974	.967
1,300	.949	.999		.981	.975
1,500	.955			.985	.982
2,000	.962			.992	.990

The maximum absolute deviation turns out to be .076, .084, .034 and .034 respectively for the Weibull(Wei), gamma, Lognormal(LN) and IG.

Thus we encounter the inverse Gaussian law again in the modelling of length of stay of patients in hospitals.

Whitmore and Eaton (1977) have studied the length of stay as a stochastic process and applied the inverse Gaussian law to schizophrenic cohort data. Among the several laws fitted to this data only the inverse Gaussian and log normal laws stand out as satisfactory. Table J.1 summarizes the several distributions fitted to the observed data. For this cohort data Whitmore and Eaton found a drift value of 0.00463 and the diffusion constant (volatility parameter) of 0.0154. The inverse Gaussian model admits the possibility that a patient may suffer a relapse during his stay in the hospital. This is in contrast to the other models examined. A relapse can be said to take place if the probability of discharge in some fixed period of time gets smaller.

K. Physiology

The tracer method is a familiar technique in physiology for studies of blood circulation and body fluids. Organic dye labels are often used with which plasma is traced. It is customary to inject a radionuclide such as ^{47}Ca or ^{45}Ca into the bloodstream and estimate the specific activity of calcium in plasma from a specific activity time curve. The injected substance goes from plasma to non-plasma several times until it settles in the bone or is eventually excreted. Sheppard and Savage (1951) considered a random walk process as a first approximation to the distribution of circulation times of the outflowing tracer. Sheppard (1968), Sheppard and Uffer (1969) proposed the inverse Gaussian law $IG(1, k^{-2})$ as the distribution of the first exit time from plasma. The physical theory is explained as follows. The label or dye is carried by the fluid with a uniform drift and is simultaneously dispersed by random interaction. At the terminus the dye is swept out of the system as if a barrier with a filter allows the fluid to pass but is absorbing for the dye.

Subsequently Wise et al (1968), Wise (1971a, 1971b, 1974, 1979) gave a detailed account of a stochastic model for turnover of radiocalcium based on observed power laws. They interpreted the hitherto accepted theory in terms of the cycle times from plasma to non-plasma and sojourn times inside and outside plasma of the injected tracer.

When physiological data are analyzed one first obtains tracer dilution curves, which describe the concentration of the tracer measured on a continuous scale. The concentration of tracer in a capillary or artery is plotted using the times from injection of the tracer in a vein. Quite often the curves are bimodal and the second component is considered as arising from the "recirculating" tracer that has passed through the heart more than once. Ignoring this part, what is left accounts for the cardiac output Q— the rate blood is pumped out by the heart. Thus

$$Q = \frac{M}{\int_0^\infty C(t)dt}$$

where M is the mass of tracer injected and $C(t)$ is mass per unit volume of the tracer that has passed only once through the heart. This primary curve $C(t)$ needs to be extrapolated and researchers have constructed an "inflection triangle" (formed by the inflection tangents) and used its parameters to fit a log-normal curve or a gamma curve.

Tracer dilation curves

In contrast to the above analysis, Wise argues that in the concentration curve, the concentration of blood is often proportional to the rate of excretion at that instant and hence the curve gives the ordinates of the probability density of the time spent within the body. In his viewpoint the typical fate of small particles of tracer is to go from plasma to non-plasma several times completing many independent cycles. Each cycle is completed with a probability p and a particle is lost by excretion with probability $1-p = q$. If N cycles are completed then the total time T of a particle from the time of injection to excretion (final departure) is approximately the sum of the time t_a spent in plasma from injection to its first exit from plasma, t_f the time spent in plasma from its last return to plasma to its final exit and the time due to the N transitions from plasma to non-plasma. Thus formally one writes

$$T = (t_a + t_f) + \sum_{i=1}^{N} t_i$$

From a physical point of view Wise claims that the density of $(t_a + t_f)$ is negligible since the cycle time densities tend to predominate. This means that we can regard T as a random sum of independent times, with N, the number of cycles considered as a geometric random variable truncated at zero.

Let $C(t)$ denote the density of T, $\Psi(s) = \mathbb{E}(e^{-Ts})$ and $\zeta(s)$ the Laplace transform of each t_i (which consists of a random sample of size n from the distribution of one complete cycle). Then

$$\psi(s) = \frac{q}{p} \sum_{n=1}^{\infty} p^n \zeta^n(s)$$
$$= \frac{q}{p} \frac{p\zeta(s)}{1 - p\zeta(s)}. \quad (K.1)$$

Wise assumes a gamma law for the total residence time T and for small q finds an approximate solution to equation (K.1). We can state this result as a proposition.

Proposition K.1 Suppose that the Laplace transforms of the total residence time T and the independent identically distributed recirculation times satisfy (K.1) for small $q < 1$. Then if $\psi(s)$ is the Laplace transform of a gamma law, namely $\Gamma(\frac{1}{2}, \beta)$, the Laplace transform of the recirculation time t_i is that of an inverse Gaussian law.

Proof Solving (K.1) for $\zeta(s)$ we have

$$\zeta(s) = \frac{\psi(s)}{q + p\psi(s)}$$
$$= \frac{1}{p + q/\psi(s)} = \frac{1}{1 - q\left\{1 - \frac{1}{\psi(s)}\right\}}.$$

Pharmacokinetics

Now take logarithms and let $\psi(s) = \left(1 - \frac{s}{\beta}\right)^{-\alpha}$. We then obtain

$$\log \zeta(s) = -\log\left[1 - q\left\{1 - \left(1 - \frac{s}{\beta}\right)^{\alpha}\right\}\right].$$

For small q the right side can be expanded to yield approximately,

$$\log \zeta(s) \doteq q\left\{1 - \left(1 - \frac{s}{\beta}\right)^{\alpha}\right\}. \tag{K.2}$$

This says that the cumulant transform of the recirculation time for small q and $\alpha = \frac{1}{2}$ is that of an inverse Gaussian distribution. Letting $q = \frac{\lambda}{\mu} < 1$, $\beta = \frac{\lambda}{2\mu^2}$, $\alpha = \frac{1}{2}$, equation (K.2) gives

$$\log \zeta(s) \doteq \frac{\lambda}{\mu}\left\{1 - \left(1 - \frac{2\mu^2 s}{\lambda}\right)^{\frac{1}{2}}\right\}$$

the cumulant transform of $IG(\mu, \lambda)$.

Wise obtains the same result except that his procedure involves equating the first few cumulants on either side of Equation (K.1). His method ususally involves fitting curves to match these cumulants. He uses numerical tables to generalize the theory to the case $\alpha \neq \frac{1}{2}$. When α is not too close to $\frac{1}{2}$ and if $W = 1 - \alpha$, Wise shows that the single recirculation time distribution can be approximated by

$$p(t) = \frac{\mu^{-\alpha}}{2K_\alpha(\phi)} t^{\alpha-1} \exp\left\{-\frac{\phi}{2}\left(\frac{\mu}{\gamma} + \frac{t}{\mu}\right)\right\}. \tag{K.3}$$

For $\alpha > +\frac{1}{2}$ the implication is that there are many cycle times, while for $\alpha \leq -\frac{1}{2}$ there is just one cycle. Wise asserts that for specific activities in plasma experimental evidence points to this conclusion. This fact has been corroborated by Weiss (1983, 1984).

Pharmacokinetics

Weiss (1982, 1983, 1984) based his analysis on a stochastic pharmacokinetical model which mirrors the topological properties of the circulatory system and reinterprets the findings of Wise and demonstrates that the distribution of circulation times of drug molecules through the body can be approximated by the inverse Gaussian law. We describe very briefly his modelling which relies heavily on the use of linear systems and transport functions.

Suppose that t_i is the time required for a drug molecule to pass through the ith tissue during systemic circulation, $F_i = \int_0^\infty h_i(t)dt \leq 1$

the fraction of drug going through the system and $p_i(t)$, the probability density of t_i is

$$p_i(t) = \frac{h_i(t)}{\int_0^\infty h_i(t)dt}.$$

Denote by μ_i, σ_i^2 the mean and variance of $p_i(t)$. Further let F_{pul} denote the fraction of drug going through the pulmonary circulation, and μ_{pul} the mean of the time distribution corresponding to this circulation. Finally let $q_i = \frac{Q_i}{Q}$ where Q_i is the blood flow volume of the ith tissue and $Q = \sum_{i=1}^N Q_i$ is the cardiac output.

Assuming linear pharmacokinetic systems Weiss shows that the density function of the transit times (assuming exponentially distributed transfer times across tissues) is

$$h(t) = \sum_{i=1}^n \left(\frac{F_{\text{pul}}}{\mu_{\text{pul}}}\right)\left(\frac{q_i F_i}{\mu_i}\right)\left(\frac{1}{\mu_{\text{pul}}} - \frac{1}{\mu_i}\right)\left(e^{-\frac{t}{t_i}} - e^{\frac{-t}{t_{\text{pul}}}}\right).$$

Weiss claims that simulations have shown an excellent agreement between the transit time densities $h(t)$ and

$$p(t) = \frac{\alpha(\alpha t)^{-\frac{3}{2}}}{\beta(2\pi)^{\frac{1}{2}}} \exp\left\{-\frac{(1-\alpha t)^2}{2\alpha\beta^2 t}\right\}. \tag{K.4}$$

If F denotes the fraction of the drug that traverses the circulation in a single pass, then Weiss approximates $h(t)/F$ by $p(t)$. The parameters α and β are given by

$$\alpha = \frac{1}{\mu_{\text{trans}}}, \quad \beta = \frac{\sigma_{\text{trans}}}{\mu_{\text{trans}}}$$

where μ_{trans} and σ_{trans} are the mean and standard deviation of the transit time distribution.

Weiss also discusses the theory leading to the analysis of concentration-time data fitted by power functions of time, namely $C(t) = At^{-a}\exp(-bt)$, which is equivalent to the assumption of gamma distributed residence times of drugs. This, in turn, implies the assumption of a random walk model of circulatory drug transport. In pharmacokinetics the mean of the residence time distribution represents a basic parameter and of particular importance is the square of the coefficient of variation. When $a = \frac{1}{2}$, the coefficient of variation is approximately equal to 2 and the density of circulation times assumes the form (K.4).

The circulatory system can be regarded as an open-loop or single pass system (the time course of drug concentration following a single passage around the circulatory system without recirculation) or the closed-loop system. In the latter system recirculation appears as a recurrent event process consisting of a sequence of independent identical recirculation times $T_1, T_2, \ldots, T_n$. Thus $\{T_n\}$ is a renewal process and if $q = 1-p$ is the probability that a molecule is eliminated and $h_c(t)$ is the density of T_i, then the renewal density is

$$h_r(t) = \frac{q}{p} \sum_{n=1}^{\infty} p^n h_c^{*n}(t), \tag{K.5}$$

$h_c^{*n}(t)$ being the density of $\sum_{i=1}^{n} T_i$. Here Weiss interprets $h_r(t)$ as the residence time distribution of the drug molecule.

Following the arguments of Wise, Weiss concludes that it appears physically reasonable to assume the generalized inverse Gaussian law as a circulation time model.

Homer and Small (1977) had used practically the renewal theory argument to determine the concentration of tracer in the face of renal excretion as

$$C(t) = \frac{M}{F} \sum_{1}^{\infty} p^n f^{*n}$$

M being the mass of tracer injected and F the average flow of blood between the injection site and the sampling site. Furthermore they assumed an $IG\left(\frac{1}{\alpha}, \frac{1}{\alpha\sigma^2}\right)$ law $(\alpha, \sigma > 0)$ for $f(t)$ and tried to estimate F and the mean circulation time $\bar{t}$ (where $lt_{t\to\infty} C(t) = \frac{1}{\bar{t}}$). Thus from the relation

$$C(t) = \frac{M}{F} \sum_{n=1}^{\infty} p^n \frac{(n/\sigma)}{\sqrt{2\pi t^3 \alpha}} \exp\left\{-\frac{(n-\alpha t)^2}{2\alpha t \sigma^2}\right\}$$

the parameters F, p, α and σ are estimated using non-linear regression to obtain the fitted curve $C(t)$.

Interspike train interval analysis

The neuron is a nerve cell and forms a basic unit of the nervous system. Its function is to process and transmit information. The output or phase of spontaneous activity of a neuron consists of a sequence of voltage impulses — a stream of point events — which possesses the characteristics of a stochastic process. Intercommunication among neurons takes places at the synapses between the axonal terminals of neurons and the dendrites of other neurons. The input signals are integrated and transmitted through its axon to the cell bodies of other neurons. The cylindrical membrane of a neuron is electrochemically sensitive and movements of ions across the membrane cause changes of membrane potential. The diffusion creates an electric field which opposes the chemical field and causes a potential difference. When there is no stimulation or input there is equilibrium and the membrane potential maintains a constant value called the resting potential. When the incoming stimuli excite the neuron, the membrane potential shifts towards positivity until this large depolarization causes the potential to cross a certain threshold. At this instant the neuron "fires" or sends an action potential along its axon. The neural signal contains coded information in a sequence of action potentials (a sequence of impulses). These impulses have a wave form and the message is conveyed by a sequence of point events in time. Owing to the spiky appearance of these sequences these waves are known as "spike trains".

Following a spike discharge the neuron undergoes an absolute refractory period and a relative refractory period before the membrane potential returns to its resting potential. When the stimulus, during a relative refractory period, is sufficiently intense another spike discharge follows. The spontaneous activity in neurons can then be likened to a stochastic process since this "firing" activity occurs at random. For an excellent review of the statistical methods used in neuronal spike train analysis the reader is referred to Yang and Chen (1978) and the references therein.

The recording of the potential difference $X(t)$ (measured in mV) of a neuron at every instant of time t over a period $[0, b]$ gives us a data set $\{X(t); t \in [0, b]\}$. The amplitude for the spikes remain nearly constant and the velocity at which it propagates along the fibre is independent of the stimulus so that spikes can be regarded as indistinguishable and instantaneous. A typical spike train data set consists of the times of occurrence of the spikes $(T_1, T_2, \ldots)$ taken over $[0, b]$ instead of the precise volts $X(t)$. Alternatively we can denote the data by $\{N(t); t \in [0, b]\}$ where $N(t)$ is the number of spikes in $(0, t)$.

Consider $\{X(t); t \in [0, \infty)\}$ as a stochastic process which represents the voltage difference between the membrane potential and the resting potential at the trigger zone of the neuron. Let $\theta(t)$ which stands for the threshold potential be a monotonic threshold function defined on $[0, \infty)$ to $[0, \infty)$. For neuronal models one is often interested in the first passage time $T_{\theta(t)}$ of the process $X(t)$ to reach the threshold $\theta(t)$, namely,

$$T_{\theta(t)} = \inf \{t \geq 0 \mid X(t) \geq \theta(t)\}$$
$$= \infty, \quad \text{if } X(t) < \theta(t) \text{ for all } t.$$

Thus $T_{\theta(t)}$ is the time that the membrane potential $X(t)$ takes to reach a critical level $\theta(T)$ in order to generate a spike.

As is customary with neuronal models we assume that spike occurrences are regenerative, which says that $\{N(t)\}$ is a renewal process. Suppose there is a spike at time 0. Then $X(t_0)$ is reset to a value x_0

(say), and when $X(t)$ reaches the threshold level $\theta(t)$ at time T, and generates a spike then $(0, T_1)$ represents one cycle of the membrane potential change. After resetting its potential to x_0, $X(t)$ starts afresh at T_1 and goes on to produce spikes in succession at times $T_2, T_3, \ldots$ Therefore, from the theory of renewal processes, the interspike intervals $T_1, T_2 - T_1, \ldots$ constitute a sequence of independent and identically distributed random variables (an argument valid only for spontaneous spike train activity and not for the simulated variety).

Denote by $F(t)$ the distribution function of the non-negative random variables $U_j = T_j - T_{j-1}$ $(j = 1, 2, \ldots, T_0 = 0)$. Then the mean of $N(t)$ is

$$R(t) = \mathbb{E}[N(t)] \qquad t \geq 0$$

and the properties of the spike train are either describable by means of $R(t)$ or $F(t)$. When $F'(t)$ exists and is equal to $f(t)$, $N(t)$ can also be described by means of the hazard rate

$$h(t) = \frac{f(t)}{1 - F(t)}.$$

An interspike interval (ISI) distribution has to be assumed in order to analyze the spike train data. Several models are employed, as for example, the gamma, the Weibull, the log normal, and the random walk model.

In the random walk model $\{\ldots, -1, 0, 1, \ldots, \theta\}$ is the state space of $X(t)$ where zero corresponds to the resting potential and θ a constant threshold function. $\{X(t) \mid t \geq 0\}$ is considered to be a time-homogeneous Markov Chain with θ as an absorbing barrier.

The diffusion model approach assumes that the membrane potential is a one-dimensional diffusion process, the value of which is reset to the resting potential x_0, at the instant corresponding to the time of the previous spike activity. To be precise let $\{W(t) \mid t \geq 0\}$ be a standard Wiener process and $\mu(\cdot)$ and $\sigma(\cdot)$ two continuous functions (known as the infinitesimal mean and variance). The membrane potential $X(t)$ is defined as the solution of the equation

$$dX(t) = \mu(X(t))dt + \sigma(X(t))dW(t), \quad t \geq 0$$
$$X_0 = x_0 < \theta.$$

The interspike intervals are independent realizations of the random variable T_θ defined by

$$T_\theta = \inf \{t \geq 0 \mid X(t) \geq \theta\}.$$

The theory of stochastic differential equations enables us to obtain (under assumptions on $\mu(\cdot)$ and $\sigma(\cdot)$) the transition probability density function $f(x,t \mid x_0, s)$ of $X(t)$ as the solution to the Fokker-Planck equation

$$\frac{\partial f}{\partial t} = -\frac{\partial}{\partial x}[\mu(x)f] + \frac{1}{2}\frac{\partial^2}{\partial x^2}[\sigma^2(x)f] \qquad (K.6)$$

with

$$p(x,s \mid x_0, s) = \delta(x - x_0), \quad p(\pm\infty, t \mid x_0, s) = 0.$$

The distributions of T_θ are known for some special cases of μ and σ.

When μ and σ are assumed positive constants we are led to the Wiener process model (in this case the neuronal model is called the perfect integrator) and the solution of (K.6) with $\mu(x(t)) = \mu$ and $\sigma(X(t)) = \sigma (\mu \geq 0, \sigma > 0)$ is

$$f(x,t \mid x_0, 0) = (2\pi\sigma^2 t)^{-\frac{1}{2}} \exp\left\{-\frac{(x - x_0 - \mu t)^2}{2\sigma^2 t}\right\}$$

and the first passage time distribution of T_θ is

$$p(t \mid \theta, x_0) = \frac{\theta - x_0}{\sigma\sqrt{2\pi t^3}} \exp\left\{-\frac{(\theta - x_0 - \mu t)^2}{2\sigma^2 t}\right\} 1_{0,\infty}(t)$$

for $\theta > x_0$, which when reparametrized by $\beta = \frac{\theta - x_0}{\sigma}$, $\nu = \frac{\mu}{\sigma}$ gives the inverse Gaussian law $IG(\frac{\beta}{\nu}, \beta^2)$. When $m = \frac{\theta - x_0}{\mu}, \lambda = \frac{(\theta - x_0)^2}{\sigma^2}$, we obtain the $IG(m, \lambda)$ law.

For details about the assumption of the Ornstein-Uhlenbeck process model the reader can consult Yang and Chen (1978) or Lánský and Smith (1989) and the references therein. A good mathematical treatment of the spike train activity and the underlying stochastic process can be found in Tuckwell (1988, 1990).

Tuckwell identifies twelve general shapes that can be fitted with ISI histograms and the basic mechanism that leads to these shapes. Even when a particular shape has been chosen there still remains the issue of parameter estimation to obtain the best fit. Levine (1991) used simulated data using three noise distributions and fitted both the simulated data and real data to four models including the log normal and the inverse Gaussian and found that they provided better fit than the hyperbolic normal and the reciprocal gamma. He found that the ISI distribution reveals the neuron's processing much more than it explains the nature of the noise.

Liao (1995) has examined both simulated data and real data in his thesis on model selection and proposes the GIG law as a larger family to model interspike interval data.

The four distributions considered by Levine and Liao are the hyperbolic normal, the reciprocal gamma, the lognormal and the inverse Gaussian. The left end point of the interspike interval distribution is considered at zero (thus the neuron's refractory period is excluded in the modelling).

1. Hyperbolic normal. Suppose that $\mathcal{L}(X) = N(\mu, \sigma^2)$ then $\mathcal{L}(X^{-1})$ given that $X^{-1} > 0$ is defined as the hyperbolic normal law. The density is given by (writing $Y = X^{-1}$)

$$f(y \mid \mu, \sigma) = \frac{\phi\left(\frac{\frac{1}{y}-\mu}{\sigma}\right)}{\Phi(\frac{\mu}{\sigma})\sigma y^2} 1_{\mathbb{R}^+}(y).$$

2. Reciprocal gamma. If $\mathcal{L}(X) = \Gamma(\alpha, 1/\beta)$ then $\mathcal{L}(X^{-1})$ is defined to be the reciprocal gamma law.
3. Lognormal. Let $\mathcal{L}(X) = N(\mu, \sigma^2)$. Then $\mathcal{L}(\exp X)$ has a lognormal law.
4. Finally we consider the $IG\left(\frac{\beta}{\nu}, \beta^2\right)$ for the inverse Gaussian model.

In his work Liao used ten data sets from the retinal ganglia of goldfish collected by Levine. The data sets include the record length, record number, run time, stimulus positions, changes of shutter state, spike numbers in different channels and spikes. The experiment times were over a period of 30 seconds and the firing rates ranged from 29 to 42 per second. Liao reports that the data sets have similar histogram type with peaks near b with a substantial number of ISI intervals with length exceeding 100. The means are different from the medians and the coefficient of variation for the actual data are larger (≈ 1.1) than that for the 3 sets of simulated data. We reproduce histograms of three data sets from the real data and artificial data in Figure K.1 - Figure K.6 The histogram is fit with the IG and log normal in Figure K.1a - Figure K.6a while the fit is made with the reciprocal gamma and the hyperbolic normal in Figure K.1b – Figure K.6b.

Levine had used the mean squared difference between the interval histogram and the theoretical curve for $0 < t \leq 200$ ms as a measure of the goodness of fit and Liao used the same measure in his work. From Table K.1 based on the calculated mean-squared errors it appears that the winner is the reciprocal gamma. Table K.2 contain the MLE of the data sets f82dat, f83dat, f87dat, f90dat and f98dat together with relevant statistics.

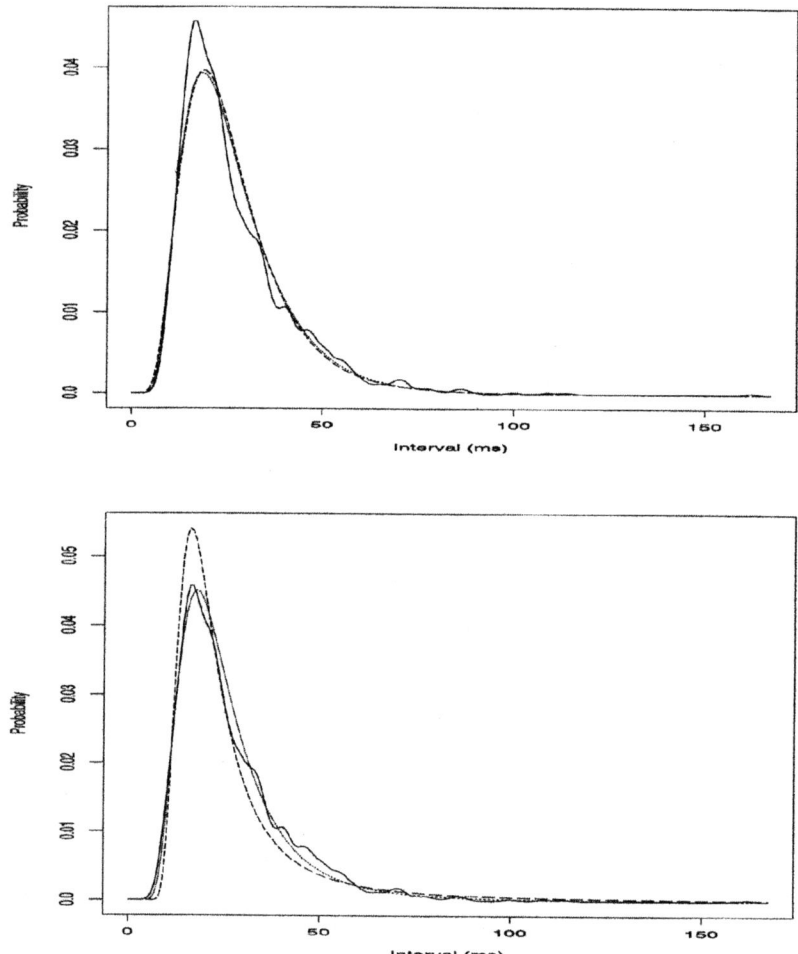

Figure K.1 *Interval histogram (solid curve) from simulated data 1 with mean firing rate=26.738 and CV=0.55. a (top): fitting with inverse Gaussian distribution (dotted curve) and lognormal distribution (dashed curve). b(bottom): fitting with reciprocal gamma distribution (dotted curve) and hyperbolic normal distribution(dashed curve)*

Interspike train interval analysis 245

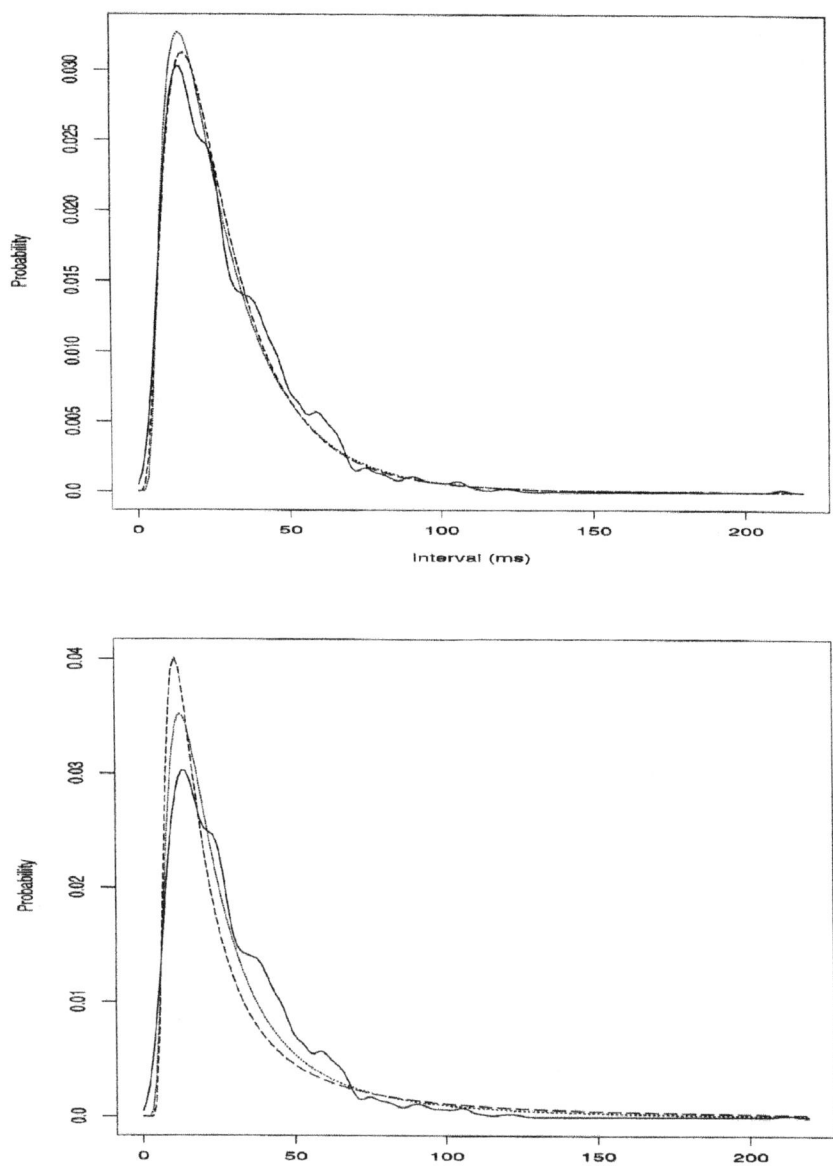

Figure K.2 *Interval histogram solid curve from simulated data 2 with mean firing rate=29.644 and CV=0.696. a (top): fitting with inverse Gaussian distribution (dotted curve) and lognormal distribution (dashed curve). b (bottom): fitting with reciprocal gamma distribution (dotted curve) and hyperbolic normal distribution (dashed curve).*

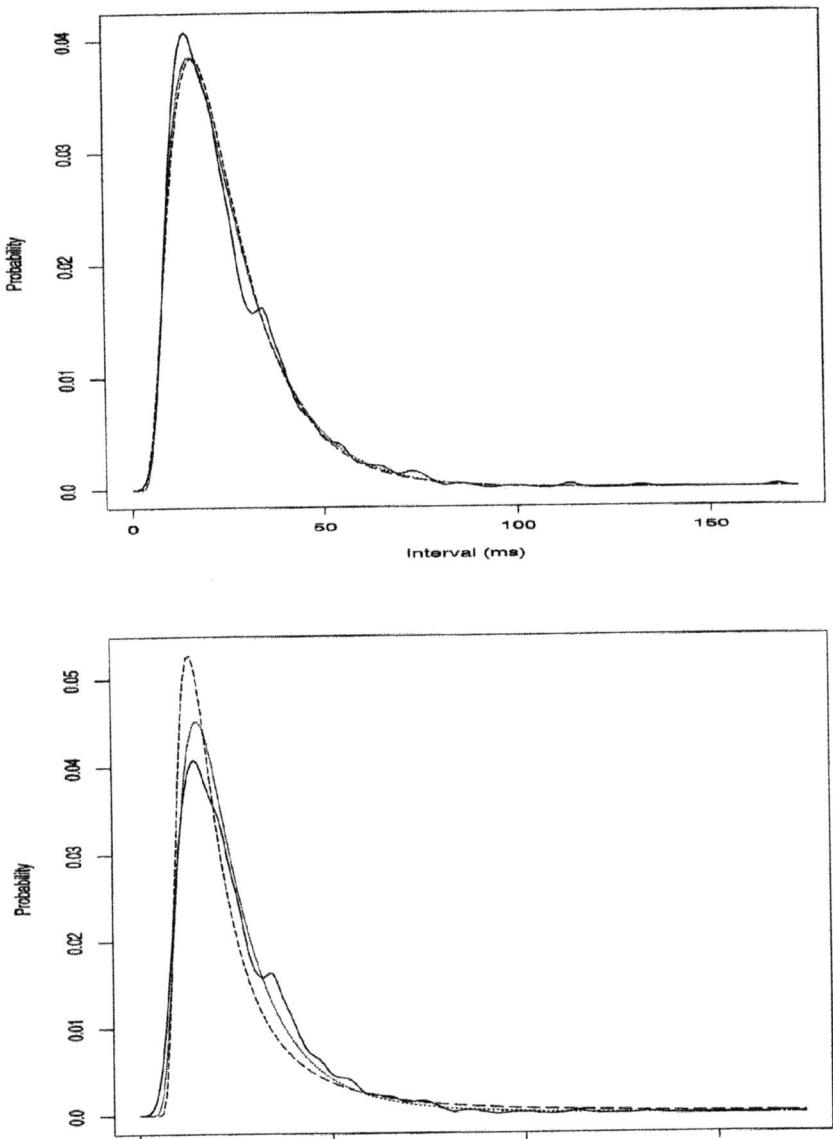

Figure K.3 *Interval histogram (solid curve) from simulated data 3 with mean firing rate=25.884 and CV=0.614. a (top): fitting with inverse Gaussian distribution (dotted curve) and lognormal distribution (dashed curve). b (bottom): fitting with reciprocal gamma distribution (dotted curve) and hyperbolic normal distribution (dashed curve)*

Interspike train interval analysis 247

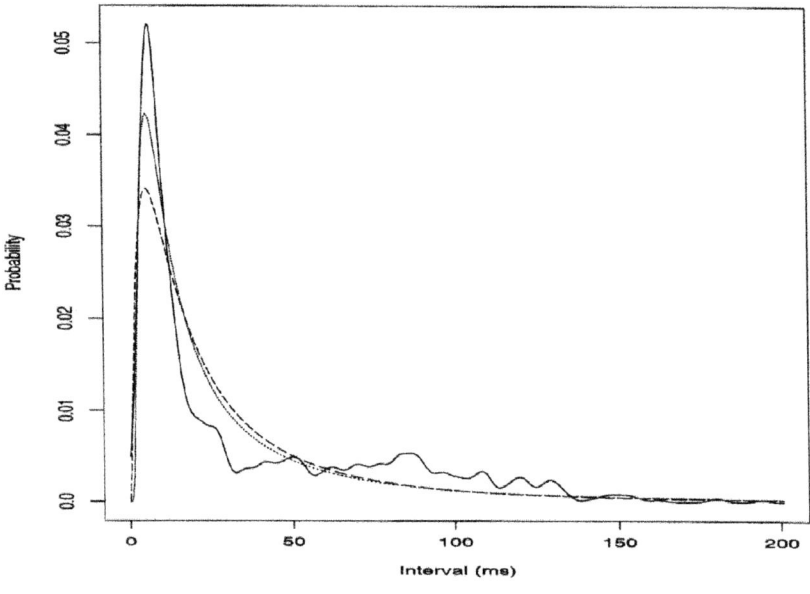

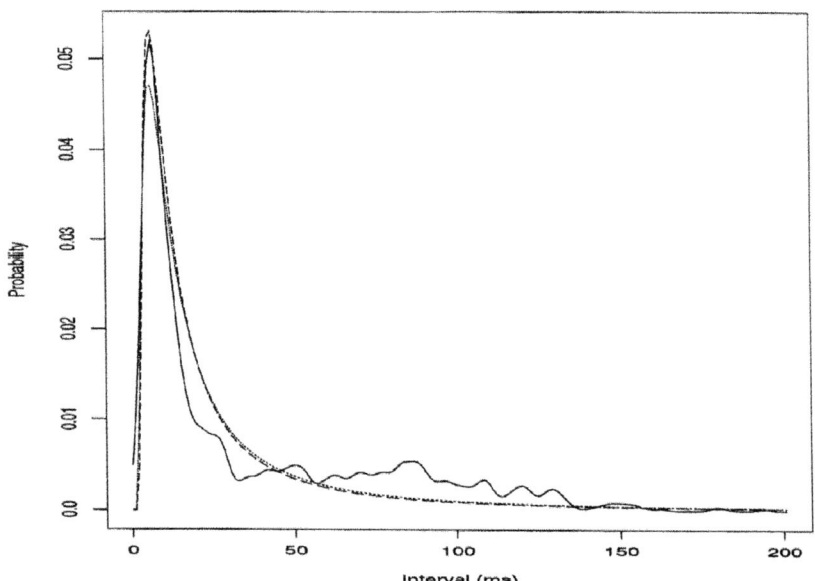

Figure K.4 *Interspike interval histogram (solid curve) from goldfish data set f68.dat with mean firing rate=36.855 and CV=1.074. a (top): fitting with inverse Gaussian distribution (dotted curve) and lognormal distribution (dashed curve). b (bottom): fitting with reciprocal gamma distribution (dotted curve) and hyperbolic normal distribution (dashed curve)*

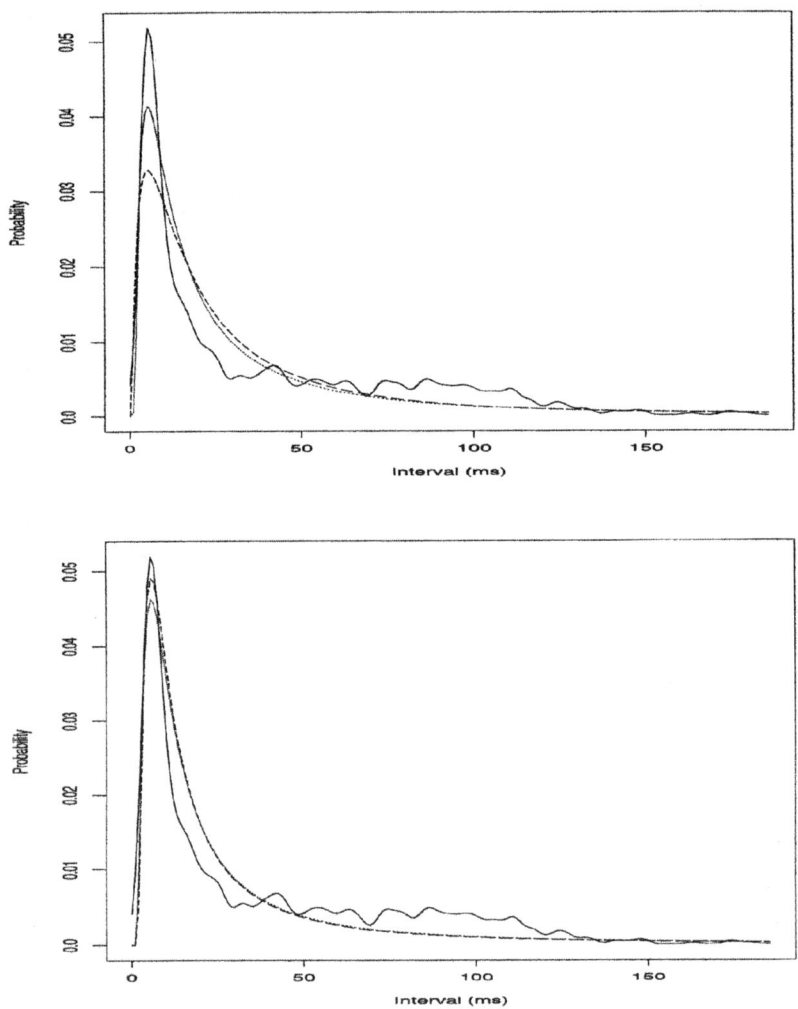

Figure K.5 *Interspike interval histogram (solid curve) from goldfish data set f72.dat with mean firing rate=36.991 and CV=1.020. a (top): fitting with inverse Gaussian distribution (dotted curve) and lognormal distribution (dashed curve) and lognormal distribution (dashed curve). b (bottom): fitting with reciprocal gamma distribution (dotted curve) and hyperbolic normal distribution (dashed curve)*

Interspike train interval analysis 249

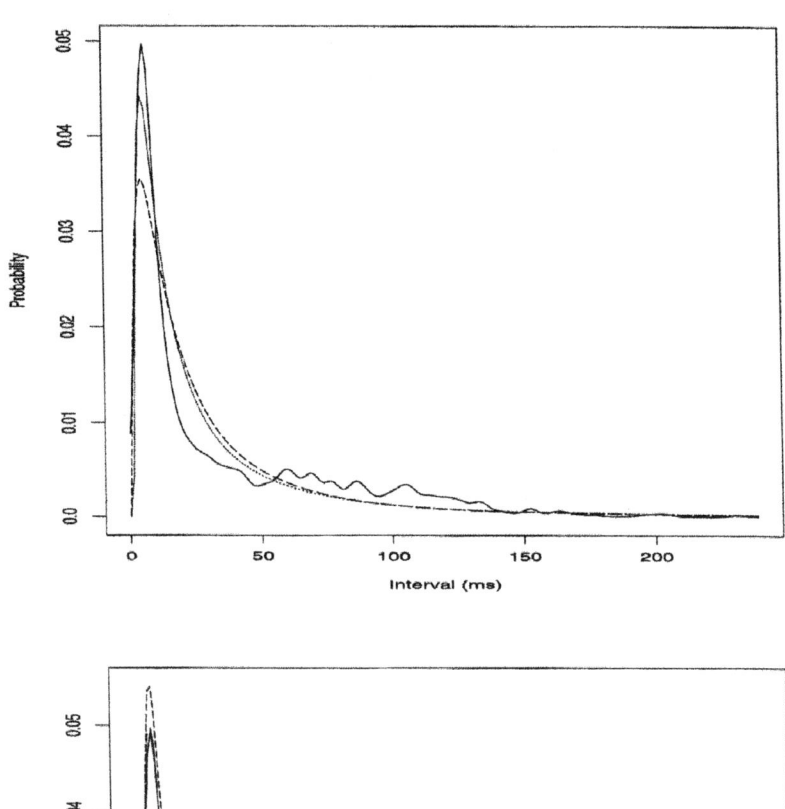

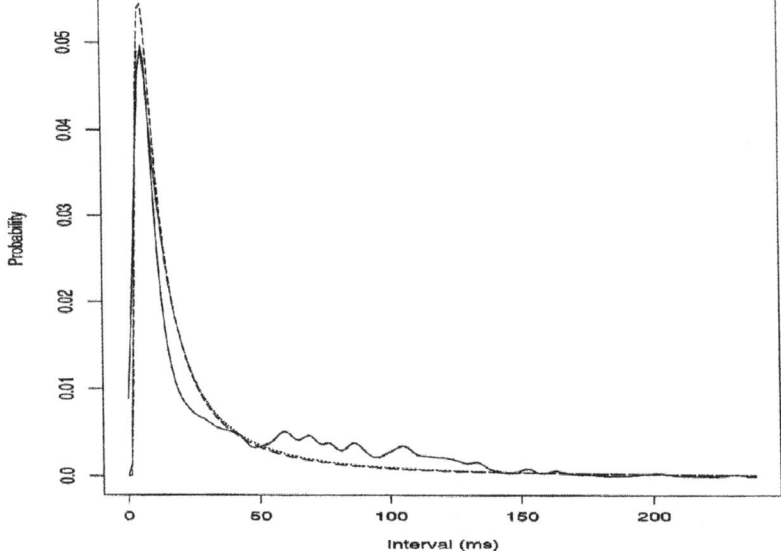

Figure K.6 *Interspike interval histogram (solid curve) from goldfish data set f75.dat with mean firing rate=36.452 and CV=1.119. a (top): fitting with inverse Gaussian distribution (dotted curve) and lognormal distribution (dashed curve). b (bottom): fitting with reciprocal gamma distribution (dotted curve) and hyperbolic normal distribution (dashed curve)*

Table K.1 *Mean squared errors (computed using mle)*

Noise distribution or data	Statistics of Fits Mean Squared Error ×10⁶			
	Hyperbolic normal	Reciprocal gamma	Log-normal	Inverse Gaussian
Normal	5.418	1.205	3.126	2.621
Gamma	7.91	2.649	0.814	1.013
Uniform	8.527	1.991	1.282	0.775
f68.dat	7.843	6.686	12.970	7.881
f72.dat	9.482	8.441	12.982	8.524
f75.dat	7.646	5.823	8.238	5.580
f78.dat	10.082	7.781	19.043	11.543
f80.dat	11.053	10.329	24.642	15.366
f82.dat	9.366	6.024	13.060	7.911
f83.dat	10.942	7.199	12.664	8.136
f87.dat	9.854	9.214	20.944	12.257
f90.dat	12.757	10.505	20.594	12.731
f98.dat	9.275	8.700	11.972	8.549

The inverse Gaussian is slightly better than the lognormal for data 3 and slightly inferior for data 2.

For the real data shown in Figures K.4 - K.6 the IG gives the best fit for data sets f72 and f75 as does the reciprocal gamma, while the reciprocal gamma is again the superior choice for f68. Over all the IG fits are better than the log normal fits and the reciprocal gamma fits are superior to the hyperbolic normal fits. Liao concludes that "while there is no overriding reason to choose models among the reciprocal gamma, the hyperbolic normal, the log normal and the inverse Gaussian for fitting the interspike intervals, it is clear that the hyperbolic normal should be eliminated because we can find a model to provide better fit for data under any circumstance". Since the GIG model includes the IG, the gamma and the reciprocal gamma as special cases, and represents the distribution of the first hitting time to level zero of a variety of time-homogeneous diffusions on $[0, \infty)$ (Barndorff-Nielsen et al. 1978), Liao suggests the use of the GIG as a model to fit the ISI data. Moreover, since the log normal and the IG, as also the lognormal and the GIG are separate families, Liao uses Cox's approach for testing hypotheses of the separate families (see Section 3.8). His thesis contains an integration of existing methods together with a new "limited" test statistic for testing

Interspike train interval analysis

separate hypotheses.

Table K.2 *The basic statistics of the data f82.dat, f83.dat, f87.dat, f90.dat, and f98.dat, and the MLEs of the parameters of the four models: inverse Gaussian, lognormal, reciprocal gamma, and hyperbolic normal.*

Models	Estimates of parameters	Data				
		f82.dat	f83.dat	f87.dat	f90.dat	f98.dat
Bs	Mean	29.42	29.13	29.41	30.04	42.65
	Median	12.00	12.00	13.00	13.00	20.00
	Mode	5.90	5.10	5.54	5.25	7.03
	SD	33.09	33.52	31.30	32.86	44.04
IG	$\hat{\beta}$	3.76	3.62	3.81	3.68	4.28
	$\hat{\nu}$	0.128	0.124	0.129	0.123	0.100
Ln	$\hat{\mu}$	2.76	2.73	2.79	2.77	3.13
	$\hat{\sigma}$	1.12	1.15	1.12	1.15	1.18
Rg	$\hat{\alpha}$	0	0	0	0	0
	$\hat{\beta}$	0.105	0.111	0.103	0.107	0.078
Hn	$\hat{\mu}$	-0.206	-0.273	-0.561	-0.321	-1.063
	$\hat{\sigma}$	0.201	0.227	0.278	0.234	0.308

Note that the values of $\hat{\alpha}$ of reciprocal gamma distribution were taken as the closest integers of the maximum likelihood estimates in both Tables K.1 and K.2.

L. Remote sensing

Photogrammetry

Detection of objects from aerial surveys often depends on the size and the detection probability is generally a function of film, season, terrain location in addition to size. In agricultural crop surveys where the size of the fields varies among farms, in forestry surveys designed to detect human destruction and insect infestation or in wildlife management surveys designed to estimate the number of ponds and lakes available for breeding waterfowl, the size plays a key note. The smallest items which cannot be captured on camera escape detection. Very large members of the population also may not be recognized if they exceed the image frame. Furthermore, even if they are detected it is necessary to capture the images of adjoining quadrats to obtain a good measure.

Of the different analytical approaches to size-dependent detection the parametric size-dependent method has been investigated in great detail both for its novelty and power and versatality by Maxim and Harrington (1982). Chief among the advantages of this method over the "scale up" technique or the discrete size-dependent detection is the ability to extract the maximum information from limited amounts ofO "ground truth" data - a jargon which applies to ground data used in remote sensing to imply that the data are error-free, the authors claim that this approach is robust to shifts in the parameters of the population, changes that could confound or nullify the calibration between imagery interpretation and ground truth.

In this approach it is assumed that the size X, of the objects within the region of interest (say fields) is a random variable with density $f(x \mid \theta)$, θ being a parameter. The region of interest could be stratified if necessary. In addition to the parent distribution $f(x \mid \theta)$, one assumes a detection function $D(x \mid \psi)$ involving a parameter ψ, which specifies the dependence of detection probability upon the field size X. $D(x \mid \psi)$ is assumed to be a monotone non decreasing function of X in $[0, 1]$. The larger the field size the greater the probability of detection. (To incorporate nonsize-dependent aspects of detection D can be multiplied by $\alpha \in (0, 1]$.) Thus one can see that $D(x \mid \psi)$ is a distribution function.

The so-called "cookie cutter" detection function defined as

$$D(x \mid c) = \begin{cases} 0 & x \leq c \\ 1 & x > c \end{cases}$$

assumes that all fields below the threshold value c will go undetected while those above are certain to be detected.

Based on this statistical model we can define the sample that arises in the imagery or aerial survey has a density

$$g(x \mid \theta, \psi) = \frac{f(x \mid \theta) \, D(x \mid \psi)}{\int_0^\infty f(x \mid \theta) \, D(x \mid \psi) \, dx}.$$

The denominator on the right side expresses the overall fraction of fields that are detected. Thus $g(x \mid \theta, \psi)$ represents the probability density of the size of the detected fields.

Maxim and Harrington recommend the use of the inverse Gaussian law $IG(\mu, \lambda)$ as an appropriate candidate for the field size distribution (parent distribution) $f(x|\theta)$. Consider then

$$f(x \mid \mu, \phi) = \sqrt{\frac{\mu\phi}{2\pi x^3}} \, \exp(\phi) \, \exp\left\{-\frac{\phi\mu}{2x} - \frac{\phi}{2\mu}x\right\} 1_{\mathbf{R}^+}(x)$$

and an analytically compatible detection function known as the extreme valve distribution, namely,

$$D(x \mid \psi) = \exp\left(-\frac{\psi}{x}\right).$$

The compatibility arises from the fact that $g(x|\mu, \phi, \psi)$ is also an inverse Gaussian law! To see this we calculate

$$\int_0^\infty f(x \mid \mu, \phi) D(x \mid \psi) dx = \frac{e^\phi \sqrt{\mu\phi}}{\phi\mu + 2\psi} \exp\left\{-\sqrt{\frac{\phi(\phi\mu + 2\psi)}{\mu}}\right\}$$

so that writing $\phi^* = \sqrt{\frac{\phi(\phi\mu + 2\psi)}{\mu}}$, $\mu^* = \sqrt{\frac{\mu(\varphi\mu+2\psi)}{\phi}}$,

$$g(x \mid \mu, \phi, \psi) = \sqrt{\frac{\mu^*\phi^*}{2\pi x^3}} \exp(\phi^*) \exp\left\{-\frac{\phi^*\mu^*}{2x} - \frac{\phi^*}{2\mu^*}x\right\} 1_{\mathbf{R}^+}(x).$$

Thus the density function of the detected field sizes is $IG(\mu^*, \lambda^*)$ where $\lambda^* = \phi^*\mu^*$, and clearly $g(x \mid \mu, \phi, \psi) = f(x \mid \mu^*, \phi^*)$.

It is interesting to note that $\frac{\phi^*}{\mu^*} = \frac{\phi}{\mu}$, and for larger field sizes, since the density function is dominated by the term $\exp\left(-\frac{\phi x}{2\mu}\right)$ so is the density of the aerial survey data.

The expected size of detected fields is μ^*, while the fraction of fields detected is

$$\sqrt{\frac{\mu\phi}{\mu\phi + 2\psi}} \, \exp\left\{\phi - \sqrt{\frac{\phi(\mu\phi + 2\psi)}{\mu}}\right\}.$$

Finally the expected fraction of the area detected is

$$\exp\left\{\phi - \sqrt{\frac{\phi(\mu\phi + 2\psi)}{\mu}}\right\}.$$

It is impossible to obtain estimates of all the three parameters without additional "ground truth" data. Indeed using the maximum likelihood principle we have from the aerial survey data

$$\mu^* = \bar{x}, \quad \lambda^{*-1} = (\phi^*\mu^*)^{-1} = \frac{1}{n}\sum_{i=1}^{n}\left(\frac{1}{X_i} - \frac{1}{\bar{X}}\right).$$

Hence we have

$$\bar{x}^2 = \frac{\mu(\mu\phi + 2\psi)}{\phi}, \quad (\mu\phi + 2\psi)^{-1} = \frac{1}{n}\sum_{i=1}^{n}\left(\frac{1}{X_i} - \frac{1}{\bar{X}}\right)$$

and

$$\frac{\mu}{\phi} = \frac{\bar{X}^2}{n}\sum_{i=1}^{n}\left(\frac{1}{X_i} - \frac{1}{\bar{X}}\right).$$

Thus $\frac{\mu}{\phi} = \frac{\mu^*}{\phi^*}$ can be estimated.

Observed detection results can be used to obtain estimates of ψ. The data consists of (X_i, U_i) from a matched ground truth and imagery experiment. (Imagery from an area of known ground truth is interpreted and the correspondence between detection and size is noted.) Here X_i is the field size of the ith field and $U_i = 1$ if the ith field is detected and $U_i = 0$ otherwise.

For a fixed ψ

$$P(U_i = 1 \mid X_i) = e^{-\frac{\psi}{X_i}}$$
$$P(U_i = 0 \mid X_i) = 1 - e^{-\frac{\psi}{X_i}}.$$

Thus if p_i denotes the probability of U_i given X_i

$$p_i = u_i e^{-\frac{\psi}{x_i}} + (1-u_i)(1 - e^{-\frac{\psi}{x_i}}).$$

The log likelihood of the outcomes $U = u_i$ $(i = 1, \ldots, n)$ is

$$\ell = \log\left(\prod_{i=1}^{n} p_i\right) = \sum_{i=1}^{n}\log(p_i)$$

and
$$\frac{\partial \ell}{\partial \psi} = -\sum_{i=1}^{n} \frac{u_i}{x_i} + \sum_{i=1}^{n} \frac{(1-u_i)e^{-\frac{\psi}{x_i}}}{x_i(1-e^{-\frac{\psi}{x_i}})}.$$

Noting that $\dfrac{\partial \ell}{\partial \psi}$ is monotone decreasing in ψ it is possible to use numerical methods to find $\hat{\psi}$ such that $\dfrac{\partial \ell}{\partial \psi}\bigg|_{\psi=\hat{\psi}} = 0$. Maxim and Harrington use a binary search technique called Bolzano's method in determing $\hat{\psi}$ based on the results of an imagery and ground truth experiment given below.

Field Size x_i	Detection outcome u_i
10	1
2	0
10	0
2	0
8	1
14	1
1	0
7	1
8	0

The maximum likelihood estimate $\hat{\psi} = 4.58$. Using this estimate $\hat{\mu}$ and $\hat{\lambda}$ can then be determined completely.

The example analyzed by Maxim and Harrington concerns the data from a study by Podwysocki (1976). In this data set the distribution of field sizes for various major grain producing countries were empirically estimated from fields detected on Landsat imagery (unmanned polar orbiting earth resources satellite). The use of the IG law seems very appropriate since all the distributions are highly peaked and skewed toward small sizes. A total of 147 grain fields detected on Landsat imagery in one sample region in Kansas yielded $\mu^* = 13.795$ and $\sigma^* = \sqrt{\dfrac{\mu^{*3}}{\lambda^*}} = \dfrac{\mu^*}{\sqrt{\phi^*}} = 10.77$. This gives $\phi^* = 1.64$, and a mode equal to 6.078.

Since Landsat's capability to detect small fields is limited, the authors resort to the use of ground truth. To this end they use a ground truth survey of sizes of agricultural fields of 10 midwestern states in the U.S as an approximation. Using the modal value of 2.5 as ground truth, they proceed to estimate all the three parameters of the model.

Use of the extreme value detection function has therefore increased the modal value from 2.5 to 6.078. Therefore this information is used in calculating μ and ϕ as follows

$$\mu\left\{\sqrt{1+\frac{9}{4\phi^2}}-\frac{3}{2\phi}\right\}=2.5.$$

Since $\frac{\mu}{\phi}=\frac{\mu^*}{\phi^*}=\frac{13.795}{1.64}=8.41$, $\mu=8.41\phi$. Hence $\phi\sqrt{1+\frac{9}{4\phi^2}}=\frac{9.5}{8.41}+1.5=1.8$ and $\phi^2=0.99$ or $\phi=0.99$. Thus $\mu=8.32$ and ψ is obtained from the relation $\mu^*=\sqrt{\frac{\mu}{\phi}(2\psi+\mu\phi)}$ as $\psi=7.19$.

The values of μ, ϕ will then completely determine the parent distribution $f(x\mid\mu,\phi)$ and the updated estimate of $\psi=7.19$ gives the best fit detection function $D(x\mid\psi)$. By constructing a graph one can verify the adequacy of the fit to the observed data and the area detection probabilities for different sizes of fields.

It is also possible to ascertain an optimal ground truth sample by studying the asymptotic variance of the maximum likelihood estimate of ψ which was obtained by numerical methods. Indeed we have

$$\text{Var}(\hat{\psi})=\left[-\mathbb{E}\left(\frac{\partial^2\ell}{\partial\psi^2}\right)\right]^{-1}.$$

By differentiating $\frac{\partial\ell}{\partial\psi}$ and simplifying we can show that

$$\text{Var}(\hat{\psi})=\left[\sum_{i=1}^{n}\frac{e^{-\frac{\psi}{x_i}}}{x_i^2(1-e^{-\frac{\psi}{x_i}})}\right]^{-1}.$$

In order to find out the field sizes X_i which minimize $\text{Var}(\hat{\psi})$, that is maximize $(\text{Var }\hat{\psi})^{-1}$ (since the denominator in the right side >0) we differentiate $(\text{Var }\hat{\psi})^{-1}$ with respect to x_i and solve for x_i. Thus

$$\frac{\partial(\text{Var }\hat{\psi})^{-1}}{\partial x_i}=\frac{e^{-\frac{\psi}{x_i}}}{x_i^4(1-e^{-\frac{\psi}{x_i}})^2}\{\psi-\psi e^{-\frac{\psi}{x_i}}-2x_i(1-e^{-\frac{\psi}{x_i}})+\psi e^{-\frac{\psi}{x_i}}\}=0$$

gives the equation

$$e^{-\frac{\psi}{x_i}}=1-\frac{\psi}{2x_i}.$$

The finite solution for x_i is given by

$$x_i=\frac{\psi}{1.594}.$$

Cookie cutter detection

Thus the optimal ground truth sample would be to select fields of size $\frac{\psi}{1.594}$ with a detection probability $D(x \mid \psi) = e^{-1.594} \doteq 0.203$

Cookie cutter detection

Suppose that the parent distribution is inverse Gaussian and the detection function $D(x \mid c)$ is the "cookie cutter". We then have

$$g(x \mid \mu, \phi, c) = \frac{f(x \mid \mu, \phi)}{\int_c^\infty f(x \mid \mu, \phi) dx} 1_{[c,\infty)}(x),$$

$$\mu^* = \int_c^\infty x \, g(x \mid \mu, \phi, c) \, dx,$$

$$\int_c^\infty f(x \mid \mu, \phi) \, dx = 1 - \Phi\left(\sqrt{\frac{\mu\phi}{c}} \left(\frac{c}{\mu} - 1\right)\right)$$
$$- e^{2\phi} \, \Phi\left(-\sqrt{\frac{\mu\phi}{c}} \left(\frac{c}{\mu} - 1\right)\right)$$

and the fraction of area detected is

$$\frac{\mu^*}{\mu} \int_c^\infty f(x \mid \mu, \phi) \, dx.$$

The maximum likelihood estimates of the parameters can be obtained using the approach of Gupta (1969) as solutions to the equations

$$\hat{\mu}^* = m'_1 + \bar{x}(1 - F(c))$$

$$\hat{\lambda}^* = \left[m'_{-1} + \left(\frac{1 - F(c)}{n}\right) \sum_{i=1}^n \frac{1}{x_i} - \frac{1}{\hat{\mu}^*}\right]^{-1}$$

where m'_r are the sample moments.

Ferguson et al. (1986) have adopted the approach of Maxim and Harrington to study the field size distributions for selected crops in the United States and Canada based on an enlarged field size data base. The data deals with field sizes measured in length width and area for ten agricultural crops across several states. Detailed tests of fit are presented to check the adequacy of the IG model for crop field size in terms of width, length and area.

An interesting aspect of the above analysis shows that both the inverse Gaussian and the reciprocal inverse Gaussian laws arise in a natural way in the area of remote sensing. To see this consider the following.

Let the field size distribution $f(x_i \mid \chi)$ be a χ^2 law with one degree of freedom so that

$$f(x_i \mid \chi) = \sqrt{\frac{\chi}{2\pi x_i}} \exp\left(-\frac{\chi}{2} x_i\right) 1_{\mathbf{R}^+}(x_i).$$

Further let the detection function $D(x_i \mid \psi)$ be defined by

$$D(x_i \mid \psi) = \exp\left(-\frac{\psi}{2x_i}\right).$$

Then

$$g(x_i \mid \chi, \psi) = \frac{f(x_i \mid \chi) D(x_i \mid \psi)}{\int_0^\infty f(x_i \mid \chi) D(x_i \mid \psi) dx_i} 1_{\mathbf{R}^+}(x_i)$$

$$= \sqrt{\frac{\chi}{2\pi x_i}} e^{\sqrt{\chi\psi}} \exp\left\{-\frac{\psi}{2x_i} - \frac{\chi x_i}{2}\right\} 1_{\mathbf{R}^+}(x_i),$$

the reciprocal inverse Gaussian law. In general the GIG (λ, χ_1, ψ) law with $D(x \mid \chi_2)$ gives us a GIG $(\lambda, \chi_1+\chi_2, \psi)$ law as the resultant density function.

M. Traffic noise intensity

In the field of transportation research it is of interest to calculate the noise-producing properties of a stream of equidistant acoustic sources moving at constant speed. Weiss (1970) developed a probabilistic model of traffic noise intensity from vehicles on a long straight highway. We discuss his derivation of the model and related articles on the subject relevant to the IG law by Kurze (1971a, 1971b, 1974) Marcus (1973, 1974, 1975), Takagi et al., (1974).

Model assumptions
(a) Traffic move in a single stream along an infinitely long straight highway.
(b) The noise-measuring apparatus is assumed to be placed at a distance d from the highway at its point of closest approach.
(c) Each car on the highway is modelled as a point source of sound.
(d) Let X denote the position of the vehicle and let $X = 0$ be the point of closest approach to the noise-measuring device.
(e) Assume that each car generates the same amount of noise and the acoustic power of a single car is a constant q.
(f) Assume vehicles travel independently of each other along the highway.
(g) Denote by $I(x)$ the intensity of noise at the measurement point generated by a car at coordinate x, where

$$I(x) = \frac{q}{x^2 + d^2}.$$

(h) Let the reference noise level be the noise due to a car ar $x = 0$ and denote the normalized intensity $I(x)/I(0) = J(x)$, so that

$$J(x) = \frac{1}{\left(1 + \left(\frac{x}{d}\right)^2\right)}.$$

(i) The respective positions X on the highway are realisations of a stationary Poisson process so that the probability density for gaps between successive cars is $\rho \exp(-\rho X)$ where ρ is the vehicular concentration.

Then if x_n represents the location of the nth car in the stream the total noise intensity generated by an infinite stream of identical sound generators is

$$J = \sum_{n=-\infty}^{\infty} J(x_n).$$

The Laplace transform of J is

$$L(\theta) = \int_{-\infty}^{\infty} e^{-\theta J} f(J) dJ = e^{k(\theta)} \quad \text{(say)},$$

where $f(J)$ is the probability density of the normalized noise intensity. It turns out that J can be represented as a shot noise process (Cox and Miller, 1965) and hence

$$k(\theta) = -\rho d \int_{-\infty}^{\infty} \left\{ 1 - \exp\left(-\frac{\theta}{1+u^2}\right) \right\} du$$

(using the change of variable $x = ud$). Weiss showed that

$$k(\theta) = -\pi \rho d e^{-\frac{\theta}{2}} \left\{ I_0\left(\frac{\theta}{2}\right) + I_1\left(\frac{\theta}{2}\right) \right\}$$

where

$$I_0(x) = \frac{1}{\pi} \int_0^{\pi} \exp(-x \cos \theta) d\theta$$

is the integral representation for the Bessel function of imaginary argument. ($I_n(x)$ is a modified Bessel function of the first kind of order n with argument x). For small J using a Tauberian argument to $k(\theta)$ an approximation to $f(J)$ can be obtained. For large $|\theta|$

$$k(\theta) \sim -2\rho d \sqrt{\pi \theta} + O(\theta^{-\frac{1}{2}})$$

and therefore

$$L(\theta) \sim \exp\left\{-2\rho d \sqrt{\pi \theta}\right\}$$

so that for small J

$$f(J) \sim \frac{\rho d}{\sqrt{J^3}} \exp\left(-\frac{\pi \rho^2 d^2}{J}\right).$$

A change of variable $\lambda = 2\pi \rho^2 d^2$ then gives

$$f(y) \sim \frac{\sqrt{\lambda}}{\sqrt{2\pi y^3}} \exp\left(-\frac{\lambda}{2y}\right) 1_{\mathbb{R}^+}(y).$$

Marcus, on the other hand, simplified the problem by assuming that the acoustic power of a single car Q is itself a random variable with a heavy tailed distribution $h(Q)$. In this case one has to calculate

$$\mathbb{E}\left\{\mathbb{E}(k(\theta \mid Q))\right\} = k^*(\theta)$$

where
$$k(\theta \mid Q) = -\pi\rho d\, e^{-\frac{\theta Q}{2d^2}} \left\{ I_0\left(\frac{\theta Q}{2d^2}\right) + I_1\left(\frac{\theta Q}{2d^2}\right) \right\}$$

(i.e, we deal with $I(x)$ instead of $J(x)$).

Under these circumstances by taking the limit as $d \to 0$ (the observer is on the highway) Marcus obtains for $\mathrm{Re}(\theta) \geq 0$

$$k(\theta) = -2\sqrt{\pi}\rho\, \sqrt{\theta}\, \mathbb{E}(\sqrt{Q}).$$

We then obtain the stable law for the noise intensity as

$$f(I) = \frac{\rho \mathbb{E}(\sqrt{Q})}{\sqrt{I^3}} \exp\left(-\frac{\rho\pi(\mathbb{E}(\sqrt{Q}))^2}{I}\right) 1_{\mathbb{R}^+}(y).$$

The distribution of Q could be modelled by the lognormal, Erlang, the improper Pareto (unnormalized) or the inverse Gaussian law viz.,

$$h(Q) = \sqrt{\frac{\lambda}{2\pi Q^3}} \exp\left\{-\frac{\lambda(Q-\mu)^2}{2\mu^2 Q}\right\} 1_{\mathbb{R}^+}(Q).$$

Let the distribution of decibel noise levels L be related to that of Q by

$$L = 10\, \log_{10}\left(\frac{Q}{I_{\mathrm{ref}}\, d_{\mathrm{ref}}^2}\right)$$

where $D_{\mathrm{ref}} = 50$ feet.

By choosing $I_{\mathrm{ref}} = 1$ and $\mu = 1$, Marcus provides a graphical comparison of emission distributions $h(Q)$. He notes that empirical distributions of emission intensity usually have values of $\sqrt{\frac{\mu}{\lambda}}$ between 0.5 and 1.0, and observed values of I also have values of $\sqrt{\frac{\mu}{\lambda}}$ between 0.5 and 1.0 and therefore the comparisons are made for $c = 0.5$ for the lognormal, Erlang (shape parameter $= 4$) and the IG and for $c = 1.0$ for the lognormal, Erlang (1) and the IG. His comparisons show that the IG approximates the lognormal when L is between L_{99} and $L_{0.5}$, particularly at important values like L_{10} and L_{50} for $0.5 \leq \sqrt{\frac{\mu}{\lambda}} \leq 1.0$. Here L_α is the 100 $\propto$ percentile of the noise level distribution.

Marcus recommends the IG as a model to use in routine analysis of traffic noise data. If the traffic is extremely heavy, or if the observer is far from the road, then the representation of the process as a linear filter of the point process of vehicle locations guarantees that the distributions of I (or of L) will be approximately inverse Gaussian with a small coefficient of variation.

N. Market Research

Whitmore (1986a) considered an inverse Gaussian ratio estimation model and applied it to a marketing survey study. In reality the ratio estimation problem involves a regression setting where for $\alpha > 0$ and $\beta > 0$, $\mathcal{L}(Y_i) = IG(\alpha x_i, \beta x_i^2)$, $i = 1, \ldots, n$. Thus the variance of Y_i is proportional to x_i. When $\alpha = \beta = 1$ the densities of Y display increasing skewness with small values of x_i. For large values of x_i the law of Y approaches a lognormal form.

Suppose that $\mathcal{L}(Y_i) = IG(\alpha x_i^a, \beta x_i^b)$ for some constants a and b, then we have $\mathcal{L}(Y_i') = IG(\alpha x_i', \beta x_i'^2)$ where $Y_i' = x_i^{b-2a} Y_i$, $x_i' = x_i^{b-a}$.

The log-likelihood based on a sample of size n is proportional to

$$\frac{n}{2} \log \beta - \frac{\beta}{2\alpha^2} \sum_{i=1}^{n} \frac{(Y_i - \alpha x_i)^2}{Y_i}$$

and the maximum likelihood estimates of α, β are given by

$$\hat{\alpha} = \frac{\bar{Y}}{\bar{X}}, \quad n\hat{\beta}^{-1} = \sum_{i=1}^{n} \frac{x_i^2}{Y_i} - \frac{n\bar{x}^2}{\bar{Y}^2}.$$

It now follows from the distributional results of Chapter 1 (Propositions 1.1 and 1.2), that

(1) $\mathcal{L}(\hat{\alpha}) = IG(\alpha, n\beta\bar{x})$

(2) $\mathcal{L}(\frac{n\beta}{\hat{\beta}}) = \chi^2_{n-1}$

(3) $\hat{\beta} \perp\!\!\!\perp \hat{\alpha}$

and

(4) $\mathcal{L}\left[\frac{(n-1)\hat{\beta}\bar{x}(\hat{\alpha}-\alpha)^2}{\alpha^2\,\hat{\alpha}}\right] = F_{1,n-1}$

Inferences on α and β can now be routinely carried out. Moreover, given a sample of n observations, one can predict the population total of N observations. Suppose that for $i = 1, \ldots, N$ the independent observations Y_i are such that $\mathcal{L}(Y_i) = IG(\alpha x_i, \beta x_i^2)$.

Let $T_2 = \sum_{i=n+1}^{N} Y_i$, $X_1 = \sum_{i=1}^{n} x_i$, $X_2 = \sum_{i=n+1}^{N} x_i$. The predictor of T_2 is $\hat{\alpha}X_2 = \hat{T}_2$, and $\mathcal{L}(\hat{T}_2) = IG(\alpha X_2, \beta X_1 X_2)$. On the other hand, since $\frac{(\alpha x_i)^2}{\beta x_i^2} = \text{constant} = \frac{\alpha}{\beta}$, $\mathcal{L}(T_2) = IG(\alpha X_2, \beta X_2^2)$. Now consider two independent samples, one of size (X_1, X_2) and the other of size X_2^2 from $IG(\alpha X_2, \beta)$. The sample mean corresponding to the sample of size (X_1, X_2) then has the law $IG(\alpha X_2, \beta X_1 X_2)$, the same as that of T_2. Similarly the sample mean from the second sample of size (X_2^2) follows the law $IG(\alpha X_2, \beta X_2^2)$ which is the same as that of T_2.

… Market research

Denote by $T = \sum_{i=1}^{N} Y_i$. Then $T = \sum_{i=1}^{n} Y_i + T_2$, so that $T = \hat{a}X_2 + T_2$.
Furthermore from the distributional results of Chapter 1 (Propositios 1.1 and 1.2),

$$\mathcal{L}\left[\frac{\hat{\beta}\,(n-1)X_1\,X_2^2\,(T_2-\hat{T}_2)^2}{n\hat{T}_2\,T_2(X_1\,\hat{T}_2+X_2T_2)}\right] = F_{1,n-1}.$$

Therefore a prediction interval for T_2 as well as T can be deduced from the above distribution.

The following example considered by Whitmore on a market survey report illustrates these ideas.

Example N.1 The projections of annual dollar sales for all products (N) of all companies in a particular consumer product industry are made by monitoring sales amounts in a panel of retail sales outlets. Denote by Y_i the actual sales for the n products of a company, and by x_i the projected amounts. It is required to make inferences on sales for the competitor's products $(Y_{n+1}, \ldots, Y_N)$ from the corresponding known projections $(x_{n+1}, \ldots, x_N)$. Table N.1 contain the figures for the projected and actual sales of 20 products of a company.

Table N.1 Projected (PS) and Actual Sales (AS) of 20 products

#i	x_i	Y_i	#i	x_i	Y_i
1	5959	5673	11	527	487
2	3534	3659	12	353	463
3	2641	2565	13	331	225
4	1965	2182	14	290	257
5	1738	1839	15	253	311
6	1182	1236	16	193	212
7	667	918	17	156	166
8	613	902	18	133	123
9	610	756	19	122	198
10	549	500	20	114	99

We then obtain

$$X_1 = 21,930,\ X_2 = 359,561$$

$$\sum_{i=1}^{20} Y_i = 22,771, \quad \sum_{i=1}^{20} \frac{x_i^2}{Y_i} = 21,463$$

$$\hat{\alpha} = 1.0383, \hat{\beta} = 0.05838.$$

A 95% confidence interval for β is $(0.9785, 1.1060)$, and a 95% prediction interval for Y at $x = 100$ is given by $(43.2, 250.8)$. Finally a 95% prediction interval for $T_2 = \sum_{i=21}^{N} Y_i$ is $(351,190 - 398,398)$.

The adequacy of the model can be assessed by computing the relative squared residuals r_i

$$r_i = \frac{\hat{\beta}(y_i - \hat{\alpha}x_i)^2}{\hat{\alpha}y_i\left[\hat{\alpha} + \left(\frac{y_i}{X_1}\right)\right]} \quad, i = 1, \ldots, n$$

and then plotting $(\frac{i}{n+1}, P(F \leq r_i))$ where $\mathcal{L}(F)$ is $F_{1,n-1}$. Another method involves plotting $(\frac{i}{n+1}, \hat{P}[Y_i \leq y_i])$ where $\hat{P}(Y_i \leq y_i)$ is the estimated distribution function.

O. Regression

Whitmore (1979b,1983) studies multiple regression methods for complete and censored data and remarks that exact inference details are unavailable although asymptotic results, when the degree of censoring is moderate, turn out to be reasonably accurate even when the sample sizes are small. The likelihood and factorization strongly resemble those of the normal model.

Consider n independent observations Y_i from $IG(\delta_i^{-1}, \sigma^{-1})$ where $\delta_i = X_i\beta$, $X_i^t = (x_{i1}, \ldots, x_{ip})^t$ is a $(p \times 1)$ vector of covariates and $\beta^t = (\beta_1, \ldots, \beta_p)$ a vector of p regression parameters. We write $X = \begin{pmatrix} X_1 \\ \cdot \\ \cdot \\ \cdot \\ X_k \end{pmatrix}$,
$e^t = (1, \ldots, 1)$ and let Y be the $n \times n$ diagonal matrix $\mathrm{diag}(y_1, \ldots, y_n)$.

The log-likelihood is proportional to

$$-\frac{n}{2} \log \sigma - \frac{\beta^t X^t Y X \beta - 2e^t X \beta + e^t Y^{-1} e}{2\sigma},$$

and the maximum likelihood estimates of β and σ can be found as

$$\hat{\beta} = (X^t Y X)^{-1} X^t e$$

and

$$\hat{\sigma} = \frac{e^t Y^{-1} e - e^t X \hat{\beta}}{n}.$$

Whitmore points out that when $\delta = X^t\beta$ is negative or zero the conditional variable $(Y \mid Y < \infty)$ is distributed as $IG(|\delta|^{-1}, \frac{1}{\sigma})$ and inference procedures do not pose any difficulties.

When $(n-r)$ observations are censored on the right at the points $a_{r+1}, \ldots, a_n$ the likelihood function is, writing $\theta = (\beta, \sigma)$,

$$\mathcal{L}(\theta) = \prod_{i=1}^{n} f(y_i, \theta) \prod_{i=r+1}^{n} P[Y_i > a_i].$$

This can be written compactly as

$$\int_{a_{r+1}}^{\infty} \cdots \int_{a_n}^{\infty} \prod_{i=1}^{n} (2\pi\sigma y_i^3)^{-\frac{1}{2}} \exp\left\{-\sum^{n} \frac{(y_i X_i^t \beta - 1)^2}{2\sigma y_i}\right\} dy_{r+1} \ldots dy_n$$

$$= \int_S \prod_{i=1}^{n} (2 \prod \sigma y_i^3)^{-\frac{1}{2}} \exp\left\{-\frac{\beta^t X^t Y X \beta - 2e^t X \beta + e^t Y^{-1} e}{2\sigma}\right\} dY$$

where
$$= C(\theta) \text{ (say)}$$

$$S = \{Y_1 = y_1, \ldots, Y_r = y_r, Y_{r+1} > a_{r+1} \ldots Y_n > a_n\}. \tag{O.1}$$

The maximum likelihood estimation of θ is obtained by solving $\frac{\partial C(\theta)}{\partial \theta} = 0$ or equivalently $\mathbb{E}_*(\frac{1}{L}\frac{\partial L}{\partial \theta}) = 0$, L being the likelihood for the complete sample and $\mathbb{E}_*(\cdot)$ denotes the expectation in the censored sample space S. Thus we have to solve

$$\mathbb{E}_*\{X^t Y X \beta - X^t e\} = 0$$

and

$$\mathbb{E}_*\{n\sigma - (\beta^t X^t Y X \beta - 2e^t X \beta + e^t Y^{-1} e\} = 0$$

resulting in the equations

$$\hat{\beta} = (X^t \mathbb{E}_*(Y) X)^{-1} X^t e$$
$$\hat{\sigma} = \frac{e^t \mathbb{E}_*(Y^{-1}) e - e^t X \hat{\beta}}{n}. \tag{O.2}$$

In the above expressions $\mathbb{E}_*(Y)$ will be a $n \times n$ diagonal matrix with the first r diagonal elements, the uncensored observations and the remaining $(n - r)$ elements, the conditional expected values $\mathbb{E}(Y_{r+1} \mid Y_{r+1} > a_{r+1}), \ldots, \mathbb{E}(Y_n \mid Y_n > a_n)$. The same applies to $\mathbb{E}_*(Y^{-1})$.

Indeed, if $F(y)$ represents the distribution function of Y,

$$\mathbb{E}(Y \mid Y > a) = \frac{F(\frac{1}{\delta^2 a})}{\delta(1 - F(a))}$$

and

$$\mathbb{E}\left(\frac{1}{Y} \mid Y > a\right) = \sigma + \delta^2 \mathbb{E}(Y \mid > a) - \frac{2a\sigma f(a)}{1 - F(a)}.$$

These relations are proved as follows

$$\mathbb{E} X \mid X > a) = \frac{\int_a^\infty x f(x \mid \mu, \lambda) dx}{1 - P[X < a \mid \mu, \lambda]}.$$

Writing $y = \frac{\mu}{x}$ in the integral we have

$$\int_a^\infty x f(x \mid \mu, \lambda) dx = \mu \int_0^{\frac{\mu}{a}} \sqrt{\frac{\phi}{2\pi y^3}} \exp\left\{\frac{\phi}{2} \frac{(y-1)^2}{y}\right\} dy$$
$$= \mu F_X\left(\frac{\mu}{a} \mid 1, \phi\right),$$

where
$$F_X(c \mid 1, \phi) = P[X \le c \mid \mathcal{L}(X) = IG(1, \phi)].$$

Moreover
$$P[X < a \mid \mu, \lambda] = P\left[\frac{X}{\mu} < \frac{a}{\mu} \mid 1, \phi\right],$$

that is
$$F_X(a \mid \mu, \lambda) = F_X\left(\frac{a}{\mu} \mid 1, \phi\right).$$

Hence
$$\mathbb{E}(X \mid X > a) = \frac{\mu F_X\left(\frac{\mu}{a} \mid 1, \phi\right)}{1 - F_X\left(\frac{a}{\mu} \mid 1, \phi\right)}.$$

Suppose now that $\mathcal{L}(Z) = IG(1, \phi)$ and $\mathcal{L}(X) = IG(\delta^{-1}, \sigma^{-1})$. Then it is easy to see that $Z = \delta X$, $\phi = \frac{\delta}{\sigma}$ $(\mu = \frac{1}{\delta}, \lambda = \frac{1}{\sigma})$ allows us to go from $\mathcal{L}(Z)$ to $\mathcal{L}(X)$ and therefore

$$P\left(Z < \frac{\mu}{a} \mid 1, \phi\right) = P\left(\delta X < \frac{1}{\delta a} \mid \delta^{-1}, \sigma^{-1}\right)$$
$$= P\left(X < \frac{1}{a\delta^2} \mid \delta^{-1}, \sigma^{-1}\right).$$

Hence
$$F_Z\left(\frac{\mu}{a} \mid 1, \phi\right) = F\left(\frac{1}{a\delta^2}\right).$$

Similarly
$$F_Z\left(\frac{a}{\mu} \mid 1, \phi\right) = P\left(Z < \frac{a}{\mu} \mid 1, \phi\right)$$
$$= P(X < a \mid \delta^{-1}, \sigma^{-1})$$
$$= F(a).$$

Finally
$$\mathbb{E}(X \mid X > a) = \frac{1}{\delta} \frac{F\left(\frac{1}{a\delta^2}\right)}{[1 - F(a)]}.$$

To obtain the second relation we need some details from conditional moments. For $x > a > 0$, the density of X is

$$g(x) = \frac{f(x)}{1 - F(a)} 1_{\mathbb{R}^+}(x).$$

Suppose now we want $\mathbb{E}(X^r \mid X > a)$ for $r > 1$.

$$\mathbb{E}(X^r \mid X > a) = \frac{1}{1 - F(a)} \int_a^\infty x^r f(x) dx = \nu_r \text{ say}.$$

When $f(x)$ is the $IG(\mu, \lambda)$ density

$$f'(x) = \left(\frac{\lambda}{2}x^{-2} - \frac{3}{2}x^{-1} - \frac{\lambda}{2\mu^2}\right) f(x)$$

so that

$$\int_a^\infty x^r f'(x)dx = \frac{\lambda}{2}\int_a^\infty x^{r-2} f(x)dx - \frac{3}{2}\int_a^\infty x^{r-1} f(x)dx$$

$$- \frac{\lambda}{2\mu^2}\int_a^\infty x^r f(x)dx.$$

Now the left side gives us

$$x^r f(x)\,|_a^\infty - r\int_a^\infty x^{r-1} f'(x)dx$$

and after multiplication of both sides by $\frac{1}{1-F(a)}$ we obtain

$$\frac{-a^r f(a)}{1-F(a)} - r\nu_{r-1} = \frac{\lambda}{2}\nu_{r-2} - \frac{3}{2}\nu_{r-1} - \frac{\lambda}{2\mu^2}\nu_r.$$

Hence we obtain

$$\frac{2a^r f(a)}{1-F(a)} + (2r-3)\nu_{r-1} = \frac{\lambda}{\mu^2}\nu_r - \lambda\nu_{r-2}$$

when $r = 1$, $\mathbb{E}(X^{r-1} \mid X > a) = 1$, i.e., $\nu_0 = 1$ and $\mathbb{E}(X \mid X > a) =$
$$\nu_1 = \frac{\mu F_Z\left(\frac{\mu}{a} \mid 1, \phi\right)}{1 - F_Z\left(\frac{\mu}{a} \mid 1, \phi\right)}.$$ (Here $\mathcal{L}(Z) = IG(1, \phi)$).
Thus

$$\nu_{-1} = \mathbb{E}(X^{-1} \mid X > a) = \frac{\frac{1}{\mu^2}\left[\mu F_Z\left(\frac{\mu}{a} \mid 1, \phi\right)\right] - \frac{2a}{\lambda}f(a)}{1 - F_Z\left(\frac{a}{\mu} \mid 1, \phi\right)} + \frac{1}{\lambda}.$$

Writing $\mu = \delta^{-1}$ and $\lambda = \sigma^{-1}$ we then obtain

$$\nu_{-1} = \frac{\delta F\left(\frac{1}{a\delta^2}\right) - 2a\sigma f(a)}{1 - F(a)} + \sigma.$$

The likelihood equations O.2 can be solved iteratively using the *EM* algorithm starting with initial estimates of β and σ.

Regression

In the absence of exact distributional results, Whitmore presents an asymptotic theory in which the information matrix I has the form (when there is no censoring)

$$I = \begin{pmatrix} \frac{X^t Y X}{\hat{\sigma}} & 0 \\ 0 & \frac{n}{2\hat{\sigma}^2} \end{pmatrix}.$$

One then uses asymptotic normal theory to perform routine statistical inference. In the presence of censoring the sample information matrix has a slightly more complicated structure since it involves taking conditional expectations, namely,

$$I = -\mathbb{E}_* \left(\frac{1}{L} \frac{\partial^2 L}{\partial \theta \partial \theta^t} \right).$$

We omit the details and refer the reader to Whitmore (1983).

Next we consider the asymptotic approach due to Bhattacharyya and Fries (1982) who also consider a reciprocal linear model.

The reciprocal linear model of Bhattacharyya and Fries (1982) assumes that the failure time Y has the distribution $IG(\mu = \frac{1}{\delta}, \lambda = \frac{1}{\sigma})$; we let $\delta_i = \frac{1}{\mu_i} = x_i^t \beta$ where x_i is a $p \times 1$ vector of covariates and β a $(p \times 1)$ vector of unknown parameters. (Thus the drift is assumed to be a linear function of the covariates.) Suppose that there are k design points $x_i \in \mathbb{R}^p (i = 1, \cdots, k)$ and that there are independent observations n_i in number, $Y_{ij}(j = 1, \cdots, n_i)$, then we may now write $Y_{ij} \sim IG(\mu_i, \sigma)$ where $\mu_i^{-1} = x_i^T \beta$. Assume further that $k \geq p + 1$. We shall use the following notation

$$N = \sum_i n_i, \quad \overline{Y}_i = n_i^{-1} \sum_j Y_{ij}, \quad \overline{Y} = N^{-1} \sum_i \sum_j Y_{ij},$$

$$R = N^{-1} \sum_i \sum_j Y_{ij}^{-1}, \quad V = \sum \sum \left(Y_{ij}^{-1} - \overline{Y}_i^{-1} \right), \quad \overline{x} = N^{-1} \sum_i n_i x_i$$

$$C = \text{diag}(n_1, \cdots, n_k), \quad D = \text{diag}(\overline{Y}_i, \cdots, \overline{Y}_k), \quad X^t = (x_1, \cdots, x_k),$$

$$e^t = (1, \cdots, 1)$$

$$S = X^t C D X, \quad Q(\beta) = \sum \sum y_{ij}^{-1} (y_{ij} x_i^t \beta - 1)^2$$
$$= (DX\beta - e)^t C D^{-1} (DX\beta - e) + V.$$

The likelihood L is proportional to

$$\sigma - \frac{N}{2} \exp(-Q(\beta)/2\sigma).$$

The partial derivatives of $\log L$ with respect to β and σ give equations

$$S\hat{\beta} = X^t Ce, \quad N\hat{\sigma} = Q(\hat{\beta})$$

whose solutions provide estimates of β and σ. When X has full rank so does S and S is invertible with probability one and the unique roots $\hat{\beta}$ and $\hat{\sigma}$ do indeed maximize L thus yielding

$$\hat{\beta} = S^{-1} X^t Ce, \quad \hat{\sigma} = R - \hat{\beta}^t \bar{x}$$
$$= N S^{-1} \bar{x}.$$

Note, however, that the roots are not necessarily the maximum likelihood estimators since $(\hat{\beta}^t x_i)^{-1}$ may be negative! Some authors refer to them as pseudo-maximum likelihood estimators. When N is not too small this is not a serious problem, and one can resort to asymptotics.

Asymptotics

The asymptotic case, as sketched by Bhattacharyya and Fries, concerns the situation where k, the number of design points x_i remains fixed and the number of replications n_i goes to infinity at a fixed rate. That is as $N \to \infty$, we let $\frac{n_i}{N} = h_i > 0$ for $i = 1, \cdots, k$. We will see later how Jogesh Babu and Chaubey (1996) treat the situation where the number of replications is small and the number of design points k is large.

Define the diagonal matrices M, Δ and H as follows.

$$M = \text{diag}\left(\delta_1^{-1} = \mu_1, \cdots, \delta_k^{-1} = \mu_k\right), \quad H = \text{diag}(h_1, \cdots, h_k)$$
$$\Delta = X^t H M X.$$

Then

$$\Delta \beta = X^t H M X \beta \quad \text{and since} \quad X\beta = M^{-1} e, \quad MX\beta = e$$

so that

$$\Delta \beta = X^t H e.$$

Also note that $N\bar{x} = X^t Ce$ by definition of $\bar{x}$. As $N \to \infty$ we note that $N^{-1} C \to H$. The next proposition shows that $\hat{\beta}$ and $\hat{\sigma}$ are strongly consistent.

Proposition O.1 *The pseudo-maximum likelihood estimators $\hat{\beta}$ and $\hat{\sigma}$ are strongly consistent.*

First we note that

$$\mathbb{E}(Y_{ij}) = \delta_i^{-1}, \quad \mathbb{E}(Y_{ij}^{-1}) = \delta_i + \sigma.$$

Asymptotics

Applying the strong law of large numbers to $\overline{Y}_i$ and $n_i^{-1}\sum_j Y_{ij}^{-1}$ we have with probability 1
i) $D \to M$
ii) $R \to e^t H M^{-1} e + \sigma$

Hence with probability 1
iii) $N^{-1}S = N^{-1}X^t C D X \to X^t H M X = \Delta$
iv) $\hat{\beta} = S^{-1}X^t C e \to S^{-1}X^t C M X \beta = \beta$
v) $X\hat{\beta} \to X\beta = M^{-1}e$
vi) $\hat{\sigma} = R - \hat{\beta}^t \overline{x} = R - N^{-1}\hat{\beta}^t X^t C e = R - N^{-1}e^t C X \beta \to \sigma$.

Writing $V_1 = \sum_i n_i \overline{Y}_i^{-1} - N\hat{\beta}^t\overline{x}$, we have $N\hat{\sigma} = V + V_1$.

The following theorem of Bhattacharyya and Fries then gives the asymptotic normality of the estimates of β and σ

Theorem O.1 *As $N \to \infty$ and $N^{-1}C \to H$*
(a) $\sqrt{N}(\hat{\beta} - \beta) \xrightarrow{D} N_p(0, \sigma\Delta^{-1})$
(b) $\dfrac{V_i}{\sigma} \xrightarrow{D} \chi^2_{k-p}$ *so that* $\sqrt{N}\left(\dfrac{\hat{\sigma}}{\sigma} - 1\right) \to N(0, 2)$
(c) *$\hat{\beta}$ and $\hat{\sigma}$ are asymptotically independent.*

Proof Letting $U_i = \overline{Y}_i \delta_i$, $i = 1, \cdots, k$, we obtain $U = DM^{-1}e = DX\beta$. This implies that

$$S^{-1}X^t C U = S^{-1}X^t C D X \beta$$
$$= \beta.$$

Since $S^{-1}X^t C e = \hat{\beta}$ we have

$$\hat{\beta} - \beta = -S^{-1}X^t C (U - e).$$

Since U_i are independent, applying the Central Limit Theorem we see that for each $i = 1, \cdots, k$

$$\sqrt{n_i}(U_i - 1) \xrightarrow{D} N_1(0, \sigma\delta_i^{-1})$$

so that

$$\sqrt{N}(U - e) \xrightarrow{D} N_k(0, \sigma M H^{-1}).$$

Now $N^{-1}S \to \Delta$ with probability 1 and $N^{-1}C \to H$. Therefore

$$\sqrt{N}(\hat{\beta} - \beta) = -\sqrt{N}S^{-1}X^t C(U - e)$$
$$= -\sqrt{N}(NS^{-1})X^t(N^{-1}C)(U - e)$$
$$\sqrt{N}(\hat{\beta} - \beta) \to -\sqrt{N}\Delta^{-1}X^t H(U - e)$$

and the limit law of $\sqrt{N}(\hat{\beta} - \beta)$ is thus seen to be $N_p(0, \sigma\Delta^{-1})$.

Next to prove (b) we have

$$e^t C X \beta = e^t C D^{-1} e$$
$$V_i = e^t C D^{-1} e - N \hat{\beta}^t \overline{x} = e^t C D^{-1} e - \hat{\beta}^t X^t C e$$
$$= e^t C D^{-1} e - e^t C X \hat{\beta}.$$

Moreover from the equation

$$\hat{\beta} - \beta = -S^{-1} X^t C(U - e)$$

we have

$$\hat{\beta} = \beta - S^{-1} X^t C(U - e)$$
$$e^t C X \hat{\beta} = e^t C X \beta - e^t C X S^{-1} X^t C(U - e)$$
$$= e^t C M^{-1} e - e^t C M^{-1}(U - e) + (U - e)^t C X S^{-1} X^t C(U - e)$$

since $U^t C X S^{-1} X^{-1} C(U - e) = e^t M^{-1} C(U - e)$.

A Taylor expansion of $\overline{Y}_i^{-1}$ around $\mu_i^{-1} = \delta_i$ gives

$$\overline{Y}_i^{-1} = \mu_i^{-1} - \mu_i^{-1}(U_i - 1) + \mu_i^{-1}(U_i - 1)^2 + O_p(N^{-1}).$$

Now multiply both sides by n_i and sum from 1 to k to obtain

$$N^{-1} \sum_i n_i Y_i^{-1} = N^{-1} \sum_i n_i \delta_i - N^{-1} \sum_i n_i \delta_i (U_i - 1)$$
$$+ N^{-1} \sum_i n_i \delta_i (U_i - 1)^2 + O_p(1).$$

This gives us

$$e^t C D^{-1} e = e^t C M^{-1} e - e^t C M^{-1}(U - e) + (U - e)^t C M^{-1}(U - e) + O_p(1).$$

Hence

$$V_1 = (U - e)^t \left[C M^{-1} - C X S^{-1} X^t C \right] (U - e) + O_p(1).$$

Observe that $C X S^{-1} X^t C = N \left[N^{-1} C X (N S^{-1}) X^t N^{-1} C \right]$
$\to N H X \Delta X^t H$ with probability 1 and we have that the limit law of V_1 is the same as that of

$$N(U - e)^t [H M^{-1} - H X \Delta X^t H](U - e).$$

Since $\sqrt{N}(U - e) \xrightarrow{D} N_k(0, \sigma M H^{-1})$, the quadratic form $N(U - e)^t$

$[HM^{-1} - HX\Delta X^t H](U-e)$ has a χ^2_{k-p} law, it being verifiable that the associated matrix of the quadratic form is symmetric idempotent with trace $(k-p)$.

To show that $\sqrt{N}\left(\frac{\hat{\sigma}}{\sigma} - 1\right) \xrightarrow{D} N(0,2)$ it suffices to note that $\mathcal{L}(\frac{V}{\sigma}) = \chi^2_{N-k}$ (Tweedie) and hence that $\sqrt{N}\left(\frac{V}{\sqrt{N}\sigma} - 1\right) \xrightarrow{D} N(0,2)$. Moreover since $\frac{V_1}{\sigma} \xrightarrow{D} \chi^2_{k-p}$ it follows that $\frac{V_1}{\sigma\sqrt{N}}$ is $O_p(1)$. As a result, from the equation

$$\sqrt{N}\left(\frac{\hat{\sigma}}{\sigma} - 1\right) = \sqrt{N}\left(\frac{V}{\sigma N} - 1\right) + \frac{V_1}{\sigma\sqrt{N}}$$

we can assert that the limit law of $\sqrt{N}\left(\frac{\hat{\sigma}}{\sigma} - 1\right)$ is indeed $N(0,2)$. Note that from Tweedie's result, V and V_1 are independent. ♣

Analysis of Reciprocals (revisited)

Based on a combination of exact theory and the asymptotic results developed so far an inverse Gaussian analogue to the normal theory of analysis of variance table (ANOVA) can be pursued.

When there is a replicated design and the model accommodates a constant term so that $n_i \geq 2$ for all i, letting $X = (e, X_1), \beta = (\beta_0, \beta_1)^t$, Bhattacharyya and Fries consider an analysis of reciprocals akin to ANOVA. Here β_1 is a $(p-1)$ vector and X_1 is a $k \times (p-1)$ matrix of covariates. In the spirit of the ANOVA setup they consider the identity

$$Y_{ij}^{-1} - \overline{Y}^{-1} = \left(x_i^t \hat{\beta} - \overline{Y}^{-1}\right) + \left(\overline{Y}_i^{-1} - x_i^t \hat{\beta}\right) + \left(Y_{ij}^{-1} - \overline{Y}_i^{-1}\right).$$

The deviation of an observed reciprocal variable from the grand mean is accounted for in terms of three components, the regression contribution due to β_1 (given β_0), and the contribution arising from pure error. Summing over i and j one obtains

$$\sum_i \sum_j \left(Y_{ij}^{-1} - \overline{Y}^{-1}\right) = Q_{\text{reg}} + Q_{\text{lof}} + Q_e$$

where

$$Q_{\text{reg}} = \sum_i \sum_j \left(x_i^t \hat{\beta} - \overline{Y}^{-1}\right) = N\left(\overline{x}^t \hat{\beta} - \overline{Y}^{-1}\right)$$

$$Q_{\text{lof}} = \sum_{i=1}^k n_i \left(\overline{Y}_i^{-1} - x_i^t \hat{\beta}\right)$$

$$Q_e = \sum_i \sum_j \left(Y_{ij}^{-1} - \overline{Y}_i^{-1}\right).$$

Here Q_{reg} measures the contribution due to regression, Q_{lof} the lack of fit and Q_e the pure error. The sum total is the total corrected sum of reciprocals analogous to ANOVA.

From inverse Gaussian theory due to Tweedie $\mathcal{L}(Q_e/\sigma) = \chi^2_{N-k}$ and $Q_e \perp\!\!\!\perp$ of Q_{reg} and Q_{lof}. As to what part is played by Q_{lof} and Q_{reg} in the analysis can be gleaned from the likelihood ratio test for certain hypotheses of interest, namely

$$H_2 : \mu_i = \delta_i^{-1} \quad \text{unrestricted}$$
$$H_1 : \mu_i^{-1} = x_i^t \beta \quad \text{the reciprocal linear model}$$
$$H_0 : \mu_i^{-1} = \beta_0 \quad \text{lack of regression.}$$

The likelihood ratio test for hypotheses H_i nested within H_j is usually given by
$$\Lambda_{ij} = -2\left[\ell_{\max}(H_i) - \ell_{\max}(H_j)\right]$$
where $\ell_{\max}(H_i)$ is the maximum of the likelihood function under H_i $i = 0, 1, 2$. We then consider efficient likelihood estimators of σ under each H_i, say $\hat{\sigma}_i$ ($i = 0, 1, 2$) and form the statistics

$$\Lambda_{ij} = N \log\left[\frac{\hat{\sigma}_i}{\hat{\sigma}_j} - 1\right].$$

Define $\hat{\sigma}_i$ $i = 0, 1, 2$ as follows

$$\hat{\sigma}_2 = \overline{Y}_- - N^{-1} \sum_i \overline{Y}_i^{-1} = N^{-1} Q_e$$
$$\hat{\sigma}_1 = \overline{Y}_- - \overline{x}^t \hat{\beta}$$
$$\hat{\sigma}_0 = \overline{Y}_- - \overline{Y}^{-1}.$$

Clearly
$$Q_l = \sum_i n_i \left(Y_i^{-1} - x_i^t \hat{\beta}\right)$$
$$= N\left(\overline{Y}_- - \sum_i n_i \overline{Y}_i^{-1}\right)$$
$$= N(\hat{\sigma}_1 - \hat{\sigma}_2) > 0.$$

Hence
$$\Lambda_{12} = N \log\left[\frac{\hat{\sigma}_1 - \hat{\sigma}_2}{\hat{\sigma}_2}\right]$$
$$= N \log\left[1 + \frac{k-p}{N-k} F_\ell\right]$$

where
$$F_\ell = \frac{Q_{\text{lof}}/(k-p)}{Q_e/(N-k)}.$$

Thus the likelihood ratio test for H_2 rejects for large values of F_ℓ.

Observe that Q_ℓ is none other than $V_1 = \sum_i n_i \overline{Y}_i^{-1} - N\bar{x}^t\hat{\beta}$ and $Q_e = V$. From the asymptotic χ^2_{k-p} distribution of Q_ℓ/σ under H_1 and the exact χ^2_{N-k} law of Q_e/σ and their independence it follows that F_ℓ has an approximate F distribution with $(k-p)$ and $(N-k)$ degrees of freedom. We remark that the only approximation involved is that of the distribution of Q_{lof}.(lof means lack-of-fit)

Suppose that we define F_{reg} as
$$F_{\text{reg}} = \frac{Q_{\text{reg}}/(p-1)}{Q_e/(N-k)},$$

we note that the numerator is > 0 and under H_0, $\mathcal{L}(Q_{\text{reg}}/\sigma)$ is asymptotically χ^2_{p-1} so that we obtain for F_{reg} an F law with $(p-1)$ and $(N-k)$ degrees of freedom.

The likelihood ratio test of H_0 under the model H_1 needs a subtler analysis as outlined cleverly by Bhattacharyya and Fries. First of all $Q_{\text{reg}} = N(\hat{\sigma}_0 - \hat{\sigma}_1)$ whereas $N\hat{\sigma}_1 = Q_{\text{lof}} + Q_e$, so that
$$\Lambda_{01} = \frac{Q_{\text{reg}}}{Q_{\text{lof}} + Q_e}$$

whereas F_{reg} is simply equal to const $\frac{Q_{\text{reg}}}{Q_e}$. But when $0 < \gamma < 1$, $\frac{N^\gamma Q_{\text{lof}}}{Q_e} \to 0$ in probability because $\frac{Q_e}{N} \to \sigma$ in probability and $\frac{Q_{\text{lof}}}{\sigma} \xrightarrow{D} \chi_{k-p}$. Thus Q_{lof} is of a smaller order of magnitude compared to Q_e and the test based on F_{reg} is asymptotically equivalent to Λ_{01}. Thus F_{reg} preserves the F law and this approximation is more effective in moderate sized samples.

The detailed analysis of reciprocals can be given as

Table O.1 *Analysis of Reciprocals*

Source	Reciprocal component	d.f	App.F ratio
Regression	$N(x^t\hat{\beta} - \bar{y}^{-1})$	$p-1$	$\frac{N-k}{p-1}\frac{Q_{\text{reg}}}{Q_e}$
Residual lof	$\sum n_i(\bar{y}_i^{-1} - \bar{x}_i^t\hat{\beta})$	$k-p$	$\frac{N-k}{k-p}\frac{Q_{\text{lof}}}{Q_e}$
error	$\sum\sum(y_{ij}^{-1} - \bar{y}_i^{-1})$	$N-k$	

Regression diagnostics

Diagnostics for checking the model assumptions pertain to the examination of residuals that measure departures of the data from the fitted equation as well as the (assumption) hypothesis that the data are indeed from an inverse Gaussian distribution.

In the one-sample case this means that $\mathcal{L}(Y_i) = IG(\mu = \frac{1}{\delta}, \sigma)$, $i = 1, \cdots, N$. Then the standardized residuals

$$\varepsilon_i = \sqrt{Y_i} \frac{|Y_i^{-1} - \mu^{-1}|}{\sqrt{\sigma}}, \quad i = 1, \cdots, N$$

have a half-normal distribution. We substitute the estimates $\overline{Y}$ and $\hat{\sigma}$ for μ and σ to form estimated residuals given by

$$\hat{\varepsilon}_i = \frac{\sqrt{NY_i} \, |Y_i^{-1} - \overline{Y}^{-1}|}{\sqrt{\sum_i (Y_i^{-1} - \overline{Y}^{-1})}}, \quad i = 1, \cdots, N.$$

The construction of a half-normal plot based on the ordered estimated residual along the lines of Daniel (1959) and Birnbaum (1959) then provides a graphical check. If the model is appropriate a plot of the ordered $\hat{\varepsilon}_i$ say $\hat{\varepsilon}_{(i)}$ against $\Phi^{-1}\left(\frac{1}{2} + \frac{i-1/2}{2N}\right)$ fall along the 45° line through the origin.

To measure the departures from the 45° line one could use the correlation coefficient to check for linearity, and the least squares estimate of the slope which should be close to one if the model holds. (Moderately large N will be needed.)

Alternatively one could plot $\hat{\varepsilon}_{(i)}$ against the expected values of $\varepsilon_{(i)}$, the ordered ε_i values. Tables of expected values provided by Govindarajulu and Eisenstat (1965) as well as Pearson and Hartley (1972) are useful in this context.

An extension to the k-sample case yields the estimated residuals as

$$\hat{\varepsilon}_{ij} = \frac{\sqrt{n_i Y_{ij}} \, |Y_{ij}^{-1} - \overline{Y}_i^{-1}|}{\sqrt{\sum_j (Y_{ij}^{-1} - \overline{Y}_i^{-1})}}$$

which are then used in an analogous fashion.

We conclude this analysis with an example considered by Bhattacharyya and Fries to illustrate the techniques used. The example is taken from Nelson (1971).

Example G.1 (continued)   Returning to the example on failure times of two batches of insulation material in a motorette test performed at elevated temperature settings, the fit of an IG reciprocal linear model yields the following analysis of reciprocals table for the first three levels (190°C, 220°C and 240°C).

Regression diagnostics

Table O.2 *Analysis of Reciprocals for failure times*

Source	d.f	Reciprocal Component	F (App)	p − value (Approx)
Regression	1	4.3268	398.4	≪ 0.001
Residual lack of fit	1	0.0042	0.3868	≈ 0.52
pure error	27	0.2932		
Total	29	4.6242		

The regression component is highly significant and there is no evidence of lack-of-fit.

Referring back to the maximum likelihood estimators given by Whitmore the estimators $\hat{\beta}$ and $\hat{\sigma}$ are called pseudo maximum likelihood estimators by Jogesh Babu and Chaubey (1996) since there is no guarantee that $x_i^t \hat{\beta} > 0$ for all i. Whitmore's model circumvents this with the assumption that when the drift is negative the density has a defective distribution, and a further assumption that the estimators are non-negative.

Jogesh Babu and Chaubey study the asymptotics of the estimators $\hat{\beta}$ and $\hat{\sigma}$ under general assumptions on the design points and show that the estimators have many desirable properties. A bootstrap procedure is suggested and it is shown that the bootstrap estimates are close to the crude estimates and provides an estimate of the bias and variance of the pseudo maximum likelihood estimators.

Recall that the parameter space is $\Theta = \{\beta = (\beta_1, \cdots, \beta_p), \bar{\nu} = \nu^{-1} \mid x_i^t \beta > 0\}$. This is replaced by the stronger assumption $\inf_i x_i^t \beta > 0$ for all i and $\beta \in \Theta$.

Let us write $Y_i^{-1} = \delta_i + \varepsilon_i$ so that it is readily seen that

$$\frac{Y_i \varepsilon_i^2}{\nu} = \frac{Y_i(Y_i^{-1} - \delta_i)^2}{\nu} = \frac{(1 - \delta_i Y_i)^2}{\nu Y_i}$$

are i.i.d. χ_1^2 random variables. The following two theorems relate to the asymptotic laws of functions of $\hat{\nu}$ and $\hat{\beta}$. Details of the proof are to be found in Jogesh Babu and Chaubey (1996).

Theorem O.2 *Suppose for each $\beta \in \Theta$, there exists a constant $c_\beta > 0$ satisfying*
  i) $x_i^t \beta \geq c_\beta$ *for all* i
  ii) $\max_{1 \leq i \leq n} \| x_i^3 \|$ *to* $((X^t X)^{-1}) \to 0$ *as* $n \to \infty$.

Further let

(a) $U_1 = \sqrt{\dfrac{n}{2}} \left(\dfrac{\hat{\nu}}{\nu} - 1 \right)$

and

(b) $U_2 = \sqrt{\dfrac{X^t M X}{\nu}} \left(\hat{\beta} - \beta \right).$

Then $\mathcal{L}(U_1, U_2^t) \to N_{p+1}(0, I_{(p+1)}).$

Here M is the diagonal matrix with diagonal entries $m_i = \delta_i^{-1}$ and by $(X^t M X)^{\frac{1}{2}}$ we mean the unique positive definite square root of the symmetric positive definite matrix $X^t M X$.

We remark that $X^t M X$ is positive definite (since $m_i > 0$ for all i) whenever $X^t X$ is positive definite.

Observe that condition (a) implies that $\dfrac{1}{\mathrm{E}(Y_i)}$ are bounded away from zero. Condition (b) holds if for some sequence of positive real numbers $q_n \to \infty$

(c) $\max\limits_{1 \leq i \leq n} \| x_i \| = O\left(q_n^{\frac{1}{3}} \right)$

and

(d) $q_n^{-1}(X^t X) \to S$, a positive definite matrix.

The proof of Theorem O.2 is based on the following lemmas which we state without proof.

Lemma O.1 Let $d_{in} = x_i^t A^{-1} x_i$ where $A = X^t M X$, and $d_n = \max\limits_{1 \leq i \leq n} d_{in}$. Then from condition (b) of the theorem $d_n \to 0$ as $n \to \infty$.

Lemma O.2 Under conditions (a) and (b) of Theorem O.2 the positive definite matrix C where $C = A(Z^t Y X)A$ in probability to I_p.

Lemma O.3 Let $Y^t = (Y_1, \cdots, Y_n)$, and $V = X^t(I - M^{-1}Y)$. Then under the hypothesis of Theorem O.2

$$\mathcal{L}\left[\sqrt{\dfrac{n}{2}} \left(\left(\dfrac{\hat{\nu}}{\nu} - 1 \right), \dfrac{V^t A}{\sqrt{\nu}} \right) \right] \to N_{p+1}(0, I_{p+1})$$

where

$$n\hat{\nu} = \sum_{i=1}^{n} (1 - \delta_i Y_i)^2 Y_i^{-1}.$$

Proof of Theorem O.2 Since $\hat{\beta} - \beta = (X^t Y X)^{-1} V$, from Lemma O.3, the vector $\{\sqrt{\dfrac{n}{2\nu}}(\hat{\nu} - \nu), (\hat{\beta} - \beta)^t (X^t Y X) A\} \to N_{(p+1)}(0, \nu I_{p+1})$. This, together with Lemma O.2 shows that $(X^t M X)^{1/2}(\hat{\beta} - \beta)$, C^{-1} and C are bounded in probability. (Note that $A(X^t Y X)(\hat{\beta} - \beta) = C(X^t M X)^{1/2}(\hat{\beta} - \beta))$. Therefore $(X^t M X)^{1/2}(\hat{\beta} - \beta) - A(X^t Y X)(\hat{\beta} -$

$\beta) = (I-C)(X^tMX^{1/2}(\hat{\beta}-\beta) \to 0$ in probability. Thus $(X^tMX)^{1/2}(\hat{\beta}-\beta)$ tends in law to
$$N_p(0, \nu I_p). \tag{O.3}$$

Furthermore since $n\tilde{\nu}\nu^{-1} = \sum_{i=1}^n Y_i\varepsilon_i^2\nu^{-1}$ (the sum of n independent χ_1^2 random variables) we see that

$$n\tilde{\nu} = (YX\beta - e)^t Y^{-1}(YX\beta - e) = n\hat{\nu} + (\hat{\beta}-\beta)^t(X^tYX)(\hat{\beta}-\beta).$$

From equation O.1 and Lemma O.2,

$$\nu^{-1}(\tilde{\beta}-\beta)^t(X^tYX)(\hat{\beta}-\beta)$$
$$= \mu^{-1}(\hat{\beta}-\beta)^t(X^tMX)^{1/2}C(X^tMX)^{1/2}(\hat{\beta}-\beta) \to \chi_1^2$$

in law. Hence $\sqrt{n}(\hat{\nu}-\hat{\nu}) \to 0$ in probability. ♣

Remarks

(1) It can be shown (using the spectral decomposition and the Cauchy-Schwarz inequality) that $tr((X^tMX)^{-1}) \leq (\max_{1\leq i \leq n} m_i^{-1}) \, tr((X^tX)^{-1})$. Since the right side tends to zero from the hypothesis of Theorem O.2,

$$\| \hat{\beta}-\beta \| \leq \| X^tMX \|^{1/2} (\hat{\beta}-\beta) \| \, tr((X^tMX)^{-1})^{1/2} \to 0$$

in probability, (Note that $X^tJX \leq \| x \|^2 trJ$, and $X^tJ^2X \leq \| X \|^2 (trJ)^2$) and we obtain the weak consistency of $\hat{\beta}$.

(2) From Theorem O.2 the weak consistency of $\hat{\nu}$ follows automatically.

(3) Suppose that $n^{-1}(X^tMX) \to T$ for a positive definite matrix then under the hypothesis of Theorem O.2

$$\mathcal{L}\left((n\hat{\nu}^{-1})^{1/2}(\hat{\beta}-\beta)\right) \to N_p(0, T^{-1}).$$

The usefulness of Theorem O.2 is restricted by the fact that M is usually unknown. However, the same limit law can be obtained if Y replaces M as shown in the next theorem.

Theorem O.3 *Under the hypothesis of Theorem O.2*

$$\mathcal{L}(U_1, U_2^{*t}) \to N_{p+1}(0, I_{p+1})$$

where

$$U_2^* = \left(\frac{X^tYX}{\nu}\right)^{\frac{1}{2}}(\hat{\beta}-\beta).$$

Proof From Theorem O.2

$$(X^tMX)^{-1}X^tYX - I = (X^tMX)^{-1}(X^t(Y-M)X) \to 0 \quad (O.4)$$

The right side tends to zero in probability. Furthermore for any two symmetric non-negative definite matrices C, D if $C^2, D^2 \to I$ it can be shown that $CD \to I$. Using this fact and Equation O.4 one can show that $((X^tMX)^{1/2})^{-1}(X^tYX)^{1/2} \xrightarrow{P} I_p$. ♣

Strong consistency and bootstrap

The strong consistency of the estimators $\hat{\beta}$ and $\hat{\nu}$ are needed to study bootstrap methodology. Jogesh Babu and Chaubey establish a Theorem using some stronger conditions than those used in Theorem O.2. We state this result in the next theorem.

Theorem O.4 *Assume that the hypothesis (a) of Theorem O.2 holds. Furthermore suppose that* $(\log n) \max_{1 \le i \le n} \| x_i \|^2 tr((X^tX)^{-1}) \to 0$ *as* $n \to \infty$. *Then* $\hat{\beta} - \beta \to 0$ *and* $\hat{\nu} - \nu \to 0$ *almost everywhere.*

The proof of the theorem rests on showing
(1) $C = A(X^tYX)A \to I_p$ a.s.,
(2) $D = (X^tMX)^{-1}(X^tYX) \to I_p$ a.s.,
(3) $C^{-1} \to I_p$ a.s.,
(4) $D^{-1} \to I_p$ a.s..

To prove (1) one first shows that $P(|\sum(Y_i - m_i)a_{in}| > \varepsilon) = O(n^{-2})$ for any set of constants a_{in}, $i = 1, \cdots, n$, provided

$$(\max_{1 \le i \le n} m_i |a_{in}|) \log n \to 0 \text{ and } (\log n) \sum_{i=1}^{n} m_i^3 a_{in}^2 \to 0. \quad (O.5)$$

Using (O.5), then the Borel-Cantelli lemma the almost sure convergence of $\sum_{i=1}^{n}(y_i - m_i)a_{in}$ to zero follows. Now take a_{in} to denote the $(k, \ell)^{\text{th}}$ element of $Ax_i x_i^t A$ to verify the above convergence results. We omit the details.

To prove (2) take a_{in} as the $(k, \ell)^{\text{th}}$ element of $(X^tMX)^{-1}x_i x_i^t$ and show that $a_{in}^2 \le \| \beta \|^2 (\max_{i \le j \le n} \| x_j \|^6)(tr(X^tX)^{-1})^2$ and $\sum^n m_i^3 a_{in}^2 \le \| \beta \| C_\beta^{-2}(\max_{1 \le j \le n} \| x_j \|^3)tr(X^tX)^{-1}$ thus verifying the above convergence results. The results in (3) and (4) follow from the fact that $|C|$ and $|D|$ converge to 1 and the $(i,j)^{\text{th}}$ cofactor goes to 1 or 0 according as $i = j$ or not.

Using these 4 results

$$\hat{\beta} - \beta = ((X^tYX)^{-1} - (X^tMX)^{-1}) X^t e$$
$$= (D^{-1} - I_p)\beta \to 0 \quad \text{a.s.}$$

The strong consistency of $\hat{\nu}$ is proved by showing that

$$(\hat{\beta} - \beta)^t(X^tYX)(\hat{\beta} - \beta)(\log n)^{-2} \to 0 \quad \text{a.s..}$$

The sampling distribution of the estimators $\hat{\beta}$ and $\hat{\nu}$ are estimated by a bootstrap procedure.

Consider Y_i^* $i = 1, \cdots, n$ which are i.i.d $IG\left(\frac{1}{x_i^t\hat{\beta}}, \hat{\nu}\right)$ and let $Y^* = $ diag $(Y_1^*, \cdots, Y_n^*)$. Given the original sample Y the pseudo maximum likelihood estimator of $\hat{\beta}$ is $\beta^* = (X^tY^*X)^{-1}X^te$.

Now

$$x_i^t(\hat{\beta} - \beta) = x_i^t A (X^tMX)^{1/2}(\hat{\beta} - \beta)$$

so that

$$\| x_i^t(\hat{\beta} - \beta) \| \leq [x_i^t(X^tMX)^{-1}x_i]^{1/2} \| (X^tMX)^{1/2}(\hat{\beta} - \beta) \|.$$

Hence under the hypothesis of Theorem O.2 we have

$$P\left\{x_i^t\hat{\beta} > \frac{1}{2}C_\beta \text{ for all } i\right\} \to 1 \quad \text{as } n \to \infty$$

and under the hypothesis of Theorem O.4 $\inf_{1 \leq i \leq n} x_i^t\beta > 0$, for all large n and for almost all sample sequences. The next theorem shows the asymptotic normality of the bootstrap estimator.

Theorem O.5 *Under the hypothesis of Theorem O.4 we have for almost all sample sequences*

$$\mathcal{L}\left(\left(X^t\hat{M}X\right)^{1/2}\left(\beta^* - \hat{\beta}\right)\right) \to N_p(0, \nu I_p)$$

where $\hat{M} = $ diag $(x_1^t\hat{\beta}, \cdots, x_n^t\hat{\beta})$. Consequently for almost all sample sequences

$$\sup_{Z \in \mathbb{R}^p} | P^*\left(\left(X^t\hat{M}X\right)^{1/2}\left(\beta^* - \hat{\beta}\right) \leq z\right)$$

$$-P\left(\left(X^tMX\right)^{\frac{1}{2}}\left(\hat{\beta} - \beta\right) \leq z\right) | \to 0.$$

Here P^* denotes the probability measure induced by the bootstrap sampling procedure given the original data.

Babu and Chaubey also remark that under general conditions of x_i it can be shown that for any $\theta > 0$ and a p-dimensional Borel set H,

$$P\left(AX^tYX\left(\hat{\beta} - \beta\right) \in \nu H\right)$$

$$-P^*\left(\left(X^t\hat{M}X\right)^{1/2}X^tY^*X\left(\beta^* - \hat{\beta}\right) \in \hat{\nu}H\right) = O\left(n^{-1/2}\right)$$
$$+ O\left(\Phi_p(\partial H)^{\theta/\sqrt{n}}\right).$$

Example G.1 (continued) As an application Babu and Chaubey consider the example studied by Bhattacharyya and Fries. Recall that these authors had found that the reciprocal linear model provided an adequate fit of the failure time data studied by Nelson. Let us recall that the x_i values are $x_i = 10^{-8}(t_i^3 - 180^3)$ and the model is

$$\delta_i = \mu_i^{-1} = \alpha + \beta x_i.$$

In analyzing the Nelson data, Babu and Chaubey used the entire data set (batches I and II) to arrive at their estimates unlike Bhattacharyya and Fries. They have

$$\hat{\alpha} = 0.037310$$
$$\hat{\beta} = 7.317285$$

and

$$\hat{\nu} = 0.040233.$$

Theorems O.3 and O.4 provide crude estimates of the covariance matrix as $\hat{\nu}(X^tYX)^{-1}$. The asymptotic normality of the estimators can be used to obtain confidence intervals for $\hat{\alpha}, \hat{\beta}$ and other related parameters.

Estimates of the variances of $\hat{\alpha}$, and $\hat{\beta}$ are

$$\widehat{\text{var}}(\hat{\alpha}) = 5.593833 \times 10^{-4}$$
$$\widehat{\text{var}}(\hat{\beta}) = 0.242709.$$

Thus a $100(1 - \alpha)\%$ confidence interval for α is $\hat{\alpha} \pm Z_{\alpha/2}\sqrt{\widehat{\text{var}}\hat{\alpha}}$ where $Z_{\alpha/2}$ is the $100(1 - \alpha)$th percentile of the standard normal.

Based on 500 bootstrap samples $(\hat{\alpha}^*_{(i)}, \hat{\beta}^*_{(i)}, \hat{\nu}_{(i)})$, $i = 1, \cdots, 50$, the averages of these give the bootstrap estimates as

$$\hat{\alpha}^* = 0.039949$$
$$\hat{\beta}^* = 7.326207$$

and

$$\hat{\nu}^* = 0.038258.$$

Estimates of the biases of the pseudo maximum likelihood estimates are, therefore

$$\hat{\alpha}^* - \hat{\alpha} = 0.002639$$
$$\hat{\beta}^* - \hat{\beta} = 0.008922$$

and
$$\hat{\nu}^* - \nu = -0.001987.$$

Also the sample variances from the bootstrap scheme gives estimates of var $\hat{\alpha}$ and var $\hat{\beta}$ as

$$\widehat{\text{var}}(\hat{\alpha}^*) = 5.214227 \times 10^{-4} \quad \text{and} \quad \widehat{\text{var}}(\hat{\beta}^*) = 0.245392,$$

values which are quite close to the crude estimates. However, the usefulness of the bootstrap technique lies in the computation of the estimates of the bias and variance of the pseudo maximum likelihood estimators.

P. Slug lengths in pipelines

A new technology termed the multiphase production system deals with a search for low cost recovery schemes for offshore hydrocarbon fields. Applications of this system are to be found in TOTAL'S development programs in Indonesia, Argentina, Egypt, Thailand and the North Sea. The past and the future developments require an estimation of slug characteristics in order to facilitate downstream process designing. Slug length data from BHRG pipelines have been analyzed carefully by Dhulesia, Bernicot and Deheuvels (1991) and the superiority of the inverse Gaussian fit over the close competitors, the lognormal and the gamma laws has been established.

Most of the multiphase pipelines encounter a slug flow regime at the outlets, caused by long distances, upstream changes in flowrate, depressurization of pipeline and the like. To obtain a reasonable estimate of the pressure drop the average slug length prediction is a key factor. The maximum possible slug length, too, enters the analysis. Dhulesia et al., propose the IG law to fit the entire slug length distribution for the data sets of a wide range of liquid and gas superficial velocities. They also present studies to predict the slug length distribution.

Continuous measurements of slug length and velocity were observed at three different locations of the pipe loops over an extended time period and upto one thousand slugs at each site were collected. The air and water were maintained at a constant flow rate. In all thirty sets of experimental data were obtained over ten series of experiments. The final analysis involved only 28 data sets (two being declared unworthy). To make the data sets homogeneous the slug lengths larger than one metre were retained. The sample sizes for these 28 data sets varied from 223 to 999.

The authors noted that an individual slug is generated at the inlet of a pipeline and moved towards the outlet. Denoting by $X(t)$ the distance of the slug from the inlet at time $T > 0$, they found it reasonable to assume that $X(t)$ followed a Brownian motion process with drift ν and diffusion constant σ^2. Thus

$$X(t) = \nu t + \sigma W(t)$$

where $\{W(t) \mid t \geq 0\}$ is a standard Wiener process. They also assume that the total liquid hold-up at $t = 0$ inside the pipeline is proportional to the cumulative slug lengths present in the pipeline at $t = 0$. Moreover, they concluded that the total liquid hold-up at $\nu = 0$ is proportional to the residence time of an individual slug at $t = 0$.

Thus if a is the length of the pipeline, $Y(a)$ the residence time of slug passing through the pipeline and $Z(a)$ is the cumulative length of all slugs present in the pipeline at time $t = 0$, then for a proportionality

constant C
$$Y(a) = CZ(a)$$
so that
$$Z(a) = C^{-1}Y(a).$$

Since the first passage through a of a slug follows $IG(\mu = \frac{a}{\nu}, \lambda = \frac{a^2}{\sigma^2})$ it follows that $Y(a)$ follows the same law and finally $Z(a)$ is distributed as $IG\left(\frac{\mu}{C}, \frac{\lambda}{C}\right)$. Now because of infinite divisibility and self decomposability of the IG law it is clear that individual slug lengths are IG distributed.

Dhulesia et al., have fitted an IG law with a threshold α (see Section 2.2), namely $IG(\mu, \alpha, \lambda)$ since the slug lengths are at least 1m long.

Table P.1 *Parameter estimates for data sets 20 and 27*

data set number	20	27
$\theta = \mu + \alpha$	4.19	10.01
$\theta = \mu^3/\lambda$	2.81	14.05
$\phi = \lambda/\mu$	13.39	23.87

The method proposed by Cheng and Amin (1981) was used to fit the $IG(\mu, \alpha, \lambda)$ law. The estimates of the parameters θ and σ^2 are given in Table P.1.

Since the lognormal and the IG fits are indistinguishable, the closeness of the fits was examined by using the Kolmogorov distance norm, viz
$$d(F, G) = \sup_x | F(x) - G(x) |.$$

The authors conclude on the basis of this statistic that at both the 1% and 5% levels the IG model was superior to the lognormal model.

Dhulesia et al., also report a non-linear transformation of the parameters θ, σ and α to a new set of parameters characterizing the fluid properties, the pipeline geometry and the operating conditions. They then find the maximum likelihood estimators by a non-linear optimization software package OPTOR developed by TOTAL and fit the $IG(\mu, \alpha, \lambda)$ model. They conclude that the slug length distributions are, in general, well fitted by the IG law (except for small slug length). With the non-linear transform, the entire slug length distribution seemed to be well predicted, as evidenced by slug length measured on a real pipeline of 42 km and 12" diameter.

Q. Ecology

A good description of stochastic population growth uses a discrete state space with population numbers taking on integer values. Demographic transitions like birth, death and growth are subject to two sources of variation. One is demographic stochasticity – in simple terms sampling randomness – and the other is environmental stochasticity or variation between individuals. Demographic stochasticity is caused by finite population size. The basic theory of population growth in fluctuating environments was first presented by Cohen (1977) and Tuljapurkar and Orzack (1980). Based on this theory Lande and Orzack (1988) derive a diffusion approximation for the logarithm of total population size in a population subject to density – independent fluctuations in vital rates. We describe, in the following discussion their development.

Time till extinction

We first digress to introduce the projection matrix first considered by Leslie (1945) which maps the state of an age structured population from one time to the next – the transition into the next state being assumed to depend only on the current state of the population.

Consider the method of computing the female population in one unit's time, given any arbitrary age distribution at time t. Let

$n_x(t)$ = the number of females in the age group x to $x+1$ at time t,

$p_x(t)$ = the probability that a female aged x to $x+1$ at time t will be alive in the age group $x+1$ to $x+2$ at time $t+1$,

$f_x(t)$ = the number of daughters in the interval t to $t+1$ per female alive aged x to $x+1$ at time t, who will be alive in the age group $0-1$ at time $t+1$.

Working from an origin of time, the age distribution at the end of one unit's interval is given by $(m+1)$ linear equations (assuming m to $m+1$ is the last age group in the life table description),

$$\sum_{i=0}^{m} f_i(t) n_i(t) = n_0(t+1)$$
$$p_0(t) n_0(t) = n_1(t+1)$$
$$p_1(t) n_i(t) = n_2(t+1)$$
$$\vdots$$
$$p_{m-1}(t) n_{m-1}(t) = n_m(t+1)$$

or in compact matrix notation

$$A(t) n(t) = n(t+1)$$

where for $i = 0, 1, \cdots$,
$$n(\cdot)^t = (n_0(\cdot), \cdots, n_m(\cdot))$$
and
$$A(t) = \begin{pmatrix} f_0(t) & f_1(t) & \cdots & f_{m-1}(t) & f_m(t) \\ p_0(t) & 0 & \cdots & 0 & 0 \\ 0 & p_1(t) & & & \\ \vdots & & & & \\ 0 & & & p_{m-1}(t) & 0 \end{pmatrix}$$
is a square matrix of size $(m+1) \times (m+1)$. This matrix is known as the Leslie matrix with age-specific fecundity rates in the first row, age-specific survival probabilities along the subdiagonal and zero elsewhere. Since the matrix projects the present state of a population into the future it is called a projection matrix in population biology. (No relation to projection operators.)

Lande and Orzack assume that the projection matrices change with time such that each of the entries (vital rates) constitutes a stationary time series. A further assumption on the life history, with non-periodic fecundities, is that no matter what the initial population size vector $n(0)$ is, the probability distribution of the natural logarithm of the total population size is asymptotically normal. In most cases it turns out that the rate of convergence to normality is fast enough to let changes in logarithmic population size, even for short periods, to be approximated by a diffusion process with constant infinitesimal mean μ and infinitesimal variance σ^2.

Since the individuals of different ages contribute unequally to the future population growth, the initial age distribution has a considerable effect on extinction probabilities. An accurate approximation of the effect of the initial age distribution on the total population size at time t is given by the initial total reproductive value in the population, $V_0 = <v, n(0)>$ where v is the dominant left eigenvector of the average projection matrix $\mathbb{E}(A(t))$.

When the fluctuations in the vital rates are "small" (assuming serially independent environments), a useful approximation for the asymptotic rates of change of mean and variance of the probability distribution of the logarithm of the population size is obtained from Tuljapurkar (1982). The infinitesimal mean μ and variance σ^2 are given by

$$\mu \approx \log \lambda_0 - \frac{\sigma^2}{2}$$
$$\sigma^2 \approx \lambda_0^{-2} \delta^t C \delta$$

where λ_0 is the dominant eigenvalue of the average projection matrix $A^* = \mathbb{E}[A(t)]$ (containing the average vital rates A_{ij}), δ is a column vector of sensitivity coefficients $\frac{\partial \lambda_0}{\partial A_{ij}} = \frac{V_i U_j}{<V,U>}$, V_i and U_j denoting the ith and jth elements of the dominant left and right eigenvectors of A^*, and C is the covariance matrix of the multivariate distribution from which the elements of $A(t)$ arise.

It is also possible to estimate μ and σ^2 fairly accurately from a single time series of observations on total population size because the growth rates form an ergodic series. (σ^2 must be calculated over segments of the sample path which are much longer than the characteristic autocorrelation time). The infinitesimal variance $\sigma^2 = 2(\ell n \lambda_0 - \mu)$ easily gives an estimate of μ.

Let $N(t)$ denote the total population size at time t and $X(t) = \ell n N(t)$ (the natural logarithm of $N(t)$). Further let $X(0) = \ell n(0)$ be the adjusted initial value at $t = 0$. Then

$$P[X(t) = x \text{ at time } t \mid X(0)] = p(x,t \mid x_0)$$

approaches the solution of the diffusion equation for the Wiener process

$$\frac{\partial p}{\partial t} = -\mu \frac{\partial p}{\partial x} + \frac{\sigma^2}{2} \frac{\partial^2 p}{\partial x^2}$$

with initial boundary condition $p(x, 0 \mid x_0) = \delta(x - x_0)$, the Dirac mass at x_0. When no extinction boundary is specified, the solution to p is $N(x_0 + \mu t, \sigma^2 t)$ where μ and σ^2 are the asymptotic results stated above. If, however, an extinction boundary (extinction below a population size of one) is imposed then

$$p(0, t \mid x_0) = 0.$$

With an absorbing barrier, the solution to the diffusion equation is

$$p(x,t) = \frac{1}{\sigma \sqrt{2\pi t}} \left[\exp \left\{ -\frac{(x - x_0 - \mu t)^2}{2\sigma^2 t} \right\} \right.$$
$$\left. - \exp \left\{ \frac{-2\mu x_0}{\sigma^2} - \frac{(x + x_0 - \mu t)^2}{2\sigma^2 t} \right\} \right].$$

If $g(t \mid x_0)dt$ denotes the probability of extinction in $(t, t+dt)$,

$$g(t \mid x_0) = -\frac{d}{dt} \int_0^\infty p(x, t \mid x_0) dx$$
$$= \frac{x_0}{\sigma \sqrt{2\pi t^3}} \exp \left\{ -\frac{(x_0 + \mu t)^2}{2\sigma^2 t} \right\},$$

and the probability of extinction before time t is $P[T \leq t]$ given by

$$G(t \mid x_0) = \int_0^t g(y \mid x_0) dy$$
$$= \Phi \left(\frac{-x_0 - \mu t}{\sigma \sqrt{t}} \right) + \exp \left(\frac{-2\mu x_0}{\sigma^2} \right) \left(1 - \Phi \left(\frac{x_0 - \mu t}{\sigma \sqrt{t}} \right) \right).$$

The probability of ultimate extinction is

$$G(\infty \mid x_0) = \begin{cases} 1 & \text{if } \mu \leq 0 \\ \exp\left\{-\frac{2\mu x_0}{\sigma^2}\right\} & \text{if } \mu > 0. \end{cases}$$

The conditional probability distribution of extinction times, considering only sample paths in which the population eventually becomes extinct and excluding those in which extinction never occurs in

$$g^*(t \mid x_0) = \frac{g(t \mid x_0)}{G(\infty \mid x_0)} = \frac{x_0}{\sigma\sqrt{2\pi t^3}} \exp\left\{-\frac{x_0 - (\mu)t)^2}{2\sigma^2 t}\right\}.$$

The conditional law of extinction times for $\mu > 0$ is exactly the same as that for $\mu < 0$ and hence the proper distribution of extinction times depends only on the magnitude and not on the sign of μ. This result follows from the theory of conditional diffusion processes (the infinitesimal variance of the conditional process for $\mu > 0$ is the same as that for unconditional processes with $\mu < 0$). The distribution is recognizable as IG whose mean and variance are

$$\mathbb{E}[T] = \frac{x_0}{|\mu|}, \quad Var[T] = \frac{x_0 \sigma^2}{|\mu|^3}.$$

The mode of the extinction time distribution is less than the mean and when $\mu = 0$, the moments do not exist. (It is assumed that the fluctuations in the projection matrix are small or moderate.)

Lande and Orzack show by computer simulations that the extinction dynamics of density-independent age-structural populations in a fluctuating environment can be modeled and predicted accurately as a diffusion process for the natural logarithm of total population size. They also note that the extinction of a population with a large value of μ is expected to be rapid, when it occurs, although eventual extinction is a low probability event. An explanation of this is given by the authors as follows. "When μ is positive there is a deterministic force acting to increase the population size. If the population goes extinct, it is most likely to happen near the beginning of the sample path (in a series of time intervals with unusually low population growth), before the population attains a size large enough that the same environmental sequence would not cause extinction. The more positive the long run growth rate, the more rapid and extreme must be the intervals of low growth in order to overcome the opposing deterministic force."

Endangered species

Survival or extinction of a population is a random phenomenon. The classical theory of demography is governed by a fixed Leslie matrix A of vital rates. Faced with an uncertain life, it is only natural that vital rates should change over time in the most unpredictable ways. When this is taken into account the specification of uncertainty becomes a key issue. For an excellent review of the development the reader should consult Tuljapurkar (1989).

Here we consider the estimation of growth and extinction parameters of an endangered species as outlined by Dennis, Munholland and Scott (1991). These authors follow the development of Lande and Orzack in their statistical analyses. Recall that Lande and Orzack showed using computer simulations based on the Wiener process (with drift), that the approximation for extinction-related quantities are reasonably accurate.

Dennis et al., consider a stochastic exponential growth model

$$N(t+1) = A(t)N(t+1)$$

where the entries of $A(t)$ are chosen randomly for each t from the same fixed multivariate population, independent of previous years. There may be correlation bewteen vital rates but there is no serial correlation between rates at different times. It is assumed that $X(t) = \log N(t)$, where $N(t)$ is the total population size at time t $(= e^t n(t), e^t = (1, 1, \cdots, 1))$ is approximately $N(x_0 + \mu t, \sigma^2 t)$ as t becomes large. Here $x_0 = \log N(0)$, $N(0)$ being the initial population size.

Now $X(t)$ could possibly cross any lower threshold size x_ℓ starting from x_0. This means that $N(t)$ reaches a lower barrier $\exp(x_\ell) = n_\ell$, say starting from $n_0 = N(0)$. For $n_\ell = 1$, $x_\ell = 0$ and the population is wiped out. As seen in the previous study $n_\ell > 1$ is often considered as a policy threshold. Denote by x_d

$$x_d = x_0 - x_\ell = \log\left(\frac{n_0}{n_\ell}\right)$$

the distance (on the logarithmic scale) from an initial population size to a lower threshold size. Then from Lande and Orzack (1988), given that the threshold is reached (conditional on all sample paths reaching the threshold), T the first time to reach the threshold is

$$f(t \mid x_d, \mu, \sigma^2) = \frac{x_d}{\sigma\sqrt{2\pi t^3}} \exp\left[-\frac{(x_d - \mid \mu \mid t)^2}{2\sigma^2 t}\right].$$

The distribution of the time to attain an upper threshold, given this threshold is attained, is also inverse Gaussian with a slight change, namely, with

$$x_d = x_u - x_0 = \log\left(\frac{n_u}{n_0}\right).$$

(In actual practice, x_d is specified by the experimenter.)

Suppose that a population is observed at times $0, t_1, \cdots, t_m$ and the recorded values are $n(0) = n_0, n(t_1) = n, \cdots, n(t_m) = n_m$. The time intervals (not necessarily equal) are $\tau_1 = t_1 - 0, \cdots \tau_m = t_m - t_{m-1}$.

Since $N(t) = \exp(X(t))$ is also a diffusion process with stationary transition probabilities, (indeed a Markov process) the probability distribution of each n_i given n_{i-1} is $p(n_i, \tau_i \mid n_{i-1})$ representing the likelihood of the system moving from n_{i-1} to n_i in a time interval τ_i is a log-normal distribution, namely

$$p(n_i, \tau_i \mid n_{i-1}) = \frac{1}{\sigma n_i \sqrt{2\pi \tau_i}} \exp\left[-\frac{(\log n_i - \log n_{i-1} - \mu \tau_i)^2}{2\sigma^2 \tau_i}\right].$$

Hence the log-likelihood of the sample observations $(n_0, \cdots, n_m)$, namely $\ell(\mu, \sigma^2)$ is

$$\ell(\mu, \sigma^2) = -\sum_{i=1}^{m} \log(n_i) - \frac{m}{2} \log \sigma^2 - \sum \log(2\pi t_i)^{1/2}$$
$$- \sum_{i=1}^{m} \frac{1}{2\sigma^2 \tau_i} (\log n_i - \log n_{i-1} - \mu \tau_i)^2.$$

The maximum likelihood estimates are found to be

$$\hat{\mu} = \frac{\sum_{i=1}^{m} \log \frac{n_i}{n_{i-1}}}{\sum_{i=1}^{m} \tau_i} = \frac{\log\left(\frac{n_m}{n_0}\right)}{t_m}$$

$$\hat{\sigma}^2 = \frac{1}{m} \sum_{i+1}^{m} \frac{1}{\tau_i} \left(\log\left(\frac{n_i}{n_{i-1}}\right) - \hat{\mu}\tau_i\right)^2.$$

Armed with these estimates it is possible to check model assumptions using a linear regression approach and estimate growth parameters, extinction parameters, or times to extinction. We consider briefly some of these issues.

The parameter representing the continuous rate of increase is denoted by r and figures in the stochastic differential equation of exponential growth. The exponential growth equation is (see Karlin and Taylor 1981 page 359)

$$\frac{dN(t)}{dt} = r(t)N(t)$$

where $r(t)$ is an instantaneous rate of growth at time t. To take into account the random environmental effects one stipulates that

$$r(t) = r + \sigma W(t)$$

where
$$W(t) = \frac{dB(t)}{dt},$$
and obtains the stochastic differential equation for the population size $N(t)$ at time t as
$$dN(t) = rN(t)dt + \sigma N(t)dB(t).$$
Here r is constant and $dB(t) \sim N(0, dt)$.

In terms of the statistical properties of the stochastic process $N(t)$, rn denotes the infinitesimal mean of $N(t)$; that is to say $rn\Delta t \approx$ average amount of change in $N(t)$ over Δt given $N(t) = n$. Now r has to be estimated from time series observations.

The quantity r is related to μ and σ^2 by the equation
$$r = \mu + \frac{\sigma^2}{2}$$
and can be estimated using the estimates of μ and σ^2 which depend on m and t_m.

Dennis et al., point out that the fundamental discrete-time nature of population growth for many vertebrate species suggests the use of the Ito interpretation of the stochastic differential equation for $N(t)$, especially when dealing with endangered species. Under the Ito calculus r is the rate constant in $\mathbb{E}(N(t)) = n_0 \exp\left[\left(\mu + \frac{\sigma^2}{2}\right)t\right]$. Under the Stratonovich Calculus r is the rate constant in the geometric mean of $N(t)$, i.e., $\exp\{\mathbb{E}(X(t))\} = \exp\{\mathbb{E}(\log N(t))\} = n_0 \exp(\mu t)$.

An estimate of the finite rate of increase denoted by $\lambda = \exp\{\mu + \sigma^2/2\}$ (the mean population size after one year divided by the initial size n_0) can also be calculated. Dennis et al., interpret λ as approximating the dominant eigenvalue of the projection matrix. They also show that as m and t_m become large
$$\tilde{\lambda} = \exp(\hat{\mu})_0 F_1\left(\frac{m-1}{2}; \frac{m-1}{4}\hat{\sigma}^2\right) \to N(\lambda, c^2)$$
where
$$_0F_1(\nu; z) = \sum_{i=0}^{\infty} \frac{z^i}{\nu)_i(i!)}, (\nu)_i = \nu(\nu+1)\cdots(\nu+i-1) \text{ with } \nu_0 = 1,$$
and $c^2 = \lambda^2 \left[\exp\frac{\sigma^2}{m} {_0F_1}\left(\frac{m-1}{2}; \frac{(m-1)^2}{4m^2}\sigma^2\right) - 1\right]$. The authors warn that convergence to normality may be slow and recommend use of the approximate normal distribution of $\tilde{r} = \hat{\mu} + \frac{m\hat{\sigma}^2}{(m-1)^2}$, namely
$$\tilde{r} \simeq N\left(r, \frac{\sigma^2}{t_m} + \frac{\sigma^4}{2(m-1)}\right)$$

for confidence interval construction.

The growth rate of the geometric mean population size is denoted by α (Tuljapurkar 1982) and equals $\exp(\mu)$, and is a better descriptor of the growth rates of the typical sample paths. Its estimate is given by

$$\hat{\alpha} = \exp(\hat{\mu})$$

or

$$\tilde{\alpha} = \exp(\hat{\mu}) {}_oF_1\left(\frac{m-1}{2}; \frac{\hat{\sigma}^2}{4}\right).$$

Given the observations $n_0, \cdots, n_m$ at times $t_0 = 0, t_1, \cdots, t_m$, it may be desirable to predict the state of the system at $t > t_m$, or predict the value of $\log N(t) = X(t)$. Using the Markovian nature of the diffusion process it can be shown that

$$\mathbb{E}[X(t) \mid X(t_m) = x_m] = x_m + \mu(t - t_m) = \gamma$$

and the mean squared prediction error (the unconditional expectation of $(X(t) - \gamma)^2$ over all possible realizations of the process) is given by

$$\sigma^2 \frac{(t - t_m)t}{t_m}.$$

For estimating the probability of reaching a lower threshold x_d we have

$$\hat{p} = \begin{cases} 1 & \text{if } \hat{\mu} \leq 0 \\ \exp\left(-\frac{2\hat{\mu}x_d}{\hat{\sigma}^2}\right) & \text{if } \hat{\mu} > 0. \end{cases}$$

Taking the latest value of n, namely n_m as the 'initial value' and writing $x_d = \log\left(\frac{n_m}{n_\ell}\right)$ we have, for $\hat{\mu} > 0$

$$\hat{p} = \left(\frac{n_\ell}{n_m}\right)^{\frac{2\hat{\mu}}{\hat{\sigma}^2}}.$$

This value can be plotted as a function of n_ℓ or even $n_m - n_\ell$ in order to asses the rate of decrease of the probability as n_ℓ decreases.

The approximate variance of $\hat{p}$ is given as

$$Var(\hat{p}) \approx (2x_d p)^2 \left[\frac{1}{t_m \sigma^2} + \frac{2\mu^2(m-1)}{(\sigma^2 m)^2}\right]$$

when $\mu > 0$ and $\hat{p}$ converges to a normal law with mean p and the above variance.

As an illustration of the principles in the discussion above we first consider the California Condor population studied by Dennis et al.

Example Q.1 One of the most endangered bird species is the California Condor. By 1987 all condors remaining in the wild had been taken into captivity and as reported in 1980 only a flock of 32 birds with 11 potential breeding pairs (now) remained. Data collected from 1965 till 1980 indicate a phenomenal decline in the condor population. The results of a survey by Wilbur (1980) and Snyder and Johnson (1985) yield the following estimates

Table Q.1 *Estimates of Condor population*

Year	Estimate	Year	Estmate
65	38	73	19
66	51	74	23
67	46	75	29
68	52	76	22
69	53	77	13
70	28	78	13
71	34	79	19
71	36	80	12

The maximum likelihood estimates of the various parameters are as follows.

$$\hat{\mu} = \frac{\log\left(\frac{n_m}{n_0}\right)}{t_m} = \frac{\log\left(\frac{12}{38}\right)}{15} = \frac{-1.15202}{15}$$
$$= -0.0768$$

$$\tilde{\sigma}^2 = \frac{1}{14}\sum_{i=1}^{15}\left(\log\left(\frac{n_i}{n_{i-1}}\right) - \hat{\mu}\right)^2$$
$$= 0.1199$$

$$\tilde{r} = \hat{\mu} + \frac{\tilde{\sigma}^2}{2}$$
$$= 0.0169$$

$$\tilde{\lambda} = \exp(\hat{\mu})\,_0F_1\left(7; \frac{7}{2}\tilde{\sigma}^2\right)$$
$$= 0.9792$$

$$\tilde{\alpha} = \exp(\hat{\mu})\,_0F_1\left(7; \frac{15}{4(14)}\tilde{\sigma}^2\right)$$
$$= 0.9297.$$

One can now plot both the IG law, namely, $IG\left(\frac{x_2}{|\hat{\mu}|}, \frac{x_d}{\hat{\sigma}}\right)$ and the distribution function

$$P(T \leq t) = \Phi\left(\frac{-x_d + |\hat{\mu}| t}{\hat{\sigma}\sqrt{t}}\right) + \exp\left(\frac{2x_d |\hat{\mu}|}{\hat{\sigma}^2}\right) \Phi\left(\frac{-x_d - |\hat{\mu}| t}{\hat{\sigma}\sqrt{t}}\right).$$

Using $n_{15} = 12$ as the starting population size and n_ℓ the lower threshold as $= 1$ for extinction we see clearly that

$$\hat{p} = 1 \text{ since } \hat{\mu} \leq 0.$$

The mean number of years to extinction is estimated as

$$\frac{x_d}{|\hat{\mu}|} = \frac{\log \frac{12}{1}}{0.0768} = \frac{248.49}{7.68} = 32.3.$$

This enabled the managers to attempt to arrest the dwindling population. The probability of extinction within twenty years is close to 0.4 and the most likely time to extinction (mode) can be shown to be under 15 years. The growth rate estimates of $\hat{\mu}, \tilde{r}, \hat{\lambda}$ and $\hat{\alpha}$ indicate a slow decline but the high value of $\tilde{\sigma}^2$ is an ominous indication of the extinction of the population.

Example Q.2 The next example deals with a Whooping Crane population – a one extremely endangered species now undergoing a promising recovery program. The Whooping Crane is a long-lived bird which stands 1.5m tall and has a wingspan of 2.1m. It becomes sexually mature at age 5 and lays 2 eggs per clutch and often only one chick is raised to fledging age.

A census of Whooping Cranes begun in 1938 has continued to the present day and Dennis et al., report that there is evidence of an overall trend of exponential growth. The Whooping Crane recovery program calls for downlisting the species from endangered to threatened status when 40 nesting pairs would correspond to a total population of 153 birds.

As a result of major programs one wild population that winters in the Aransas National Wildlife refuge on the Gulf Coast of Texas increased from 18 birds in 1938 to 146 birds in 1989. The data analyzed by Dennis et al.,(first considered by Binkley and Miller in the Canadian Journal of Zoology(1983)vol.61,p.2769 and later updated by Boyce) are given in Table Q.2.

Table Q.2 *Whooping Crane (Aransas, Texas)*

Year	j	Population size n_{j-1}	n_j	$y_j = \ell n(n_j/n_{j-1})$
38	1	18	22	0.20067070
39	2	22	26	0.16705408
40	3	26	16	−0.48550782
41	4	16	19	0.17185
42	5	19	21	0.10009
43	6	21	18	−0.15415
44	7	18	22	0.20117
45	8	22	25	0.12783
46	9	25	31	0.21511
47	10	31	30	−0.03279
48	11	30	34	0.12517
49	12	34	31	−0.09238
50	13	31	25	−0.21511
51	14	25	21	−0.17435
52	15	21	24	0.13353
53	16	24	21	−0.13353
54	17	21	28	0.31468
55	18	28	24	−0.15415
56	19	24	26	0.08004
57	20	26	32	0.20764
58	21	32	33	0.03077
59	22	33	36	0.08701
60	23	36	39	0.08005
61	24	39	32	−0.19783
62	25	32	33	0.03077
63	26	33	42	0.24116
64	27	42	44	0.04652
65	28	44	43	−0.01298
66	29	43	48	0.11000
67	30	48	50	0.04082
68	31	50	56	0.11333
69	32	56	57	0.01770
70	33	57	59	0.03448

Endangered species

Table Q.2 (ctd.)

Year	j	Population size n_{j-1}	n_j	$y_j = \ell n(n_j/n_{j-1})$
71	34	59	51	−0.14571
72	35	51	49	−0.04000
73	36	49	49	0.00000
74	37	49	57	0.15123
75	38	57	69	0.19105
76	39	69	72	0.04256
77	40	72	75	0.04082
78	41	75	76	0.01325
79	42	76	78	0.02597
80	43	78	73	−0.06625
81	44	73	73	0.00000
82	45	73	75	0.02703
83	46	75	86	0.13686
84	47	86	97	0.12037
85	48	97	110	0.13126
86	49	110	134	0.19736
87	50	134	138	0.02941
88	51	138	146	0.05635 ⋯
89	52	146	.	.

The 1940-1941 transition was deleted as an outlier. Thus the calculations based on 50 transitions give

$$\hat{\mu} = \frac{\log\left(\frac{26}{18}\right) + \log\left(\frac{146}{16}\right)}{50} = 0.05156$$

$$\tilde{\sigma}^2 = 0.1475$$

$$\tilde{r} = 0.5895$$

$$\tilde{\lambda} = 1.061$$

$$\tilde{\alpha} = 1.053.$$

Starting from a population size of 110 birds the mean time to reach 153 birds is estimated as 7.34 years; thus the threshold is reachable by 1993, and in this case $\hat{p} = 1$.

R. Entomology

Biological models for describing the life history of an insect population through the various stages of its life-cycle have been constructed by several authors, as for example, Read and Ashford (1968), Kempton (1979), Munholland (1988), Munholland and Kalbfleish (1991) and Munholland and Dennis (1992). These statistical models are characterized by semi-Markov processes with discrete state space corresponding to the various stages of the life-cycle of an individual and are known as microscopic models. We follow this development in the spirit of the work due to Munholland and Kalbfleish and illustrate it with some examples considered by Munholland and Dennis.

A stochastic model

The development of an insect can be regarded as an aggregation of small increments of growth over time measured on a variety of scales like weight and age, and possibly some losses of development. The net growth accumulated upto a time $t > 0$ of an individual will be denoted by $X(t)$. The process $\{X(t) \mid t \geq 0\}$ is assumed to be a Brownian motion process with positive drift parameter ν (the development rate) and variance parameter $\sigma^2 = 1$. It is assumed that ν does not change with the various life stages. When $t = 0$ (the origin) the growth level $X(0) = 0$. In actual practice $X(t)$ is not observable and instead a discrete development stage is observed and the number of adults and any other insects in immature stages is recorded. The growth level $X(t)$ is related to the observed life stage by requiring that an amount α_i is necessary for molting into stage i. Let there be $(k+1)$ stages $\alpha_0 < \alpha_1 < \cdots < \alpha_k < \infty$ where $\alpha_0 = 0$, and suppose that the first time T_i for the development process $X(t)$ to attain α_i in stage i has an inverse Gaussian law with parameters μ_i and λ_i, i.e., $IG(\mu_i, \lambda_i)$, $i = 1, \cdots, k$. Thus an individual is alive in stage i at time t if and only if

$$T_i \leq t < T_{i+1}.$$

The parameters μ_i and λ_i are stage dependent and

$$\mu_i = \frac{\alpha_i}{\nu}, \quad \lambda_i = \alpha_i^2 \quad i = 1, \cdots, k.$$

The molting threshold $\alpha_0 = 0$ implies that $T_0 = 0$ with probability 1. The entry time T_i into stage i is also referred to as recruitment time or molting time to stage i.

We remark that some investigators have a different definition of stage occupancy as for example, $\alpha_i \leq X(t) < \alpha_{i+1}$ if and only if an individual is in stage i. This definition permits a biological inconsistency in that an individual in stage i at time t can possibly enter an earlier

A stochastic model

stage at time t. By defining stage occupancy in terms of entry times the inconsistency is avoided.

The time spent in stage i (sojourn or residence time) is defined as

$$S_i = T_{i+1} - T_i,$$

and the net development accumulated in stage i is $\beta_i = \alpha_{i+1} - \alpha_i$.

The above definition of sojourn time in stage i clearly shows that

$$\mathcal{L}(S_i) = IG(\beta_i/\nu, \beta_i^2).$$

So far we have assumed that there is no mortality in any of the stages. In real life it is meaningful to introduce an age-dependent mortality rate $\theta(t)$ or stage-dependent mortality rate $\theta_i(s)$. Stage frequency data on live insects, however, seem to be less informative for the estimation of mortality parameters, besides adding to the complexity.

As a simple approximation, the mortality rate $\theta(t)$ can be assumed to be a constant θ, and this then implies that the stage duration times of an individual are exponentially distributed with parameter θ. Thus with this assumption of a constant death rate we can calculate the probability of an individual insect being in stage i at time t as

$$p_i(t) = P \text{ (survival to time } t) \, P(T_i \leq t \leq T_{i+1} \mid \text{survival to time } t).$$

Now

$$P(\text{survival to time } t) = \exp(-\theta t)$$

while

$$\begin{aligned} & P(T_i \leq t \leq T_{i+1} \mid \text{survival to time } t) \\ =& P\left[(T_i \leq t) \cap (T_{i+1} > t) \mid \text{survival to time } t\right] \\ =& P\left[(T_i + S_i > t) \cap (T_i \leq t) \mid \text{survival to time } t\right] \\ =& G_i(t) - G_{i+1}(t) \end{aligned}$$

where $G_i(t) = P(T_i \leq t)$.

Thus for $i = 0, 1, \cdots, k$ we obtain

$$p_i(t) = [G_i(t) - G_{i+1}(t)] \exp(-\theta t), \tag{R.1}$$

an equation which describes the structure of the insect population at arbitrary points of time besides giving the probabilities at various stages.

It is worth mentioning that Kempton (1979) advanced three distributions for $G_i(t)$, namely, the normal, gamma and the inverse Gaussian due to the additive properties associated with these laws. But the motivation behind the use of the IG law as proposed by Munholland and Kalbfleish is based on biological factors and the resulting homogeneous semi-Markov model for the life history process thus represents a continuous process in continuous time, allowing for biological interpretation of the parameters of the model. We now turn to the problem of estimation.

Estimation and model adequacy

A population under study deals with individuals randomly dispersed over a confined homogeneous area with independent life histories described by $p_i(t)$ as in equation (R.1). Assume that sampling is conducted at times $t_i < t_2 \cdots < t_m$ and that at time t_j a proportion, c, of the area is sampled and n_{ij}, the number of insects in stage i is observed in the jth sample, $i = 1, \cdots, k$, $j = 1, \cdots, m$. Furthermore it is assumed that when c is small, the n_{ij} have independent Poisson distributions, i.e.,

$$\mathcal{L}(n_{ij}) = P_o(\eta p_i(t_j)),$$

where η denotes the expected abundance over the sampling area, c and the $p_i(t)$ are defined as in equation R.1.

Letting $\psi = (\nu, \beta_0, \beta_1, \cdots, \beta_{k-1}, \theta, \eta)$, the log-likelihood function $\ell(\psi)$ is

$$\ell(\psi) = \sum_{j=1}^{m}\sum_{i=1}^{k} n_{ij} \log[\eta p_i(t_j)] - \eta p_i(t_j) - \log(n_{ij}!). \qquad \text{(R.2)}$$

Since stage frequency data usually carry little information about t_0, the likelihood function tactitly assumes that chronological time and development time have coincidental origins at $t = 0$. Strictly speaking t_0 should enter into the likelihood and t_j should be adjusted as $d_j = t_j - t_0$. The analysis performed by Munholland and Kalbfleish assumes that the development time origin is a known constant.

Nonlinear regression packages like the NLIN procedure of SAS (SAS institute 1985) can be used in determining the maximum likelihood estimates, the asymptotic or approximate correlation matrix and standard errors of the parameter estimates.

As a rule, initial estimates of the parameters are required for iterative calculations. These are usually guessed by the entomologists.

A particularly useful approximation of $p_i(t)$ for $i = 1, \cdots, k-1$, is given by

$$p_i(t) = \left\{\Phi\left(\frac{\alpha_{i+1} - \nu t}{\sqrt{t}}\right) - \Phi\left(\frac{\alpha_i - \nu t}{\sqrt{t}}\right)\right\} \exp(-\theta t).$$

The adequacy of the model can be tested either by a goodness of fit test or residual analysis. Pearson's chi-square is the yardstick for the former, namely

$$X^2 = \sum_{i=1}^{k}\sum_{j=1}^{m} [n_{ij} - \hat{\eta}\hat{p}_i(t_j)]^2 / [\hat{\eta}\hat{p}_i(t_j)].$$

Estimation and model adequacy 301

An alternative method uses the statistic

$$-2 \ln \Lambda = 2 \sum_{i=1}^{k} \sum_{j=1}^{m} n_{ij} \log\left[n_{ij}/(\hat{\eta}\hat{p}_i(t_j))\right].$$

Munholland and Dennis remark that these two statistics are special instances of the family of power-divergence statistics of Read and Cressie (1988), namely

$$H_\lambda = 2 \left[\sum_{i=1}^{k} \sum_{j=1}^{m} n_{ij} \left\{ \left(\frac{n_{ij}}{\hat{\eta}\hat{p}_i(t_j)}\right)^\lambda - 1 \right\} \right] / \lambda(\lambda+1).$$

Observe that when $\lambda = 1$, $H_1 = X^2$ while for $\lambda \to 0$, $H_\lambda \to -2 \ln \Lambda$.

All the three statistics have an approximate chi-square law with $mk - (k+3)$ degrees of freedom when null hypothesis of model adequacy holds. Because of the prevalence of a large number of cells with zero frequencies and small expected frequencies the X^2 statistic turns out to be superior to $-2 \ln \Lambda$. Read and Cressie recommend using $H_{2/3}(\lambda = 2/3)$ to get the best approximation Munholland and Dennis propose the use of the residual

$$r_{ij} = \frac{3}{2} \left[n_{ij}^{\frac{2}{3}} - (\hat{\eta}\hat{p}_i(t_j))^{2/3} \right] / (\hat{\eta}\hat{p}_i(t_j))^{\frac{1}{6}}$$

which, under the null hypothesis of model adequacy is asymptotically normal.

Other parameters of biological interest that can be estimated are
a) π_i – the probability an individual enters stage i
b) p_i – the expected recruitment to stage i per area sampled
c) m_i – the probability that an individual in stage i will die before entry to stage $i+1$ (wrongly referred to as stage specific mortality rate)
d) τ_i – the mean residence time in stage i
e) $\tau_{i,i+1}$ – the conditional residence time in stage i given entry to stage $i+1$

These parameters depend on the distribution of $S_i = T_{i+1} - T_i$.

When the mortality rate is constant, so that stage duration times are exponentially distributed with parameter θ, we can calculate π_i as follows

$$\pi_i = \int_0^\infty g_i(t) \exp(-\theta t) dt.$$

Taking $g_i(t)$ as the inverse Gaussian density $IG\left(\frac{\alpha_i}{\nu}, \alpha_i^2\right)$ it is easy to see that π_i is the Laplace transform of $g_i(t)$, i.e.,

$$\pi_i = \exp\left[\alpha_i \nu \left\{ 1 - \left(1 + \frac{2\theta}{\nu^2}\right)^{1/2} \right\} \right].$$

From this it follows that $\rho_i = \eta p_i$. Now $\frac{\pi_{i+1}}{\pi_i}$ represents the conditional probability that an individual enters stage $i+1$ given that it enters stage i so that $m_i = 1 - \frac{\pi_{i+1}}{\pi_i}$. This is then the probability that the individual dies in stage i given entry to stage i.

In the absence of mortality, the sojourn time $S_i = T_{i+1} - T_i$ in stage i has the inverse Gaussian law $IG\left(\frac{\beta_i}{\nu}, \beta_i^2\right)$. Denoting its density by $f_i(x)$ and the survivor function of S_i by $\overline{F}_i(x)$ we have

$$\tau_i = \int_0^\infty \overline{F}_i(x) \exp(-\theta x) dx$$

$$= \frac{1}{\theta} - \frac{1}{\theta} \int_0^\infty e^{-\theta x} f_i(x) dx$$

$$= \frac{1 - \exp\left[\beta_i \nu \left\{1 - \left(1 + \frac{2\theta}{\nu^2}\right)^{1/2}\right\}\right]}{\theta}$$

$$= \frac{m_i}{\theta}.$$

Finally $\pi_{i,i+1}$ is obtained from the following integral

$$\pi_{i,i+1} = \int_0^\infty \int_x^\infty f_i(y) \exp(-\theta y)/(1 - m_i) dy.$$

The expected proportion $q_i(t)$ of individuals in stage i at time is given by

$$q_i(t) = \frac{G_i(t) - G_{i+1}(t)}{G_i(t)}.$$

The time t_{im}^* at which $q_i(t)$ attains its maximum value is the solution of

$$\frac{\partial q_i(t)}{\partial t} = 0.$$

Estimates of both the quantities can be obtained by using the maximum likelihood estimates of the parameters involved in the equations.

Consider the subpopulation of individuals that survive to enter stage i, and let ψ_{iy} denote the time at which $100y\%$ of the subpopulation has been recruited to stage i, we can determine ψ_{iy} as the value of t such that

$$y = \Phi\left(\frac{ct - \alpha_i}{\sqrt{t}}\right) + \exp(2c\alpha_i) \Phi\left(-\frac{ct + \alpha_i}{\sqrt{t}}\right)$$

where $c = (\nu^2 + 2\theta)^{1/2}$.

Estimation and model adequacy

Finally if $\lambda(t)$ represents the expected number of individuals at time t in the population, per area sampled, and $\lambda_i(t)$ the corresponding expected number in stage i

$$\lambda(t) = \eta \sum_{i=1}^{k} p_i(t) \quad \text{and} \quad \lambda_i(t) = \eta p_i(t).$$

Estimates and standard errors of these parameters are calculated using the δ- method.

Table R.1 *Comparison of observed (O) and expected (E) numbers of grasshoppers*

Instar 1		Instar 2		Instar 3		Instar 4		Adult	
Obs	Exp	Obs	Exp	Obs	Exp	Obs	Exp	Obs	Exp
5	5.2±2.7		<0.5						
6	7.5±2.1		0.7±0.5						
14	11.3±2.7		2.0±0.9		<0.5				
10	10.8±2.6	1	3.4±1.1		0.5±0.3				
7	8.2±2.1	5	4.8±1.4	1	1.3±0.6		<0.5		
1	6.2±1.7	10	5.2±1.5	0	2.0±0.8		1.0±0.5		
1	5.1±1.5	8	5.1±1.4	1	2.3±0.8	1	1.4±0.6		
3	3.8±1.2	8	4.7±1.3	4	2.5±0.9	2	1.9±0.7		<0.5
7	3.0±1.0	12	4.3±1.2	6	2.6±1.0	0	2.4±0.8		0.5±0.3
0	2.1±0.8	7	3.7±1.1	6	2.6±1.0	6	2.4±0.9		0.8±0.4
1	1.6±0.7	1	3.2±1.0	6	2.5±0.9	4	3.2±1.0	1	1.1±0.5
	0.8±0.4	1	2.1±0.7	3	2.1±0.8	2	3.6±1.0	1	1.9±0.7
	0.5±0.3	4	1.6±0.6	4	1.7±0.7	4	3.6±1.0	5	2.5±0.9
	<0.5	0	1.3±0.5	1	1.5±0.6	3	3.4±1.0	2	2.8±1.0
		1	0.9±0.4	1	1.2±0.5	5	3.2±0.9	6	3.3±1.1
		1	0.7±0.3	1	0.9±0.4	2	2.9±0.9	5	3.7±1.1
			0.5±0.3		0.7±0.4	0	2.6±0.8	6	4.0±1.1
			<0.5	1	0.6±0.3	0	2.2±0.8	6	4.3±1.1
					<0.5	1	1.7±0.7	6	4.5±1.1
						1	1.2±0.5	1	4.5±1.1
							1.0±0.5	3	4.4±1.0
							0.7±0.4	3	4.3±1.0
							0.5±0.3	5	4.0±1.0
							<0.5	3	3.8±1.1
								4	3.5±1.1
								2	3.1±1.2
								2	2.9±1.3
								2	2.5±1.4
								1	2.1±1.4

Munholland and Dennis have analyzed data sets first studied by Qasrawi (1966) and reported in Ashford et al. (1970). These pertain to the study of the grasshopper *Chorithippus parallelus* over a site 3,500 square metres in area over which the environmental conditions, soil types, and vegetation were reasonably homogeneous. The grasshopper overwinters in the egg stage and hatching begins the following spring. In addition to the egg and adult stages, the insect goes through four instar during its life cycle. The adult female lays eggs in the late summer and all adults die prior to the winter season. Mortality may occur at any time during development.

In all 29 samples were collected at 3-4 day intervals between May 20, 1964 and September 23, 1964. The data consist of the number of observed insects in each of the four instars and the number of adults. Egg counts are not recorded and individuals are recruited to the population via the first instar. The data set is given in Table R.1. A constant mortality rate is assumed. Using the model given by Equation R.1 the observed likelihood yields $-2 \ln \Lambda = 95.2622$. Based on $(5 \times 29 - (5+3))$ 137 degrees of freedom the model seems reasonably adequate with a p-value larger than 0.95. The X^2 statistic has the value 26.544 based on 20 degrees of freedom with a P-value larger than 0.1.

The model is fitted simultaneously to all stages (as opposed to stage-wise estimation) Munholland and Dennis report that residual analysis revealed no apparent outliers or influential observations. Table R.2 gives the parameter estimates, the standard errors and the correlation matrix for the fitted model.

Table R.2 *Parametric estimates(SE) and correlation matrix*

	β_0	β_1	β_2	β_3	β_4	ν	θ	η
	38.730	12.620	12.420	7.540	9.520	0.796	0.021	71.200
	2.850	1.820	1.460	1.290	1.600	0.053	0.003	18.230
β_0	1.000	0.033	0.340	0.248	0.270	0.840	0.288	0.318
β_1		1.000	0.103	0.128	0.152	0.496	-0.422	-0.480
β_2			1.000	-0.100	0.052	0.495	-0.095	-0.111
β_3				1.000	-0.154	0.311	-0.024	-0.030
β_4					1.000	0.329	-0.016	-0.021
ν						1.000	-0.030	-0.042
θ							1.000	0.961
η								1.000

S. Small area estimation

Inference techniques for unbalanced and balanced two-factor experiments under an inverse Gaussian model can be used in the area of estimation for small regions. Chaubey et al., (1996) consider the analysis of variance methodology for the inverse Gaussian law and adapt it for estimation of small area parameters in finite populations. We now turn to the details of this development as outlined by Chaubey et al.

A finite population $U = \{1, 2, \cdots k \cdots N\}$ is divided into D non-overlapping domains $U_1., U_2., \cdots, U_D.$, the size of domain i being $N_i..$ The population is also divided along another line into G non-overlapping regions called groups denoted by $U._1, U._2, \cdots, U._G$, the size of group j being $N._j$. This leads to a cross-classification of domains and groups resulting in DG cells denoted by U_{ij} the size of which is N_{ij} (say). Thus the entire population size $N = \sum_i N_i. = \sum_j N._j = \sum_i \sum_j N_{ij}$. The domains may represent geographical areas being sampled whereas the groups may be less numerous than the domains and represent age or sex.

A probability sample s of size n, being a subset of U is drawn according to a simple random sampling plan. Denote the parts of s that fall within $U_i., U._j$ and U_{ij} by $s_i., s._j$ and s_{ij}, respectively. Their respective sizes are denoted by $n_i., n._j$ and n_{ij}.

Associated with the kth population unit is the value y_k of a variable of interest y. For the sample s, one observes y_k for $k \in s$. The domain totals are denoted by T_i where

$$T_i = \sum_{k \in U_i.} y_k \quad \text{for } i = 1, \cdots, D.$$

The problem is to estimate T_i.

In the regression context, (we have seen in the section on accelerated life tests) suppose that $\mathcal{L}(y_k) = IG(\mu_k, \lambda)$ where $\mu_k^{-1} = x_k^t \beta$, an estimator of β akin to that proposed by Särndal (1984) is

$$\hat{\beta} = \left(\sum_s \frac{x_k x_k^t}{\pi_k} y_k \right)^{-1} \sum_s \frac{x_k}{\pi_k}$$

where $\pi_k = P(k \in s)$. In the case of simple random sampling π_k is a constant $\frac{n}{N}$ for all k.

Writing $\hat{y}_k = x_k^t \hat{\beta}$ and $e_k = y_k - \hat{y}_k$ the ith domain total T_i has the following modified regression estimator

$$\hat{t}_{i(IG)} = \sum_{k \in U_i.} \hat{y}_k + \frac{N}{n} \sum_{k \in s_i.} e_k.$$

The additive effects model for the data based on an inverse Gaussian population is

$$\mu_{ij}^{-1} = \mu + \alpha_i + \beta_j \tag{S.1}$$

with $\sum \alpha_i = \sum \beta_j = 0$, and α_i, β_j representing the domain effect and group effect respectively and μ is an overall effect. Of course we have $\mu_{ij} > 0$ for all cells (i,j) and $\lambda > 0$. Recall that with $\sigma = \frac{1}{\lambda}$

$$f(j;\mu,\sigma) = (2\pi\sigma y^3)^{-\frac{1}{2}} \exp\left\{-(y\mu^{-1}-1)^2/2\sigma y\right\}.$$

With the parametrization arising from Equation S.1, we note that the parameter space Ω is

$$\Omega = \{(\mu,\alpha_i,\cdots,\alpha_D,\beta_1,\cdots,\beta_G,\sigma) \mid \sum \alpha_i = \sum \beta_j = 0, \mu+\alpha_i+\beta_j > 0,$$
$$\sigma > 0\}.$$

Conditional on the population and sample sizes n_{ij}, the log-likelihood ℓ is

$$\ell = -\frac{1}{2}\log\sigma \sum_{ij} n_{ij} - (2\sigma)^{-1} \sum_{i,j,k} y_{ijk}^{-1}[y_{ijk}(\mu+\alpha_i+\beta_j-1)]^2.$$

Now with the n_{ij} known the dimension of the parameter space is $(2+(D-1)+G-1)$. Routine computation of the derivative of ℓ with respect to μ, α_i, β_i now yields

$$\hat{\mu} y_{\cdot 1\cdot} + \sum_{i=1}^{D-1} \hat{\alpha}_i(y_{i\cdot} - y_{D\cdot}) + \sum_{j=1}^{G-1} \hat{\beta}_j(y_{\cdot j} - y_{\cdot G}) = n_{\cdot\cdot},$$

$$\hat{\mu}(y_{i\cdot} - y_{D\cdot}) + \hat{\alpha}_i y_{i\cdot} + \sum_{i=1}^{D-1} \alpha_i y_D.$$

$$+ \sum_{j=1}^{G-1} \hat{\beta}_j \{(y_{ij} - y_{Dj}) - (y_{iG} - y_{DG})\} = n_{i\cdot} - n_{D\cdot},$$

$$\hat{\mu}(y_{\cdot j} - y_{\cdot G}) + \sum_{i=1}^{D-1} \hat{\alpha}_i \{(y_{ij} - y_{iG}) - (y_{Dj} - y_{DG})\}$$

$$+ \hat{\beta}_j y_{\cdot j} + \sum_{i=1}^{G-1} \hat{\beta}_j(y_{\cdot G} = n_{\cdot j} - n_{\cdot G}. \tag{S.2}$$

In equation S.2 the totals and means follow the standard notations.

Small area estimation

The solutions for $(\hat{\mu}, \hat{\alpha}_i, \hat{\beta}_j)$ are known as the pseudo-maximum likelihood estimators and may not be nonnegative. But as $n_{ij} \to \infty$ they coincide with the MLE.

It is not difficult to show that for a model with interaction term γ_{ij} the estimator of the ith domain total T_i is

$$\hat{t}_i^* = \sum_j N_{ij}\overline{y}_{ij},$$

the post-stratified estimator which is not of great interest in small area estimation. In the absence of interaction the corresponding estimator is

$$\tilde{t}_i = \sum_j N_{ij}\hat{\mu}_{ij} + \sum_j \hat{N}_{ij}(\overline{y}_{ij} - \hat{\mu}_{ij})$$

where

$$\hat{N}_{ij} = \frac{n_{ij}N}{n_{..}}.$$

Särndal and Hidiroglou (1989) proposed a modified regression estimator for small area estimation. Chaubey et al. have compared the performances of both t_i^* and $\tilde{t}_i$ (suitably modified) with their estimator given by

$$\bar{t}_i = \sum_j N_{ij}\overline{y}_{.j} + \sum F_i \hat{N}_{ij}(\overline{y}_{ij} - \overline{y}_{.j})$$

where

$$F_i = \begin{cases} N_{i.}/\hat{N}_{i.} & \text{if } \hat{N}_{i.} \geq N_{i.} \\ \hat{N}_{i.}/N_{i.} & \text{otherwise} \end{cases}$$

and

$$\hat{N}_{i.} = \frac{n_{i.}N}{n_{..}}.$$

The comparison is based on the mean absolute error and absolute relative bias. The modified estimator $\tilde{\bar{t}}_i$ is

$$\tilde{\bar{t}}_i = \sum_j N_{ij}\hat{\mu}_{ij} + \sum_j F_i \hat{N}_{ij}(\overline{y}_{ij} - \hat{\mu}_{ij}).$$

Household income data for Canadians in 1986 obtained from Statistics Canada (1987) was used for generating the values of the parameters in a simulation study. The data was divided into 10 provinces (for domains) and 6 educational groups. An IG model is first fitted and then using the estimates of the parameters an inverse Gaussian super population model is considered. Values of D, G and N_{ij} are then chosen from this population. Sets of values of μ_{ij} and σ are obtained by varying the parameters μ_{ij} by $10^{-c_1}\mu_{ij}$ and σ by $c_2\sigma$ for selected combinations of (c_1, c_2). Large c_1 values correspond to small means while large c_2 values are indicative of an increased dispersion. One thousand random samples are chosen from the IG values generated using a sampling fraction of 1% and 5% with replacement. For each sample the totals are estimated over the domains using $\tilde{t}_i, \bar{\tilde{t}}_i$ and $\bar{t}_i$.

The mean absolute error of an estimator $\hat{\theta}$ is based on

$$\frac{1}{1000} \sum_{j=1}^{1000} |\hat{\theta}_{ij} - \theta_i| / \theta_i$$

while the absolute relative error is calculated using

$$\|\frac{1}{1000} \sum_{j=1}^{100} |\hat{\theta}_{ij} - \theta_i| / \theta_i\|.$$

Chaubey et al., note that reductions in biases are pronounced for many samples and that the mean absolute relative error as well as the absolute relative bias tends to diminish with decreasing values of the mean and dispersion parameter. Larger gains in both these areas seem to go with small values of the coefficient of variation.

T. CUSUM

Cusum charts

Optimum cumulative sum control charts for location and shape parameters of the IG distribution have been suggested by Hawkins and Olwell(1997) and these charts are used for detecting step changes in both parameters. These authors have also examined a data set which involves the task completion times of crews of workers at a General Motors Plant in Oshawa,Ontario,first considered by Desmond and Chapman(1993). One example involved data which had been well modelled by an inverse Gaussian law. Increases in the mean μ or decreases in λ will tend to decrease the service rate and cause a slowdown in the overall process in an assembly line. On the other hand a decrease in μ and increase in λ allows the management in determining factors that will help in improving service rates and diminish the variation. This is a typical example of the Cusum chart at work.

Cusum charts- construction

From the discussion found in Hawkins (1992) and the references therein the upward Cusum for a member of the general exponential family

$$f(x) = exp\{a(x)b(\theta) + c(x) + d(\theta)\}$$

is defined by

$$S_o^+ = 0$$
$$S_n^+ = max\left(0, S_{n-1}^+ + T_n - k^+\right)$$
$$T_n = a(X_n)$$
$$k^+ = \left\{\frac{d(\theta_0) - d(\theta_1)}{b(\theta_1) - b(\theta_0)}\right\}.$$

Here θ_0 and θ_1 are the values of the parameter in control and out ofcontrol respectively. The downward Cusum is defined by

$$S_o^- = 0$$
$$S_n^- = max\left(0, S_{n-1}^- + T_n - k^-\right)$$
$$T_n = a(X_n)$$
$$k^- = \left\{\frac{d(\theta_1) - d(\theta_0)}{b(\theta_1) - b(\theta_0)}\right\}.$$

Cusum charts for location

We assume that for the inverse Gaussian law with parameters μ, λ .the λ parameter is fixed. Thus we have from the exponential family representation $b(\mu) = -(\lambda/2\mu^2)$, $d(\mu) = (\lambda/2)$, and $T_n = \overline{X}_n$. Now the reference values k^+ and k^- simplify to (the plus sign indicates the in-control value and the minus sign the out-of-control value)

$$k^+ = 2\mu\mu^+/(\mu + \mu^+) \text{ ,and } k^- = -2\mu\mu^-/(\mu + \mu^-).$$

The Cusum scheme for upward detection of location consists of a chart given by equations involving S_o^+, S_n^+, and k^+ and likewise the scheme for downward detection involves the equations given by S_o^-, S_n^-, and k^-. The k values depend on the in-control and the out-of-control values to which the Cusum is to be " tuned " in order to obtain the maximum sensitivity.

Thus the scheme signals when either $S_n^+ > h^+$ or $S_n^- < h^-$ where the h-values are set by considering the average run lengths (ARL),namely the average number of samples until $S_n^+ > h^+$ or $S_n^- < h^-$, resulting in a false signal when the process remains in control.

Tables of the average run lengths for the inverse Gaussian distribution law require four entries involving the specification of the values of μ, λ, k and h. If we use the transformation $Y_n = \lambda X_n/\mu^2 = \rho X_n$, then since the random variable Y_n follows the $IG(\phi, \phi^2)$ law where $\phi = \lambda/\mu$, the table now requires only three entries,namely those of ϕ, μ and k. Thus the Cusum of the X_n values will involve a rescaling of the transformed Y_n values and we then have the Cusum

$$S_n^+ = max\left(0, S_{n-1}^+ + T_n - k\right).$$

The rule - signal if $S_n^+ > h^+$ - for the above Cusum has the same meaning as the rule - signal if $T_n^+ > \rho h^+$ for the Cusum

$$T_n^+ = max\left(0, T_{n-1}^+ + \rho X_n - \rho k\right).$$

This means that the desired Cusum of X_n can be designed by dividing the reference value k by ρ, then finding the the interval ρh that gives the needed ARL for the transformed Cusum and then dividing this value by ρ to obtain the h-value for the X_n Cusum. Hawkins and Olwell (1997) provide tables of the transformed h-values (for the Y_n variables) corresponding to a few values of k and ϕ for selected decreasing choices of ϕ (i.e.,increasing values of the mean)while another table shows the h-values for selected increasing choices of ϕ. For more detailed informationconcerning the computational procedure the reader is advised to contact Olwell.

The authors also have published another table which presents ARL values for various shifts in the mean μ. The Cusum parameters have been selected for maximum power in order to detect changes to the indicated out-of-control state value.

Cusum charts for shape parameter

When the mean is fixed one can treat the IG law as a one-parameter exponential family with $a(x) = (x-\mu)^2/(x\mu^2)$, $b(\lambda) = -\lambda/2$ and $d(\lambda) = ln(\lambda/2)$.

The scheme for the Cusum for λ then involves the equations

$$S_o^+ = 0$$
$$S_n^+ = max\left(0, S_{n-1}^+ + a(X_n) - k\right)$$
$$T_n = a(X_n)$$
$$k = \left\{\frac{d(\lambda_0) - d(\lambda)}{b(\lambda_0) - b(\lambda)}\right\}$$
$$= \left\{\frac{ln(\lambda_0/\lambda)}{\lambda_0 - \lambda}\right\}.$$

We now let $Y_n = \{\lambda(X-\mu)^2 X\mu^2\} = \lambda a(X)$ so that Y_n has a Chi-squared law with one degree of freedom and therefore a shift in λ is equivalent to a scale shift of a gamma law with parameters $(1/2, 2)$. A scale shift for the gamma (α, β) law uses the upward Cusum scheme which has the equations

$$S_o^+ = 0$$
$$S_n^+ = max\left(0, S_{n-1}^+ + Y_n - k\right)$$
$$k = \left\{\frac{(\alpha\beta\beta^+) ln(\beta/\beta^+)}{\beta^+ - \beta}\right\}.$$

This scheme now signals $S_n^+ > h^+$ or $S_n^- < h^-$, the h-values being determined by the desired in-control ARL. Since $\beta = 2$ for the Chi-squared law, when λ shifts to λ^+ the law of Y_n shifts to the $\Gamma\left(\frac{1}{2}, \frac{2\lambda}{\lambda^+}\right)$ law. Thus the out-of control value of the gamma scale parameter β shifts to the value $(2\lambda/\lambda^+)$. Finally the Cusum scheme becomes

$$S_o^+ = 0$$
$$S_n^+ = max\left(0, S_{n-1}^+ + \lambda a(X_n) - k\right)$$
$$k = \left\{\frac{\lambda ln(\lambda^+/\lambda)}{\lambda - \lambda^+}\right\}.$$

Downward shifts are defined analogously.

Cusums for λ thus involve(using the above scale changes) the use of a Chi-squared one law. Tables and software for these Cusums are available from Hawkins (1992). Hawkins and Olwell also provide tables of h-values and corresponding ARLs for upward as well as downward shifts in λ for selected values of k. They apply the Cusum scheme to a data set analyzed by Desmond and Chapman (1993) involving the task completion times of crews of workers at the G.M plant in Oshawa, Ontario. The task completion times are accurately described as inverse Gaussian random variables (after screening for outliers and removing the"outliers" or bogus readings). For this data set they found μ to be estimated by 42.6257 and λ to be estimated by 66.282. The units of time were in seconds. Table T.1 and T.2 give the ARL values for the Cusum for the lambda and the mean parameters respectively. In both tables an in-control ARL of 100 is used for $\mu = 42.6257$ and $\lambda = 66.282$. Moreover Table T.1 is a Cusum of individual observations. Hawkins and Olwell point out that these schemes provide for a quick response to shift in μ and λ allowing quick identification and remedy of out-of-control conditions. The Cusum of individual observations does not display increased ARLs for out-of-control states with increased λ as was noted by Olwell using the Shewhart scheme.

Table T.1 *Average run lengths for the Cusum for the lambda parameter for the General Motors data*

ARLs out-of-control

λ	ARL
10.000	3.05
20.000	5.79
30.000	10.27
40.000	18.05
50.000	32.52
60.000	62.59
66.282	100.00
70.000	80.82
80.000	52.60
90.000	38.92
100.000	31.06
110.000	26.01
120.000	22.51
130.000	19.95
140.000	18.06
150.000	16.50

Table T.2 *Average run lengths for the Cusum for μ for the General Motors data set*

ARLs out-of-control

μ	ARL
20	6.93
25	11.32
30	18.83
35	33.16
40	62.59
42.6257	100.00
45	64.42
50	34.28
60	16.54
70	10.86
80	8.23

U. Plutonium Estimation
Model development

A probabilistic model that incorporates the nature of radioactive contamination of a nuclear weapons site has been proposed by Wise and Kabaila (1989).The interest has focussed on regions of significant contamination that have resulted from local fallout from nuclear weapons tests and from explosive fragmentation of plutonium in safety trials. Plutonium here refers to a mixture of three isotopes of plutonium with mass numbers 239,240 and 241. These are initially formed in a nuclear reactor and they may be detected when freshly deposited on the soil surface but are difficult to be detected when the isotopes begin migrating through the soil surface. Several techniques exist for determining the concentrations of plutonium in soil. The Australian Radiation Laboratory has done a number of radiological surveys of nuclear explosions which the United Kingdom Atomic Weapons Research Establishment conducted between 1953 and 1963 at the Monte Bello Islands, Mu and the Maralinga test range in Australia. The model developed by Wise and Kabaila involves direct sampling of the soil and indirect observation of emitted radium using the NaI detector.

The model

In the simplest version of the model- called the one-component model- it is assumed that activity is confined to particles which are uniformly and randomly distributed throughout a volume at a rate λ and that the paricles have activities which are randomly distributed with a probability density $K(\alpha) = \beta \exp(-\beta \alpha)$. A general expression for the Laplace transform of the probability density for the measurement results as well as a numerical analysis of the transform has revealed that it approximates the transform for the inverse Gaussian law.

Suppose that U denotes the set of position vectors of particles uniformly distributed with rate λ per unit volume in a three dimensional region S. Let the activity of the particle located at $\vec{u}$ be denoted by α_u a random variable with probability density $k(\alpha_u)$. Further let $f(\vec{u})$ denote the response of the measuring instrument for a particle of unit activity with position vector $\vec{u}$. Then it turns out that the total contribution to the measurement result, $V(S)$ is

$$V(S) = \sum_{\vec{u}} f(\vec{u}) \alpha_u.$$

Using a method due to Karlin and Taylor (1981, Section 16.2) the transform of the density of $V(S)$. using the circular symmetry of $f(.)$, has been shown to be

$$\phi_X(\tau; S) = \exp\left(-\lambda m_S \tau / \rho(\tau + \beta)\right)$$

Model development

where $m_S = d^2 t \rho \pi/4$, the mass of the soil sample, ρ, the density of the soil sample and X is a random variable denoting the soil sample activity. Here d is the diameter of the corer and S is the cylinder (r $< d/2$). The distribution of X is obtained from inverting the Laplace transform $\phi_X(\tau, S)$ and will be denoted by $p_X(x; m_S, \lambda, \beta)$.

Let Y be a random variable denoting the number of counts observed. Then the Laplace transform, $\phi_Y(\tau, S)$ of the density of Y has an excellent approximation which is that of an inverse Gaussian law $IG(\mu, \theta)$ and will be denoted by $f(y; \mu, \lambda)$. In terms of the parameters of the inverse Gaussian law we have the first two cumulants given by
$\mu = KT_1(h)\lambda/\beta$ and $\theta = KT_1^3(h)\theta^2/(2T_2(h)\beta)$. The moment estimates of the rate λ and the exponential parameter β are
$\hat{\beta} = 2\bar{T}/\sigma_T^2$, $\hat{\lambda} = 2\bar{T}^2\rho/m_S\sigma_T^2$ where $\bar{T}$ is the mean of the sampled activity and σ_T^2 is the sample variance.

In practical applications, in order to use maximum likelihood methods of estimation Wise and Kabaila approximate the probability density of the NaI detector data by

$$p_Y(y; \lambda, \beta) = f(y - \psi; \lambda, \beta)$$

where ψ is a threshold parameter. As pointed out by the authors, this model ignores the Poisson nature of the NaI detector counting process. Had this counting process been included in the model then the probability density of the NaI detector data would be that of a Poisson-inverse Gaussian mixture.

Given n sample masses m_i and activities x_i, i = 1,...,n, the corresponding loglikelihood is

$$L_X(\lambda, \beta) = \sum_{i=1}^{n} \log p_X(x_i, m_i, \lambda, \beta).$$

Likewise, given N NaI detector counts y_i, the parameters can be estimated from the associated loglikelihood function, namely

$$L_Y(\lambda, \beta) = \sum_{i=1}^{N} \log p_Y(y_i; \lambda, \beta).$$

Estimates of the parameters λ and β are calculated from the equations

$$z_i = y_i - \psi \quad i = 1, ..., n$$
$$N\bar{z} = \sum_{i=1}^{N} z_i$$
$$W = N/\bar{z} \sum_{i=1}^{N} [z_i^{-1} - \bar{z}^{-1}]$$
$$\hat{\lambda} = 2T_2(h)W/T_1^2(h)$$
$$\hat{\beta} = T_1(h)K\hat{\lambda}/\bar{z}.$$

Wise and Kabaila found that the one-parameter model was not adequate for their purpose and suggested the use of a two-component model. A low activity component of a two-component model is one that has a larger β. Estimates of the parameters of this model is obtained by maximizing the sum of the loglikelihoods for the soil sampling data and the NaI detector data.

The density for the NaI detector data in a two-component model is approximated as follows. The NaI detector data is considered as the sum of two independent counts from (i) a high activity component (ii) a low activity component and (iii) a constant background-ψ. Their detailed analysis which included the Poisson nature of the radiation detector counting process showed that the high activity component is approximated by the law $f_Y(y, \lambda_1, \beta_1)$ while the combined factors arising from (ii) and (iii) could be described by a density $f_2(y; \lambda_2, \beta_2)$ which was approximately Gaussian with mean μ and variance σ^2 given by

$$\mu = \psi + 1238\lambda_2(K/\beta_2)$$
$$\sigma^2 = \mu + 277\lambda_2(K/\beta_2)^2.$$

For the data analysed by Wise and Kabaila they found that the two-component model was quite satisfactory as evidenced by a sufficient reduction in the loglikelihood.

REFERENCES

Ahmad, M., Chaubey, Y.P., and Sinha, B.K. (1991). Estimation of a common mean of several univariate inverse Gaussian populations. *Annals of the Institute of Statistical Mathematics*, **43**, (2), 357-67.

Ahmed, A.N. and Abouammoh, A.M. (1993). Characterizations of the gamma, inverse Gaussian and negative binomial distributions via their length-biased distributions. *Statistical papers*, **34**, 167-73.

Ahsanullah, M. and Kirmani, S.N.U.A. (1984). A characterization of the Wald distribution. *Naval Research Logistics Quarterly*, **31**, 155-8.

Aitkin, M. and Clayton, D. (1980). The fitting of exponential, Weibull and extreme value distributions to complex survival data using GLIM. *Applied Statistics*, **29**, 156-63.

Akman, O. and Gupta, R.C. (1992). A comparison of various estimators of the mean of an inverse Gaussian distribution. *Journal of Statistical Computation and Simulation.* **40**, 71-81.

Al-Hussaini, E.K. and Abd-El-Hakim, N.S.(1981). Bivariate inverse Gaussian distribution. *Annals of the Institute of Statistical Mathematics*, **33**, 57-66.

Al-Hussaini, E.K. and Ahmad, K.E. (1984). Information matrix for a mixture of two inverse Gaussian distributions. *Communications in Statistics*, B**13**, 785-800.

Al-Hussaini, E.K., Nagi,S., and Abd-El-Hakim. (1989). Failure rate of the inverse Gaussian-Weibull mixture model. *Annals of the Institute of Statistical Mathematics*, **41**, 617-22.

Aminzadeh, M.S. (1993). Statistical quality control via the inverse Gaussian distribution. *ASA proceedings of the Quality and Productivity Section.* (AmerStatVA), 168-70.

Amoh, R.K. (1983). Classification procedures associated with the inverse Gaussian distribution. Unpublished Ph.D. dissertation, University of Manitoba, Winnipeg.

Amoh, R.K. (1984). Estimation of parameters in mixtures of inverse Gaussian distributions. *Communications in Statistics*, **13**, 1031-43.

Amoh, R.K. (1985). Estimation of a discriminant function from a mixture of two inverse Gaussian distributions when sample size is small. *Journal of Statistical Computing and Simulation*, **20**, 275-86.

Amoh, R.K. (1986). Asymptotic results for a linear discriminant function estimated from a mixture of two inverse Gaussian populations. *Journal of Statistical Planning and Inference*, **14**, 233-43.

Amoh, R.K. and Kocherlakota, S. (1986). Errors of misclassification associated with the inverse Gaussian distribution. *Communications in Statistics*, A**15**, 589-612.

Ang, A.H-S. and Tang, W.H. (1975). *Probability concepts in Engineering Planning and Design.* Volume I. New York, Wiley.

Anscombe, F.J. (1950). Sampling theory of the negative binomial and logarithmic series distributions. *Biometrika*, **37**, 358-382.

Anscombe, F.J. (1961). Estimating a mixed exponential response law. *Journal of the American Statistical Association*, **56**, 493-502

Ashford, J.R., Read, K.L.Q., and Vickers, G.G. (1970). A system of stochastic models applicable to studies of animal population dynamics. *Journal of Animal Ecology*, **37**, 29-50.

Athreya, K.B. (1986). Another conjugate family for the normal distribution. *Statistics and Probability Letters*, **4**, 61-4.

Atkinson, A.C. (1979). The simulation of generalized inverse Gaussian, generalized hyperbolic, gamma and related random variables. Research report No.52, Department of Theoretical Statistics, University of Aarhus.

Atkinson, A.C. (1982). The simulation of generalized inverse Gaussian and hyperbolic random variables. *SIAM Journal of Scientific Statistical Computing*, **3**(4), 502-15.

Atkinson, A.C. and Lam Yeh. (1982). Inference for Sichel's compound Poisson distribution. *Journal of American Statistical Association*, **77**(377), 153-8.

Babu, G.J., and Chaubey, Y.P. (1996). Asymptotics and bootstrap for inverse Gaussian regression. *Annals of the Institute of Statistical Mathematics*, **448**(1), 75-88.

Bachelier, L. (1900). Théorie de la spéculation. *Annales des Sciences de l'Ecole Normale Superieure*, Paris, **17**(3), 21-86.

Bahadur, R.R. (1967). Rates of convergence of estimates and test statistics. *Annals of Mathematical Statistics*, **38**, 303-24.

Bahadur, R.R. (1971). *Some limit theorems in statistics*. SIAM, Philadelphia.

Baker, R. (1979). The inverse Gaussian distribution. *GLIM Newsletter* **1**, 36-7.

Balakrishnan, N. and Chen,W.S. (1997). CRC *Handbook of Tables for the order statitics from inverse Gaussian distributions with applications*. CRC Press. Florida.

Banerjee, A.K. and Bhattacharyya, G.K. (1974). A Bayesian study of the inverse Gaussian distribution. Technical report No.399, Department of Statistics, University of Wisconsin (Madison).

Banerjee, A.K. and Bhattacharyya, G.K. (1976). A purchase incidence model with inverse Gaussian interpurchase times. *Journal of the American Statistical Association*, **71**, 823-9.

Banerjee, A.K. and Bhattacharyya, G.K. (1979). Bayesian results for the inverse Gaussian distribution with an application. *Technometrics*, **21**(2), 247-51.

Barbour, C.D. and Brown, J.H. (1974). Fish species diversity in lakes. *The American Naturalist*, **108**, 473-89.

Bardsley, W.E. (1980). Note on the use of the inverse Gaussian distribution for wind energy applications. *Journal of Applied Meteorology*, **19**, 1126-30.

Bar-Lev, S.K. (1981). Asymptotic properties of the maximum conditional likelihood estimate for exponential subfamilies of distributions. Unpublished manuscript.

Bar-Lev, S.K. (1983). A characterization of certain statistics in exponential models whose distributions depend on a sub-vector of parameters only. *Annals of Statistics*, **11**, 746-52.

Bar-Lev, S.K. and Enis, P. (1986). Reproducibility and natural exponential families with power variance functions. *Annals of Statistics*, **14**, 1507-22.

Bar-Lev, S. and Reiser, B. (1982). An exponential sub-family which admits UMPU tests based on a single test statistic. *Annals of Statistics*, **10**(3), 979-89.

Bar-Lev, S.K., Bshouty, D., and Letac, G. (1992). Natural exponential families and self-decomposability. *Statistics and Probability Letters*, **13**(2), 147-52.

Bar-Lev, S.K. (1994). A derivation of conditional cumulants in exponential families. *The American Statistician*, **48**(2), 126-9.

Barndorff-Nielsen, O.E. (1978). *Information and exponential families in statistical theory*. Wiley, Chichester.

Barndorff-Nielsen, O.E. (1988). *Parametric statistical models and likelihood*. Lecture notes in Statistics **50**, Springer Verlag, New York.

Barndorff-Nielsen, O.E. (1994). A note on electrical networks. *Advances in Applied Probability*, **26**, 63-67.

Barndorff-Nielsen, O.E. and Blaesild, P. (1983a). Exponential models with affine dual foliations. *Annals of Statistics*, **11**, 753-69.

Barndorff-Nielsen, O.E. and Blaesild, P. (1983b). Reproductive exponential families. *Annals of Statistics*, **11**, 770-82.

Barndorff-Nielsen, O.E. and Halgreen, C. (1977). Infinite divisibility of the hyperbolic and generalized inverse Gaussian distributions. *Zeitschrift für Wahrscheinlichkeitstheorie und verwandte Gebiete*, **38**, 309-11.

Barndorff-Nielsen, O.E. and Jørgensen, B. (1991). Some parametric models on the simplex. *Journal of Multivariate Analysis*, **39**, 106-16.

Barndorff-Nielsen, O.E. and Koudou, A.E. (1997). Trees with random conductivities and the (reciprocal)inverse Gaussian distribution. (to appear).

Barndorff-Nielsen, O.E., Blaesild, P., and Halgreen, C. (1978). First hitting time models for the generalized inverse Gaussian distribution. *Stochastic Processes and Applications*, **7**, 49-54.

Barndorff-Nielsen, O.E., Blaesild, P., and Seshadri, V. (1991). Multivariate distributions with generalized inverse Gaussian marginals and associated Poisson mixtures. *Canadian Journal of Statistics* **20**(2), 109-20.

Barr, A. (1991). On the linear combinations of inverse Gaussian distributions. *Communications in Statistics* A, **20**, 2891-2905.

Bartlett, M.S. (1937). Properties of sufficiency and statistical tests. *Proceedings of the Royal Statistical Society, London*, A**160**, 268-82.

Basu, A.K. and Wasan, M.T. (1971). On the first passage time processes of Brownian motion. *Queen's Mathematical Preprint* No. 1971-38. Queen's University, Kingston, Ontario.

Beekman, J.A. (1985). A series for infinite time ruin probabilities. *Insurance mathematics and Economics*, **4**, 129-34.

Berg, P.T. (1994). Deductibles and the inverse Gaussian distribution. *Astin Bullettin*, **24**,319-23.

Betró, B. and Rotondi, R. (1991). On Bayesian inference for the inverse Gaussian distribution. *Statistics and Probability Letters*, **11**, 219-24.

Bhattacharyya, G.K. and Fries, A. (1982). Fatigue failure models-Birnbaum-Saunders vs. inverse Gaussian. *IEEE Transactions in Reliability* R**31** (5), 439-40.

Bhattacharyya, G.K. and Fries, A. (1982). Inverse Gaussian regression and accelerated life tests, in *Survival Analysis*, edited by John Crowley and Richard A. Johnson, IMS Lecture Notes, Monograph Series, 101-18.

Bhattacharyya, G.K. and Fries, A. (1983). Analysis of two factor experiments under an inverse Gaussian model. *Journal of the American Statistical Association*, **78**, 820-6.

Bhattacharyya, G.K. and Fries, A. (1986). On the inverse Gaussian multiple regression model checking procedures. *Reliability and Quality Control*, (ed. A.P. Basu) Elsevier, North Holland, 87-100.

Bhattacharyya, G.K. and Soejeti, Z. (1989). A tampered failure rate model for stepstress accelerated life tests. *Communications in Statistics-Theory and Methods*, **18**, 1627-44.

Bhattacharya, S.K. (1987). Bayesian normal analysis with an inverse Gaussian prior. *Annals of the Institute of Statistical Mathematics*, **39**, A, 623-6.

Bhattacharya, S.K. (1986). E-IG model in life testing. *Calcutta Statistical Association Bulletin*, **35**, 85-90.

Bhattacharya, S.K. and Kumar, S. (1986). Bayesian life estimation with an inverse Gaussian prior. *South African Statistical Journal*, **20**, 37-43.

Bickel, P.J. and Doksum, K.A. (1977). *Mathematical Statistics: Basic ideas and selected topics*. Holden Day. San Francisco.

Birnbaum, A. (1959). On the analysis of factorial experiments without replication. *Technometrics*, **1**, 347-57.

Birnbaum, Z.W. and Saunders, S.C. (1969). A new family of life distributions. *Journal of Applied Probability*, **6**, 319-27.

Blaesild, P. and Jensen, J.L. (1985). Saddlepoint formulas for reproductive exponential models. *Scandinavian Journal of Statistics*, **12**, 193-202.

Bobée, B., Ashkar, F., and Perrault, L. (1993). Two kinds of moment ratio diagrams and their applications in hydrology. *Stochastics and Hydraulics*, **7**, 41-65.

Bobée, B., Rasmussen, P., Perrault, L., and Ashkar, F. (1994) Risk analysis of hydrologic data: review and new developments concerning the Halphen distributions. L. Duckstein & E. Parent (eds). *Engineering Risk in Natural Resources Management*, 177-190

Box, G.E.P. (1949) A general distribution theory for a class of likelihood criteria. *Biometrika*, **36**, 317-46. .

Bravo, G. (1986). Unpublished Ph.D. Thesis submitted to University of Sherbrooke.

Bravo, G. and MacGibbon, B. (1988). Improved estimators for the parameters of an inverse Gaussian distribution. *Communications in Statistics*, **17**(12), 4285-99.

Bulmer, M.G. (1974). On fitting the Poisson lognormal distribution to species-abundance data. *Biometrics*, **30**(1), 101-10.

Burbeck, S.L. and Luce, R.D. (1982). Evidence from auditory simple reaction times for both change and level detectors. *Perception and Psychophysics*, **32**(2), 117-33.

Capocelli, R.M. and Ricciardi, L.M. (1972). On the inverse of the first passage time probability problem. *Journal of Applied Probability*, **9**, 270-87.

Chan, M.Y. (1982). Modified moment and maximum likelihood estimators for the parameters of the three-parameter inverse Gaussian distribution. Unpublished Ph.D. thesis, University of Georgia, Athens, Georgia.

Chan, M.Y., Cohen, A.C., and Whitten, B.J. (1983). The standardized inverse Gaussian distribution tables of the cumulative probability function. *Communications in Statistics — Simulation and Computation*, **12**, 423-42.

Chan, M.Y., Cohen, A.C., and Whitten, B.J. (1984). Modified maximum likelihood and modified moment estimators for the three-parameter inverse Gaussian distribution. *Communications in Statistics – Simulation and Computation*, **13**, 47-68.

Chandra, N.K. (1990). On the efficiency of a testimator for the mean of an inverse Gaussian distribution. *Statistics and Probability Letters*. **10**,431-7.

Chang, D.S. (1994). Graphical analysis for Birnbaum - Saunders distribution. *Microelectronics and Reliability*, **34**(1), 17-22.

Chang, D.S. (1994). A note on constructing the confidence bounds of the failure rate for the inverse Gaussian distribution. *Microelectronics and Reliability*, **34**(1), 187-90.

Chatfield, C., Ehrenberg, A.S.C. and Goodhardt, G.J. (1966) Progress on a simplified model of stationary purchasing behaviour. *Journal of the Royal Statistical Society*, Part 3, 317-69

Chatfield, C. and Goodhardt, G.J. (1977). A consumer purchasing model with Erlang interpurchase times. *Journal of American Statistical Association*, **68**, 828-35. .

Chaturvedi, A. (1985). Sequential estimation of an inverse Gaussian parameter with prescribed proportional closeness. *Calcutta Statistical Association Bulletin*, **34**, 215-9.

Chaturvedi, A. (1996). Correction to sequential estimation of an inverse Gaussian parmeter with prescribed proportional closeness. *Calcutta Statistical Assiciation Bullettin*. **35** , 211-2.

Chaturvedi, A., Pandey, S.K., and Gupta, M. (1991). On a class of asymptotically risk-efficient sequential procedures. *Scandinavian Acturial Journal*, **1**, 87-96.

Chaubey, Y.P. (1991). A study of ratio and product estimators under a superpopulation model. *Communications in Statistics-Theory and Methods*, **20**(5& 6), 1731-46.

Chaubey, Y.P., Nebebe, F., and Chen, P. (1996). Small area estimation under an inverse Gaussian model in finite population sampling, *Survey Methodology*, **22**(1), 33-41.

Cheng, R.C.H. (1984). Generation of inverse Gaussian variates with given sample mean and dispersion. *Applied Statistics*, **33**, 309-16.

Cheng, R.C.H. and Amin, N.A.K. (1981). Maximum likelihood estimation of parameters in the inverse Gaussian distribution, with unknown origin. *Technometrics*, **23**, 257-63.

Cheng, R.C.H. and Amin, N.A.K. (1983). Estimating parameters in continuous univariate distributions with a shifted origin. *Journal of the Royal Statistical Society*, **B** **45**(3), 394-403.

Cheng, R.C.H. and Iles, T.C. (1989). Significance tests for embedded models in three parameter distributions. Technical report 89-SI. School of Mathematics, University of Wales College of Cardiff.

Cheng, R.C.H. and Iles, T.C. (1990). Embedded models in three parameter distributions and their estimation. *Journal of the Royal Statistical Society*, B**52**(1), 135-49.

Chernoff, H. (1954). On the distribution of the likelihood ratio. *Annals of Mathematical Statistics*, **25**, 573-9.

Chhikara, R.S. (1972). Statistical inference related to the inverse Gaussian distribution. Unpublished Ph.D. thesis, Oklahoma State University, Stillwater.

Chhikara, R.S. (1975). Optimum tests for the comparison of two inverse Gaussian distribution means. *Australian Journal of Statistics*, **17**, 77-83.

Chhikara, R.S. and Folks, J.L. (1974). Estimation of the inverse Gaussian distribution function. *Journal of the American Statistical Association*, **69**, 250-4.

Chhikara, R.S. and Folks, J.L. (1975). Statistical distributions related to the inverse Gaussian. *Communications in Statistics*, **4**, 1081-91.

Chhikara, R.S. and Folks, J.L. (1976). Optimum test procedures for the mean of first passage time in Brownian motion with positive drift (inverse Gaussian distribution). *Technometrics*, **18**, 189-93.

Chhikara, R.S. and Folks, J.L. (1977). The inverse Gaussian distribution as a lifetime model. *Technometrics*, **19**, 461-8.

Chhikara, R.S. and Folks, J.L. (1989). *The inverse Gaussian distribution, theory, methodology and applications*. Marcel Dekker Inc. New York.

Chhikara, R.S. and Guttman, I. (1982). Prediction limits for the inverse Gaussian distribution. *Technometrics*, **24**, 319-24.

Cohen, A.C. and Whitten, B.J. (1985). Modified moment estimation for the three parameter inverse Gaussian distribution. *Journal of Quality Technology*, **17**(3), 147-53.

Cohen, J.E. (1977). Ergodicity of age structure in populations with Markovian vital rates III: finite state moments and growth rate-an illustration, *Advances in Applied Probability*, **9**, 462-75.

Cox, D.R. (1961). Tests of separate families of hypotheses. Proc. Fourth Berkeley Symposium, **1**, 105-23.

Cox, D.R. (1962). Further results on tests of separate families of hypothesis. *Journal of the Royal Statistical Society*, **B,24**, 406-24.

Cox, D.R. and Miller, H.D. (1965). *The theory of stochastic processes*. Chapman and Hall, New York.

Cox, D.R. and Lewis, P.A.W. (1966). *The statistical analysis of series of events*. Methuen & Co. Ltd. London.

Cramèr, H. (1946). *Mathematical Methods of Statistics*. Princeton. New Jersey.

Currie, I.D. and Stephens, M.A. (1986). Gurland-Dahiya statistics for the inverse Gaussian and gamma distributions. Research Report No. 88-14, Department of Mathematics and Statistics, Simon Fraser University.

Dagpunar, J.S. (1989). An easily implemented generalized inverse Gaussian generator. *Communications in Statistics B*, **18**,703-10.

Damianou, C., and Agrafiotis, G.K. (1988). An inverse Gaussian model for divorce by marriage cohort. *Biometrics Journal*, **30**, 607-13.

Daniel, C. (1959). Use of half-normal plots in interpreting factorial two-level experiments. *Technometrics*, **1**, 314-41.

Davis, A.S. (1977). Linear statistical inference as related to the inverse Gaussian distribution. Unpublished Ph.D. thesis. Oklahoma State University, Stillwater.

Davis, A.S. (1980). Use of the likelihood ratio test on the inverse Gaussian distribution. *American Statistician*, **34**, 108-10.

Dean, C., Lawless, J.F., and Willmot, G.E. (1989). A mixed Poisson inverse Gaussian regression model. *Canadian Journal of Statistics*, **17**, 171-81.

De Groot, M.H. and Goel, P.K. (1979). Bayesian estimation and optimal designs in partially accelerated life testing, *Naval Research Logistics Quarterly*, **26**, 223-35.

Dennis, B., Munholland, P.L., and Scott, M.J. (1991). Estimation of growth and extinction parameters for endangered species. *Ecological Monographs*, **61**, 115-143.

Desmond, A.F. (1985). Stochastic models of failure in random environments. *Canadian Journal of Statistics*, **13**, 171-83.

Desmond, A.F. (1986). On the relationship between two fatigue-life models. *IEEE Transactions in Reliability*, R**35**(2), 167-9.

Desmond, A.F. and Chapman,G.R. (1993) Modelling task completion data with inverse Gaussian mixtures. *Applied Statistcs*, **42**, 603-13.

Dhulesia, H., Bernicot, M., and Deheuvels, P. (1991). Statistical analysis and modelling of slug lengths. In: Multi-Phase Production (A.P. Burns, ed.) Elsevier, London, 88-112.

Diaconis, P. and Ylvisaker, D. (1979). Conjugate priors for exponential families. *Annals of Statistics*, **7**, 269-81.

Diekmann, A. and Mitter, P. (1984). A comparison of the 'sickle function' with alternative models of divorce rates in; Diekman, A. and Mitter, P. (eds), *Stochastic Modelling of Social Processes*. London, Academic Press.

Doksum Kjell A. and Arnljot Hoyland. (1992). Models for variable-stress accelerated life testing experiments based on Wiener processes and the inverse Gaussian distribution. *Technometrics*, **34**, No.1, 74-82.

Dugué, D. (1941). Sur un nouveau type de courbe de fréquence. *Comptes Rendus de l'Academie des Sciences*, Paris, Tome 213, 634-5.

Durairajan, T.M. (1985). Bahadur efficient test for the parameter of inverse Gaussian distribution. *Journal of Agricultural Statistics*, **37**, 192-7.

Eaton, W.W. and Whitmore, G.A. (1977). Length of stay as a stochastic process: A general approach and application to hospitalization for schizophrenia. *Journal of Mathematical Sociology*, **5**, 272-92.

Edgeman, R.L. (1989a). Inverse Gaussian control charts. *Australian Journal of Statistics*, **31**, 78-84.

Edgeman, R.L. (1989b). Control of inverse Gaussian processes. *Quality Engineering*, **2**, 265-76.

Edgeman, R.L. (1990). Assessing the inverse Gaussian distribution assumption. *IEEE Transactions on Reliability*, R**39**, 352-5.

Edgeman, R.L. (1996). SPRT & CUSUM results for inverse Gaussian processes. *Communications in Statistics-Theory and Methods*, **25**(11), 2797-2806.

Edgeman, R.L. and Salzburg,P.M. (1991). A sequential sampling plan for the inverse Gaussian mean. *Statistical Papers*, **32**, 45-53.

Edgeman, R.L., Scott, R.C., and Pavur, R.J. (1988). A modified Kolmogorov-Smirnov test for the inverse Gaussian distribution with unknown parameters. *Communications in Statistics* B**17**(4), 1203-12.

Ehrenberg, A.S.C. (1959). Pattern of consumer purchases. *Applied Statistics*, **8**(1), 26-41.

Embrechts, P. (1983). A property of the generalized inverse Gaussian distribution with some applications. *Journal of Applied Probability*, **20**, 537-44.

Engelhardt, M., Bain, L.J., and Wright, F.T. (1981). Inferences on the parameters of the Birnbaum-Saunders fatigue life distribution based on maximum likelihood estimation. *Technometrics*, **23**(3), 251-6.

Essenwanger, O.M. (1959). Probleme der windstatistik. Meteor, Rundsch, **12**, 37-47.

Ferguson, M., Badhwar, G., Chhikara, R.S., and Pitts, D. (1986). Field size distributions for selected agricultural crops in the United States and Canada. *Remote sensing and environment*, **19**, 25-45.

Fisher, R.A., Corbet, A.S., and Williams, C.B. (1943). The relation between the number of species and the number of individuals in a random sample from an animal population. *Journal of Animal Ecology*, **12**, 42-58.

Fletcher, H. (1911). *Physical Review*, **33**(2), 82.

Folks, J.L. and Chhikara, R.S. (1978). The inverse Gaussian distribution and its statistical application — a review. *Journal of the Royal Statistical Society*, B**40**, 263-89.

Folks, J.L., Pierce, D.A., and Stewart, C. (1965). Estimating the fraction of acceptable product. *Technometrics*, **7**, 43-50.

Foster, H.A. (1924). Theoretical frequency curves and their application to engineering problems. *American Society of Civil Engineering*.

Fries, A. and Bhattacharyya, G.K. (1983). Analysis of two-factor experiments under an inverse Gaussian model. *Journal of the American Statistical Association*, **78**(384), 820-6.

Fries, A., and Bhattacharyya, G.K. (1986). Optimal design for an inverse Gaussian regression model. *Statistics and Probability Letters*. **4**, 291-4.

Gacula, M. C. Jr. and Kubala, J.J. (1975). Statistical models for shelf life failures. *Journal of Food Science*, **40**, 404-9.

Gani, J. and Prabhu, N.U. (1963). A storage model with continuous infinitely divisible inputs. *Proceedings of the Cambridge Philosophical Society*, **59**, 417-30.

Gendron,M., and Crepeau,H.(1989). On the computation of the aggregate claim distribution when individual claims are inverse Gaussian. *Insurance mathematics and Economics*. **8**,251-8.

Gerber,H.U. (1992).On the probability of ruin for infinitely divisible claim amount distributions. *Insurance Mathematics and Economics*, **11**, 163-6.

Ghosh, B.K. (1970). *Sequential tests of statistical hypothesis*. Reading, Massachusetts: Addison-Wesley Publishing Co.

Goh, C.J., Tang, L.C., and Lim, S.C. (1989). Reliability modelling of stochastic wear-out failure. *Reliability Engineering and Systems Safety*, **25**, 303-14.

Good, I.J. (1953). The population frequencies of species and the estimation of population parameters. *Biometrika*, **40**, 237-64.

Gossiaux, A., and Lemaire, J. (1981). Methodes d'ajustement de distributions de sinistres. *Bulletin of the Association of Swiss Actuaries*, **81**, 87-95.

Govindarajulu, Z., and Eisenstat, S. (1965). Best estimates of location and scale parameters of a chi(1 d.f.) distribution using ordered observations. *Reports of Statistical Applied Research*, Union of Japanese Scientists and Engineers, **12**, 150-63.

Gradshteyn, I.S. and Rhyzhik, I.M. (1963). *Table of Integrals, Series and Products.* Academic Press, Inc. California.

Grimmet, G. (1993). Random graphical networks. In Chaos and Networks. Statistical and probablistic aspects. (eds). O.E. Barndorff-Nielsen, J.L. Jensen and W.S. Kendall, Chapman and Hall, London. 288-301.

Gumbel, E.J. (1941). The return period of flood flows. *Annals of Mathematical Statistics,* **12,** 163-190.

Gumbel, E.J. (1958). *Statistics of extremes.* Columbia University Press. New York, N.Y.

Gupta, R.C., and Akman, H.O. (1995). On the reliability studies of a weighted inverse Gaussian model. *Journal of Statistical Planning and Inference,* **48,** 69-83.

Gupta, R.P. (1969). Maximum likelihood estimate of the parameters of a truncated inverse Gaussian distribution. *Metrika,* 51-3.

Gydesen, H. (1984). A stochastic approach to models for the leaching of organic chemicals in soil. *Ecological Modelling,* **24,** 191-205.

Hadwiger, H. (1940a). Naturliche Ausscheidefunktionen fur Gesamtheiten und die Losung der Erneurungsgleichung,*Mitteilunggen der Vereinigung Schweizerischer Versicherungsmathematiker,*bf 40,31-49.

Hadwiger, H. (1940b). Eine analytische reproductions funktion für biologische gesamtheiten. *Skandinavisk Aktuarietidsskrift,* **23,** 101-13.

Hadwiger, H. (1942). Wahl einer Naherungsfunktion fur Verteilungen auf Grund einer funktionalgleichung. *Blatter fur Versicherungsmathematik,* 5345-352.

Hadwiger, H., and Ruchti, W. (1941). Darstellung der Fruchtbarkeit durch eine biologische Reproduktionsformel, *Archiv fur mathematische Wirtschaftsund Sozialforschung,* **7,** 30-34.

Harvey, A.C. (1976). Estimating regression models with multiplicative heteroscedasticity. *Econometrica,* **44,** 461-5.

Hasofer, A.M. (1964). A dam with inverse Gaussian input. *Proceedings of the Cambridge Philosophical Society,* **60,** 931-3.

Hassairi, A. (1993). Les$(d + 3)G$-orbites de la classe de Morris-Mora des familles exponentielles de R^d. *Comptes Rendus de l'Academie des Sciences,* Paris, t. 317 Série 1, 887-890.

Hawkins, D.M. (1992). Evaluation of average run lengths of cumulative sum charts for an arbitrary data distribution. *Communications in Statistics. Simulation and Computation* **21,** 1001-20.

Hawkins, D.M., and Olwell, D.H. (1997). Inverse Gaussian cumulative sum control charts for location and shape.*The Statistician* , **46,**3,323-35.

Hazen, A. (1914). Storage to be provided in impounding reservoirs for municipal water supply. *Transactions of the American Society of Civil Engineering,* Pap. 1308, **77,** 1547-50.

Hirano, K. and Iwase, K. (1989). Minimum risk scale equivariant estimator: estimating the mean of an inverse Gaussian distribution with known coefficient of variation. *Communications in Statistics,* **18**(1), 189-97.

Hoem, J.M. (1976). The statistical theory of demographic rates: a review of current developments (with discussion). *Scandinavian Journal of Statistics,* **3,** 169-85.

Hoem, J.M. and Berge, E. (1975). Some problems in Hadwiger fertility graduation. *Scandinavian Actuarial Journal*, 129-44.

Holden, A.V. (1975). A note on convolution and stable distributions in the nervous system. *Biological Cybernetics*, **20**, 171-3.

Holden, A.V. (1976). *Models of the stochastic activity of neurons. Lecture Notes in Biomathematics*, **12**, Springer-Verlag, Berlin.

Holgate, P. (1970). The modality of some compound distributions. *Biometr- ika*, **56**, 666-7.

Holla, M.S. (1966). On a Poisson-inverse Gaussian distribution. *Metrika*, **11**, 115-21.

Homer, L.D., and Small, A. (1977). A unified theory for estimation of cardiac output, volumes of distribution and renal clearances from indicator dilution curves. *Journal of Theoretical Biology*, **64**, 535-50.

Horton, R.E. (1913). Frequency of recurrence of Hudson River floods. *U.S. Weather Bureau Bulletin*, **Z**, 109-12.

Hougaard, P. (1984). Life table methods for heterogeneous populations: Distributions describing the heterogeneity. *Biometrika*, **71**(1), 75-83.

Hougaard, P. (1986). Survival models for heterogeneous populations derived from stable distributions. *Biometrika*, **73**(2), 387-96.

Howlader, H.A. (1985). Approximate Bayes estimation of the reliability of two parameter inverse Gaussian distribution. *Communications in Statistics*, **A14**, 937-46.

Hsieh, H.K. (1990). Inferences on the coefficient of variation of an inverse Gaussian distribution. *Communications in Statistics*, **19**(5), 1589-605.

Hsieh, H.K. (1990). Estimating the critical time of the inverse Gaussian hazard rate. *IEEE Transactions on Reliability*, **39**, 342-5.

Hsieh, H.K. and Korwar, R.M. (1990). Inadmissibility of the uniformly minimum variance unbiased estimator of the inverse Gaussian variance. *Communications in Statistics*, **19**(7), 2509-16.

Hsieh, H.K., Korwar, R.M., and Rukhin, A.L. (1990). Inadmissibility of the maximum likelihood estimator of the inverse Gaussian mean. *Statistics and Probability Letters*, **9**(1), 83-90.

Huff, B.W. (1974). A comparison of sample path properties for the inverse Gaussian and Bessel processes. *Scandinavian Actuarial Journal*, 157-66.

Huff, B.W. (1975). The inverse Gaussian distribution and Root's barrier construction. *Sankhyā*, **37**A, 345-53.

Iliescu, D.V. and Vodă, V.G. (1981). On the inverse Gaussian distribution. *Bulletin Mathématique de la Société des Sciences Mathématiques de la République Socialiste de Roumanie*, Tome **25**(73), No.4, 381-92.

Isogai, T., and Uchida,H. (1989). Some measurement models related to an inverse Gaussian distribution (STMA V32 3280).*Reports of Statistical Application Research* (Union of Japanese Scientists and Engineers),**36**(3),1-11.

Iwase, K. (1989). Linear regression through the origin with constant coefficient of variation for the inverse Gaussian distribution. *Communications in Statistics*, **18**(10), 3587-93.

Iwase, K. and Setô, N. (1983).Uniformly minimum variance unbiased estimation for the inverse Gaussian distribution. *Journal of the American Statistical Association*, **78**, 660-3.

Iyengar, S. (1985). Hitting lines with two-dimensional Brownian motion. *SIAM Journal of Applied Mathematics*, **45**, 983-9.

Iyengar, S., and Patwardhan, G. (1988). Recent developments in the inverse Gaussian distribution. Handbook of Statistics 7, (eds. P.R. Krishnaiah and C.R. Rao), 479-90, Elsevier, North-Holland, Amsterdam.

Jain, G.C., and Khan, M.S.H. (1979) On an exponential family. *Mathematisch Operationforschung und Statistik Series*, **10**(1), 153-68.

Jaisingh, L.R., Dey, D. K., and Griffith, W.S. (1993). Property of a multivariate survival distribution generated by a Weibull and inverse Gaussian mixture. *IEEE Transactions on Reliability*, **42**(4), 618-22.

Johnson, L.G. (1951). The median ranks of sample values in their population with an application to certain fatigue studies. *Industrial Mathematics*, **2**, 1-6.

Jones, G. and Cheng, R.C.H. (1984). On the efficiency of moment and maximum likelihood estimators in the three-parameter inverse Gaussian distribution. *Communications in Statistics*, A **13**, 2307-14.

Jørgensen, B. (1982a). *Statistical properties of the generalized inverse Gaussian distribution*, Springer Verlag, Heidelberg.

Jørgensen, B. (1982b). Identifiability problems in Hadwiger fertility graduation. *Scandinavian Actuarial Journal*, 103-9.

Jørgensen, B. (1989). *The theory of exponential dispersion models and analysis of deviance*, 1^a Escola de modelos lineares, Associacào Brasileira de Estatistica, Universidade de Sao Paolo.

Jørgensen, B., Seshadri, V., and Whitmore, G.A. (1991). On the mixture of the inverse Gaussian distribution with its complementary reciprocal. *Scandinavian Journal of Statistics*, **18**, 77-89.

Joshi, S. and Shah, M. (1990). Sequential analysis applied to testing the mean of an inverse Gaussian distribution with known coefficient of variation. *Communications in Statistics*, **19**(4), 1457-66.

Kabe, D.G. and Laurent A.G. (1981). On some nuisance parameter free uniformly most powerful tests. *Biometrical Journal*, **23**, No.3, 245-50.

Kappenman, R.F. (1979). On the use of a certain conditional distribution to obtain unconditional distributional results. *The American Statistician*, **33**(1), 23-4.

Kappenman, R.F. (1985). A testing approach to estimation. *Communications in Statistics*, A 2365-77.

Karan Singh, R., and Chaturvedi, A. (1989). Sequential procedures for estimating the mean of an inverse Gaussian distribution. *Journal of the Indian Society of Agricultural Statistics*, **41**,300-8.

Karlin, S., and Taylor, H.M. (1981). *A second course in Stochastic Processes*. Academic Press.New York.

Kemp, C.D., and Kemp, A.W. (1965). Some properties of the Hermite distribution. *Biometrika*, **52**, 381-94.

Kempton, R.A. (1979). Statistical analysis of frequency data obtained from sampling an insect population grouped by stages, Statistical distributions in Ecological work, 401-18, (eds. J.K. Ord, G.P. Patil, and C. Taille).

Kendall, D.G. (1957). Some problems in the theory of dams. *Journal of the Royal Statistical Society*, B**19**, 207-12.

Keyfitz, N. (1968). *Introduction to the mathematics of population with revisions.* Reading, Mass.Addison-Wesley.

Khan, M.S.H. and Jain, G.C. (1978). A class of distributions in the first emptiness of a semi-infinite reservoir. *Biometrics Journal,* **20**(3), 243-52.

Khatree, R. (1990). Characterization of inverse Gaussian and gamma distributions through their length-biased distributions. *IEEE Transactions on Reliability,***38**, 610-1.

Khatri, C.G. (1962). A characterization of the inverse Gaussian distribution. *Annals of Mathematical Statistics,* **33**, 800-3.

Kirmani, S.N.U.A. and Ahsanullah, M. (1987). A note on weighted distributions. *Communications in Statistics,* **16**(1), 275-80.

Kobayashi, Y. (1975). A simple derivation of some properties of the inverse Gaussian distribution. (Japanese). *MFEdShig,* **25**, 4-8.

Kocherlakota, S. (1986). The bivariate inverse Gaussian distribution: An introduction. *Communications in Statistics,* **15**(4), 1081-112.

Kokonendji, C.C. and Seshadri, V. (1994). The Lindsay transform of natural exponential families with cubic variance functions. *Canadian Journal of Statistics,* **22**(2), 259-72.

Konstantinowsky, D. (1914). *Wien Ber,* **123**, 1717.

Korwar, R.M. (1980). On the uniformly minimum variance unbiased estimators of the variance and its reciprocal of an inverse Gaussian distribution. *Journal of the American Statistical Association,* **75**, 734-5.

Kourouklis, S. (1996). Improved estimation under Pitman's measure of closeness.*Annals of the Institute of Statistical Mathematics,* **48**(3),509-18.

Kourouklis,S.(1997). A new property of the inverse Gaussian distribution with applications.*Statistics and Probability Letters,***32**,161-6.

Koziol, J.A. (1989). *CRC Handbook of percentage points of the inverse Gaussian distribution.* CRC Press, Florida.

Krapivin, V.F. (1965). *Tables of Wald's distribution.* Nauka, Moscow. (In Russian).

Kurze, U.J. (1971a). Statistics of road traffic-noise. *Journal of Sound and Vibration,* **18**, 171-95.

Kurze, U.J. (1971b). Noise from complex road traffic. *Journal of Sound and Vibration,* **19**, 167-77.

Kurze, U.J. (1974). Frequency curves of road traffic noise. *Journal of Sound Vibration,* **33**, 171-85.

Lancaster, T. (1972). A stochastic model for the duration of a strike. *Journal of the Royal Statistical Society,* A**135**(2), 257-71.

Lande, R. and Orzack, S.H. (1988). Extinction dynamics of age-structured populations in a fluctuating environment. *Proceedings of the National Academy of Sciences (USA),* **85**, 7418-21.

Lánsky, P. (1983). Inference for the diffusion models of neuronal activity. *Mathematical Biosciences,* **67**, 247-260.

Lánsky, P. and Radil, T. (1987). Statistical inference on spontaneous neuronal discharge patterns. *Biological Cybernetics,* **55**, 299-311.

Lánsky, P. and Smith, C.E. (1989). The effect of a random initial value in neural first-passage time models. *Mathematical Biosciences,* **93**, 191-215.

Le Cam L. and Morlat, G. (1949): Les lois des débits des rivières françaises. *La Houille Blanche No. spécial B.*

Lehmann, E.L. (1959). *Testing statistical hypothesis.* Wiley, New York.

Leslie, P.H. (1945). On the use of matrices in certain population mathematics. *Biometrics*, **33**, 182-212.

Letac, G. (1991). *Lecture notes on natural exponential families and their variance functions.* I.M.P.A. Rio de Janeiro.

Letac, G. and Mora, M. (1986). Sur les fonctions-variance des familles exponentielles naturelles sur **R**. *Comptes Rendus de l'Academie des Sciences*, Paris, 302 Série I, No.15, 551-4.

Letac, G. and Mora, M. (1990). Natural real exponential families with cubic variances. *Annals of Statistics*, **18**, No.1, 1-37.

Letac, G. and Seshadri, V. (1983). A characterization of the generalized inverse Gaussian by continued fractions. *Zeitschrift für Wahrscheinlichkeitstheorie und verwandte Gebiete*, **62**, 485-9.

Letac, G. and Seshadri, V. (1985). On Khatri's characterization of the inverse Gaussian distribution. *Canadian Journal of Statistics*, **13**(3), 249-52.

Letac, G. and Seshadri, V. (1986). On a conjecture concerning inverse Gaussian regression. *International Statistical Review*, **54**, 187-90.

Letac, G. and Seshadri, V. (1989). The expectation of X^{-1} as a function of $\mathbf{E}(\mathbf{X})$ for an exponential family on the positive line. *Annals of Statistics*, **17**, No.4, 1735-41.

Letac, G., Seshadri, V., and Whitmore, G.A. (1985). An exact chi-squared decomposition theorem for inverse Gaussian variates. *Journal of the Royal Statistical Society*, B**47**, 476-81.

Levine, M.W. (1991). The distribution of intervals between neural impulses in the maintained discharges of retinal ganglion cells. *Biological Cybernetics*, **65**, 459-67.

Levine, M.W. (1987). Variability in the maintained discharges or retinal ganglion cells. *Journal of the Optical Society of America A*, **4**, 2308-2320.

Levine, M.W., Saleh, E.J., and Yarnold, P. (1988). Statistical properties of the maintained discharge of chemically isolated ganglion cells in goldfish retina. *Visual Neurosciences*, **1**, 31-46.

Liao, Q. (1995). Model selection with applications to interspike interval data. Unpublished Ph.D. thesis. University of Pittsburgh.

Lieblein, J. and Zelen, M. (1956). Statistical investigation of the fatigue life of deep-groove ball bearings. *Journal of Research of the National Bureau of Standards*, **57**, 273-316.

Lingappaiah, G.S. (1983). Prediction in samples from the inverse Gaussian distribution. *Statistica*, **43**, 259-65.

Lombard, F. (1978). A sequential test for the mean of an inverse Gaussian distribution. *South African Statistical Journal*, **12**(2), 107-15.

Lotka, A.J. (1939). Theorie analytique des associations biologiques. Part II. analyse demographique avec application particuliere a l'espece humaine. *Actualites Scientifiques et Industrielles*, **780**, Paris, Hermann et Cie.

Lukacs, E. and Laha, R.G. (1964). *Applications of characteristic functions.* Charles Griffin and Co., Ltd, London.

MacGibbon, K.B. and Shorrock, G.E. (1995). Estimation of the lambda parameter of an inverse Gaussian distribution. (to appear in Statistics and Probability Letters).

Mahmoud, M. (1991). Bayesian estimation of the 3-parameter inverse Gaussian distribution. *Trabjos da Estadistica*, **6** (1),45-62.

Mann, N.R., Schafer, R.E., and Singpurwalla, N.D. (1974). *Methods for Statistical Analysis of Reliability and Life Data*. Wiley, New York.

Marcus, A.H. (1975). Some exact distributions in traffic noise theory. *Advances in Applied Probability*, **7**, 593-606.

Marcus, A.H. (1975). Power sum distributions: an easier approach using the Wald distribution. *Journal of American Statistical Association*, **71**, 237-8.

Marcus, A.H. (1975). Power laws in compartmental analysis. Part I: a unified stochastic model. *Mathematical Biosciences*, **23**, 337-50.

Maxim, L.D. and Harrington, L. (1982). "Scale up" estimators for aerial surveys with size-dependent detection. *Photogrammetric Engineering and Remote Sensing*, **48**(8), 1271-87.

McCool, J.I. (1979). Analysis of single classification experiments based on censored samples from the two-parameter Weibull distribution. *Journal of Statistical Planning and Inference*, **3**, 39-68.

McCullagh, P. and Nelder, J.A. (1983). *Generalized linear models*. Chapman and Hall, New York.

Mehta, J.S. (1969). Estimation in inverse Gaussian distribution. *Trabajos de Estadistica* (Madrid), XX(1), 103-11.

Michael, J.R., Schucany, W.R., and Haas, R.W. (1976). Generating random variates using transformations with multiple roots. *American Statistician*, **30**(2), 88-90.

Miura, C.K. (1978). Tests for the mean of the inverse Gaussian distribution. *Scandinavian Journal of Statistics*, **5**, 200-4.

Mora, M. (1986a). Classification des fonctions variance cubiques des familles exponentielles sur **R**. *Comptes Rendus de l'Academie des Sciences*, Paris 302, Série I, **16**, 587-90.

Mora, M. (1986b). Familles exponentielles et fonctions variances, Thèse de 3e cycle, Université Paul-Sabatier.

Morlat, G. (1956). Les lois de probabilités de Halphen. *Revue de Statistique Appliquée*, **4**, 21-46.

Munholland, P.L. (1988). Statistical aspects of field studies on insect populations. Ph.D. thesis. University of Waterloo, Waterloo, Ontario.

Munholland. P.L., and Dennis, B . (1992). Biological aspects of a stochastic model for insect life history data. *Environmental Entomology*,**21**,6, 1229-38.

Munholland, P.L. and Kalbfleisch, J.D. (1991). A semi-Markov model for insect life history data. *Biometrics*, **47**, 1117-26.

Munholland, P.L., Kalbfleisch, J.D., and Dennis B. (1989). A stochastic model for insect life history data. 136-44, In: L. McDonald, B.F. Manly, J. Lockwood and J. Logan (eds.) Estimation and analysis of insect populations. Springer Verlag, Berlin.

Nádas, A. (1971). Times to intermittent and permanent failures as Brownian crossing times. *Journal of Applied Probability*, **8**, 838-40.

Nádas, A. (1973). Best tests for zero drift based on first passage times in Brownian motion. *Technometrics*, **15**(1), 125-32.

Nath, G.B. (1977). Estimation of the parameter of the inverse Gaussian distribution from generalized censored samples. *Metrika*, **24**(1), 1-6.

Nelder, J.A. and Wedderburn, R.W.M.(1972). Generalized linear models. *Journal of the Royal Statistical Society.* A**135**,(33),370-84.

Nelson, W. (1969). Hazard plotting for incomplete failure data. *Journal of Quality Technology*, **1**, 27- 52.

Nelson, W.B. (1971). Analysis of accelerated life test data.*IEEE Transactions on Electrical Insulation* EI-6,165-81.

Nelson, W.B. (1980). Accelerated life testing - step stress models and data analysis. *IEEE Transactions of Reliability*, **29**, 103-8.

Nelson, W. (1982). *Applied life data analysis.* Wiley, New York.

Nelson, W. (1990). *Accelerated Testing: Statistical Models, Test Plans, and Data Analysis.* New York: John Wiley.

Newby, M. and Winterton, J. (1983). The duration of industrial stoppages. *Journal of the Royal Statistical Society*, A**146** (1), 62-70.

Newby, M. (1985). A further note on the distribution of strike duration. *Journal of the Royal Statistical Society*, A**148**, 350-6.

Oberhofer, W. and Kmenta, J. (1974). A general procedure for obtaining maximum likelihood estimates in generalized regression models. *Econometrica*, **42**, 579-90.

O'Brien, C.M. (1986). A graphical method of checking the goodness of-fit of the inverse Gaussian distribution. *GLIM Newsletter*, No.13, 32-5.

Ord, J.K. and Whitmore, G.A. (1986). The Poisson-inverse Gaussian distribution as a model of species abundance. *Communications in Statistics*, **15**, 853-71.

O'Reilly, F.J. and Rueda, R., (1992). Goodness-of-fit for the inverse Gaussian distribution. *The Canadian Journal of Statistics*, **20**(4), 387-97.

Ostle, B. (1963). *Statistics in Research.* Ames , Iowa State University Press.

Padgett, W.J. (1978). Comment on inverse Gaussian random number generation. *Journal of Statistical Computing and Simulation*, **8**, 78-9.

Padgett, W.J. (1979). Confidence bounds on reliability for the inverse Gaussian model. *IEEE Transactions in Reliability*, R**28**, 165-8.

Padgett, W.J. (1981). Bayes estimation of reliability for the inverse Gaussian distribution. *IEEE Transactions of Reliability*, **R-30**(4), 384-5. (see also Supplement: NAPS document No. 03794-C).

Padgett, W.J. (1982). An approximate prediction interval for the mean of future observations from the inverse Gaussian distribution. *Journal of Statistical Computing and Simulation*, **14** (3), 191-9.

Padgett, W.J. and Tsoi, S.K. (1986). Prediction intervals for future observations from the inverse Gaussian distribution. *IEEE Transactions on Reliability*, R**35** (4), 406-8.

Padgett, W.J. and Wei, L.J. (1979). Estimation for the three-parameter inverse Gaussian distribution. *Communications in Statistics - Theory and Methods*, A**8** (2), 129-37.

Padmanabhan, P. (1978). Applications of the inverse Gaussian distribution in valuation and in estimation of conversion probabilities of convertible debentures:

An empirical study. Unpublished M.B.A. thesis, Faculty of Management, McGill University, Montreal.

Pakes, A.G.(1974). Some limit theorems for Markov chains with applications to branching processes in Studies in Probability and Statistics. Papers in honour of Edwin G. Pitman.E.J.Williams ed.,*Jerusalem Academic Press*,21-39.

Pal, N. and Sinha, B.K. (1989). Improved estimators of an inverse Gaussian distribution. 215-22. Statistical data analysis and inference. (Y. Dodge, ed.), Elsevier Science Publishers BV (North-Holland).

Palmer, J. (1973). Bayesian analysis of the inverse Gaussian distribution. Unpublished Ph.D. thesis, Oklahoma State University.

Pandey, B.N. and Malik, H.J. (1988). Some improved estimators for a measure of dispersion of an inverse Gaussian distribution. *Communications in Statistics*, **17**(11), 3935-49.

Pandey, B.N., and Malik, H.J. (1969). Estimation of the mean and the reciprocal of the mean of the inverse Gaussian distribution. *Communications in Statistics-Simulation and Computation*, B **18**,1187-1201.

Pandey, M., Ferdous, J., and Uddin, Md.B. (1991). Effect of a new selection procedure on adaptive estimation. *Communications in Statistics. Theory and Methods*, **20**(10), 3365-79.

Park, E., Bar-Lev, S.K., and Enis, P. (1988). Minimum variance unbiased estimation in natural exponential families generated by stable distributions. *Communications in Statistics*, **17**(12), 4301-13.

Parsian, A., and Sanjarifarsipour, N. (1988). Optimum tests for the comparison of two inverse Gaussian shape parameters. *Calcutta Statistical Association Bullettin*,**37**,233-6.

Patel, J.K. (1966). Tolerance limits-a review. *Communications in Statistics-Theory and Methods*, **15**, 2719-62.

Patel, R.C. (1965). Estimates of parameters of truncated inverse Gaussian distribution. *Annals of the Institute of Statistical Mathematics*, (Tokyo) **17**, 29-33.

Patil, S.A. and Kovner, J.L. (1976). On the test and power of zero drift on first passage times in Brownian motion. *Technometrics*, **18**, 341-2.

Patil, S.A. and Kovner, J.L. (1979). On the power of an optimum test for the mean of the inverse Gaussian distribution. *Technometrics*, **21**, 379-81.

Pavur, R.J., Edgeman, R.L., and Scott, R.C. (1992). Quadratic statistics for the goodness-of-fit test of the inverse Gaussian distribution. *IEEE Transactions on Reliability*, **41**(1), 118-23.

Pearson, E.S. and Hartley, H.O. (1972). *Biometrika Tables for Statistics*, **11**, Cambridge University Press.

Pericchi, L.R., and Rodriguez-Iturbe. (1985). On the statistical analysis of floods. *International Statistical Institute Celebration of Statistics* 511-41.

Perng, S.K. and Littell, R.C. (1976). A test of equality of two normal population means and variances. *Journal of the American Statistical Association*, **71**, 968-71.

Perrault, L., Fortin, V., and Bobée, B. (1994). The Halphen family of distributions for modeling maximum annual flood series: new developments. Envirosoft conference.

Podwysocki, M.H. (1976). An estimate of field size distributions for selected sites in the major grain producing countries. *Goddard Space Flight Center X*, **923**, 76-93, Greenbelt, Maryland.

Proschan, F. (1963). Theoretical explanation of observed decreasing failure rate. *Technometrics*, **5**, 375-83.

Qasrawi, H. (1966). A study of the energy flow in a natural population of the grasshopper. *Chorthippus parallelus*, Zett (Orthopetra: Acrididae). Unpublished Ph.D. thesis. University of Exeter.

Ragab, A., and Green, J. (1987). Optimal predictive method for estimating the inverse Gaussian density. *Metron*, **45**, 51-61.

Ramachandran, B. and Seshadri, V. (1988). On a property of strongly reproductive exponential families on R. *Statistics and Probability Letters*, **6**(3), 171-4. Erratum. *Statistics and Probability Letters* (1988), **7**(1), 87.

Rao, C.R. (1973). *Linear statistical inference and its applications*. Second ed., New York. John Wiley & Sons.

Read, K.L.Q. and Ashford, J.R. (1968). A system of models for the life cycle of a biological organism. *Biometrika*, **55**, 211-21.

Read, K.L.Q. and Cressie, N.A.C. (1988). *Goodness-of-fit statistics for discrete multivariate data*. Springer-Verlag, Berlin.

Rodieck, R.W., Kiang, N.Y.S., and Gerstein, G.L. (1962). Some quantitative methods for the study of spontaneous activity of single neurons. *Biophysics Journal*, **2**, 351-68.

Roy, L.K. (1967). Some properties and applications of the inverse Gaussian distribution. Unpublished M.Sc. thesis, Queen's University, Kingston, Ontario.

Roy, L.K. (1970). Estimation of the parameter of the inverse Gaussian distribution from a multi-censored sample. *Statistica*, **30**, 563-7.

Roy, L.K. and Wasan, M.T. (1968a). Properties of the time distribution of standard Brownian motion. *Revista Trabajos de Estadistica*, 67-82.

Roy, L.K. and Wasan, M.T. (1968b). The first passage time distribution of Brownian motion with positive drift. *Mathematical Biosciences*, **3**, 191-204.

Roy, L.K. and Wasan, M.T. (1969). A characterization of the inverse Gaussian distribution. *Sankhyā*, **31**A, 217-8.

Rubinstein, G.Z. (1985). Models for count data using the Sichel distribution. Unpublished Ph.D. thesis. University of Cape Town.

Sahin, I. and Hendrick, D.J. (1978). On strike durations and a measure of termination. *Applied Statistics*, **27**(3), 319-24.

Samanta, M. (1983). Estimating the mean of an inverse Gaussian distribution with known coefficient of variation. Technical report #140. Department of Statistics, University of Manitoba, Winnipeg.

Samanta, M. (1985). On tests of equality of two inverse Gaussian distributions. *South African Statistical Journal*, **19**, 83-95.

Sankaran, M. (1968). Mixtures by the inverse Gaussian distribution. *Sankhyā*, **30**B, 455-8.

Sarndal, C. (1984). Design consistent versus model dependent estimation for small domains. *Journal of the American Statistical Association*, **79**, 624-631.

Sarndal, C. and Hiridoglou, M.A. (1989). Small domain estimation. A conditional analysis. *Journal of the American Statistical Association*, **84**, 266-75.

Saunders, S.C. (1974). A family of random variables closed under reciprocation. *Journal of the American Statistical Association*, **69**, (346), 533-9.

Schrödinger, E. (1915). Zur theorie der fall und steigversuche an teilchen mit Brownscher bewegung. *Physikalische Zeitschrift*, **16**, 289-95.

Schwarz, C.J. and Samanta, M. (1991). An inductive proof of the sampling distributions for the maximum likelihood estimators of the parameters in an inverse Gaussian distribution. *The American Statistician*, **45**(3), 223-5.

Seal, H.L. (1969). *Stochastic Theory of a Risk Business*, New York: John Wiley & Sons Inc.

Seal, H.L. (1978). From aggregate claim distributions to probability of ruin. *Astin. Bulletin*, **10**, 47-53.

Seshadri, V. (1981). A note on the inverse Gaussian distribution. *Statistical Distributions in Scientific work* (eds C. Taillie et al.), D. Reidel, Dordrecht, **4**, 99-103.

Seshadri, V. (1983). The inverse Gaussian distribution: some properties and characterizations. *Canadian Journal of Statistics*, **11**, 131-6.

Seshadri, V. (1987). Discussion of Jørgensen's paper, *Journal of the Royal Statistical Society*, B**49**(2), 156.

Seshadri, V. (1988). Exponential models, Brownian motion and independence. *Canadian Journal of Statistics*, **16**(3), 209-21.

Seshadri, V. (1989). A U-statistic and estimation for the inverse Gaussian distribution. *Statistics and Probability Letters*, **7**(1), 47-9.

Seshadri, V. (1992). General exponential models on the unit simplex and related multivariate inverse Gaussian distributions. *Statistics and Probability Letters*, **14**(5), 385-91.

Seshadri, V. (1993). *The inverse Gauusian distribution-a case study in exponential families*. Oxford University Press.

Seshadri, V. and Shuster, J.J. (1974). Exact tests for zero drift based on first passage times in Brownian motion. *Technometrics*, **16**(1), 133-4.

Shaban, S.A. (1981a). On the discrete Poisson-inverse Gaussian distribution. *Biometrics Journal*, **23**, 297-303.

Shaban, S.A. (1981b). Computation of the Poisson–inverse Gaussian distribution. *Communications in Statistics – Theory and Methods*, A**10**(14), 1389-99.

Shabhan, S.A. (1991). Testing inverse Gaussian vs lognormal. *Egyptian Statistical Journal*, **35**,1-12.

Shapiro, S.S. and Wilk, M.B. (1965). An analysis of variance test for normality (complete samples). *Biometrika*, **52**, 591-611.

Sheppard, C.W. (1962). *Basic principles of the tracer method*. Wiley, New York.

Sheppard, C.W. and Uffer, M.B. (1969). Stochastic models for tracer experiments in the circulation II. Serial random walks. *Journal of Theoretical Biology*, **22**, 188-207.

Sherif, Y.S. and Smith, M.L. (1980). First-passage time distribution of Brownian motion as a reliability model. *IEEE Transactions on Reliability*, R**29**(5), 425-6.

Shoukri, M.M. (1986). K-sample likelihood ratio test on the inverse Gaussian distribution. *Statistician*, **35**, 27-32.

Shuster, J.J. (1968). On the inverse Gaussian distribution function. *Journal of the American Statistical Association*, **63**, 1514-6.

Shuster, J.J. and Miura, C. (1972). Two way analysis of reciprocals. *Biometrika*, **59**, 478-81.

Sichel, H.S. (1971). On a family of discrete distributions particularly suited to represent long-tailed frequency data. *Proceedings of the Third Symposium on Mathematical Statistics*, (ed. N.F.Laubscher) Pretoria, South Africa, 51-97.

Sichel, H.S. (1973a). Statistical valuation of diamondiferous deposits. *Journal of the South African Institute of Mining Metallurgy*, **73**, 235-43.

Sichel, H.S. (1973b). The density and size distribution of diamonds. *Bulletin of the International Statistical Institute*, **45**, 420-7.

Sichel, H.S. (1974). On a distribution representing sentence-length in written prose. *Journal of the Royal Statistical Society*, Ser. A **137**, 25-34.

Sichel, H.S. (1975). On a distribution law for word frequencies. *Journal of the American Statistical Association*, **70**, 542-7.

Sichel, H.S. (1982a). Repeat-buying and the generalized inverse Gaussian–Poisson distribution. *Applied Statistics*, **31**, 193-204.

Sichel, H.S. (1982b). A bibliometric distribution which really works. *Journal of the American Society for Information Science*, **36**, 314-21.

Sichel, H.S. (1986). Word frequency distributions and type-token characteristics. *Mathematical Sciences*, **11**, 45-72.

Silcock, H. (1954). The phenomenon of labour turnover. *Journal of the Royal Statistical Society*, A**117**, 429-40.

Sinha, S.K. (1986). Bayesian estimation of the reliability function of the inverse Gaussian distribution. *Statistics and Probability Letters*, **4**, 319-23.

Smith, C.E., and Lánský, P. (1994). A reliability application of a mixture of inverse Gaussian distributions. *Applied Stochastic models and Data Analysis*, **10**, 61-9.

Smoluchowsky, M.V. (1915). Notiz über die Berechning der Brownschen Molkularbewegung bei des Ehrenhaft-millikanchen Versuchsanordnung. *Physikalische Zeitschrift*, **16**, 318-21.

Snyder, N.F.R. and Johnson, E.V. (1985). Photographic censusing of 1982-1983 California Condor population. *Condor*, **87**, 1-13.

Stark, G.E. and Chhikara, R.S. (1988). Kolmogorov-Smirnov statistics for the inverse Gaussian distribution. Presented at the IASTED International Conference on Applied Simulation and Modelling, held in Galveston, Texas, May 18-20, 1988.

Stein, G.Z. (1988). Modelling counts in biological populations. *Mathematical Scientist*, **13**, 56-65.

Stein, G.Z., and Juritz, J.M. (1987). Bivariate compound Poisson distributions. *Communications in Statistics*, **16**(12), 3591-607.

Stein, G.Z., and Juritz, J.M. (1988). Linear models with an inverse Gaussian-Poisson error distribution. *Communications in Statistics-Theory and Methods*, **17**(1), 557-71.

Stein, G.Z., Zucchini, W., and Juritz, J.M. (1987). Parameter estimation for the Sichel distribution and its multivariate extension. *Journal of the American Statistical Association*, **82**, 938-44.

Stewart, D.A. and Essenwanger, Ol.M. (1978). Frequency distribution of wind speed near the surface. *Applied Meteorology*, **17**, 1633-42.

Sundarier, V. H (1996). Estimation of a process capability index for inverse Gaussian distribution. *Communications in Statistics-Theory and Methods*, **25**(10), 2381-93.

Tadikamala, P.R.(1990). Kolmogorov-Smirnov type test statistics for the gamma Erlang-2 and the inverse Gaussian distributions when the parameters are known. *Communications in Statistics*,B, **19**,305-14.

Takagi, K., Hiramatsu, K., Yamamoto, T. and Hashimoto, K.(1974). Investigations on road traffic noise based on an exponentially distributed vehicles model - single line flow of vehicles with the same acoustic power. *Journal of Sound and Vibration*, **36**(3), 417-31.

Takhashi, R. (1987). Normalizing constants of a distribution which belongs to the domain of attraction of the Gumbel distribution. *Statistics and Probability Letters*, **5**,197-200.

Tang, L.C. and Chang, D.S. (1994). Reliability bounds and tolerance limits of two inverse Gaussian models. *Microelectronics and Reliability*, **34**(2), 247-59.

Tang, L.C. and Chang, D.S. (1994). Tolerence limits for inverse Gaussian distribution. *Journal of Statistical Computing and Simulation*, **51**, 21-9.

Tang, L.C. and Chang, D.S. (1994). Confidence interval for the critical time of inverse Gaussian failure rate. *International Journal of Reliability, Quality and Safety Engineering*, **1**(3), 379-89.

Tang, L.C., Lim, S.C., and Goh, C.J. (1988). On the reliability of components subject to sliding wear - a first report. *Scripta Metallurgica*, **22**, 1177-81

Tapiero, C.S., and Nina Toren.(1987). A Brownian motion model of return migration. *Applied Stochastic models and Data Analysis*,New York, Wiley,151-60.

Teugels, J. and Willmot, G. (1987). Approximations for stop - loss premiums - *Insurance mathematics and Economics*, **6**, 195-202.

Thompson, J.R. (1978). Some shrinkage techniques for estimating the mean. *Journal of the American Statistical Association*, **63**, 113-23.

Travadi, R.J., and Ratani, R.T. (1990). On estimation of reliability function for IG distribution with known CV.(STMA V32 2292). *Journal of the Indian Association for productivity,quality and reliability.*(IAPQR Transactions),**15**, 29-37.

Tremblay, L. (1992). Using the Poisson inverse Gaussian in bonus-malus systems. *Astin Bullettin*, **22**, 97-106.

Tuckwell, H. (1988). *Introduction to theoretical neurobiology.* Vols. 1 & 2. New York: Cambridge University Press.

Tuljapurkar, S.D. (1989). An uncertain life. Demography in random environments. *Theoretical Population Biology*, **35**, 227-94.

Tuljapurkar, S.D. and Orzack, S.H. (1980). Population dynamics in variable environments I. Long-run growth rates and extinction. *Theoretical Population Biology*, **18**, 314-42.

Tweedie, M.C.K. (1941). A mathematical investigation of some electrophoret- ic measurements on colloids. Unpublished M.Sc. Thesis, University of Reading, England.

Tweedie, M.C.K. (1945). Inverse statistical variates. *Nature*, **155**, 453.

Tweedie, M.C.K. (1946). The regression of the sample variance on the sample mean. *Journal of London Mathematical Society*, **21**, 22-8.

Tweedie, M.C.K. (1947). Functions of a statistical variate with given means, with special reference to Laplacian distributions. *Proceedings of the Cambridge Philosophical Society*, **43**, 41-9.

Tweedie, M.C.K. (1956). Some statistical properties of inverse Gaussian distributions. *Virginia Journal of Science*, **7**(3), New Series, 160-5.

Tweedie, M.C.K. (1957a). Statistical properties of inverse Gaussian distributions-I. *Annals of Mathematical Statistics*, **28**, 362-77.

Tweedie, M.C.K. (1957b). Statistical properties of inverse Gaussian distributions-II. *Annals of Mathematical Statistics*, **28**, 696-705.

Upadhyay, S.K., Agrawal, R., and Singh, U. (1994). Bayesian prediction results for the inverse Gaussian distribution utilizing guess values of parameters. *Microelectronics and Reliability*, **34**(2), 351-5.

Vilmann, H., Kirkeby, S., and Kronberg, D. (1990). Histomorphometrical analysis of the influence of soft diet on masticatory muscle development in the muscular dystrophic mouse. *Archives of oral Biology*, **35**, No. 1, 37-42.

Vodă, V.Gh. (1973). A note on the one-parameter inverse Gaussian distribution. *Revista de Ciencias Matematicas*, IV, Serie A, 47-55.

Von Alven, W.H. (ed), (1964). *Reliability Engineering* by ARINC, Englewood Cliffs, New Jersey, Prentice Hall.

Wald, A. (1944). On cumulative sums of random variables. *Annals of Mathematical Statistics*, **15**, 283-96.

Wald, A. (1947). *Sequential analysis*. John Wiley, New York.

Wani, J.K. and Kabe, D.G. (1973). Some results for the inverse Gaussian process. *Skandinavisk Aktuarietidsskrift*, 52-7.

Wasan, M.T. (1968). On an inverse Gaussian process. *Skandinavisk Aktuarietidsskrift*, **60**, 69-96.

Wasan, M.T. (1969a). Sufficient conditions for a first passage time process to be that of Brownian motion. *Journal of Applied Probability*, **6**(1), 218-23.

Wasan, M.T. (1969b). First passage time distribution of Brownian motion with positive drift (Inverse Gaussian Distribution). Queen's Papers in Pure and Applied Mathematics,No.19, Queen's University, Kingston, Ontario.

Wasan, M.T. and Buckholtz, P.G. (1973). Differential representation of a bivariate inverse Gaussian process. *Journal of Multivariate Analysis*, **3**, 243-7.

Wasan, M.T. and Roy, L.K. (1969). Tables of the inverse Gaussian percentage points. *Technometrics*, **11**(3), 591-604.

Washio, Y., Morimoto, H., and Ikeda, N. (1956). Unbiased estimation based on sufficient statistics. *Bulletin of Mathematical Statistics*, **6**, 69-94.

Watson, G.N. (1944). *A treatise on the theory of Bessel functions*, University Press, Cambridge.

Weiss, G.H. (1970). On the noise generated by a stream of vehicles. *Transportation Research*, **4**, 229-33.

Weiss, M. (1982). Moments of physiological transit time distributions and the time course of drug disposition in the body. *Journal of Mathematical Biology*, **15**, 305-18.

Weiss, M. (1983). Use of gamma distributed residence times in Pharmacokinetics. *European Journal of Clinical Pharmacology*, **25**, 695-702.

Whitmore, G.A. (1975). The inverse Gaussian distribution as a model of hospital stay. *Health Services Research*, **10**, 297-302.

Whitmore, G.A. (1976). Management applications of the inverse Gaussian distribution. *International Journal of Management Science*, **4**(2), 215-23.

Whitmore, G.A. (1978). Discussion of a paper by J.L. Folks and R.S.Chhikara, *Journal of the Royal Statistical Society*, B**40**, 263-89.

Whitmore, G.A. (1979). An inverse Gaussian model for labour turnover. *Journal of the Royal Statistical Society.* A**142**(4), 468-78.

Whitmore, G.A. (1983). A regression method for censored inverse-Gaussian data. *Canadian Journal of Statistics*, **11**(4), 305-15.

Whitmore, G.A. (1986a). First passage time models for duration data regression structures and competing risks. *The Statistician*, **35**, 207-19.

Whitmore, G.A. (1986b). Normal-gamma mixtures of inverse Gaussian distributions. *Scandinavian Journal of Statistics*, **13**, 211-20.

Whitmore, G.A. (1986c). Inverse Gaussian ratio estimation. *Applied Statistics*, **35**, 8-15.

Whitmore, G.A. (1990). On the reliability of stochastic systems: a comment. *Statistics and Probability Letters*, **10**, 65-7.

Whitmore, G.A. and Lee, M.T. (1991). A multivariate survival distribution generated by an inverse Gaussian mixture of exponentials. *Technometrics*, **33**, 39-50.

Whitmore, G.A. and Neufeldt, A.H. (1970). An application of statistical models in mental health research. *Bulletin of Mathematical Biophysics*, **32**, 563-79.

Whitmore, G.A. and Seshadri, V. (1987). A heuristic derivation of the inverse Gaussian distribution. *The American Statistician*, **41**, 280-1.

Whitmore, G.A. and Yalovsky, M. (1978). A normalizing logarithmic transformation for inverse Gaussian random variables. *Technometrics*, **20**(2), 207-8.

Wicksell, S.D. (1931). Nuptiality, fertility, and reproductivity. *Skandinavisk Aktuarietidsskrift*, 125-157

Wilbur, S.R. (1980). Estimating the size and trend of the California Condor population 1965-1978. *California Fish and Game*, **66**, 40-8. .

Willmot, G.E. (1986). Mixed compound Poisson distributions. *Astin Bulletin*, **16**, S59-S89.

Willmot, G.E. (1987). The Poisson-inverse Gaussian distribution as an alternative to the negative binomial. *Scandinavian Actuarial Journal*, 113-127.

Willmot, G.E. (1988). Parameter orthogonality for a family of discrete distributions. *Journal of the American Statistical Association*, **83**(402), 517-521.

Willmot, G.E., and Panjer, H.H. (1987). Difference equation approximations in evaluation of compound distributions. *Insurance mathematics and Economics* **6**,Elsevier, North Holland, 43-56.

Wise, K and Kabaila, P. (1989). The estimation of plutonium concentrations in soil. *Australian Journal of Statistics*, **31**, 25-41.

Wise, M.E. (1966). Tracer dilution curves in cardiology and random walk and log-normal distributions. *Acta Physiologica Pharmacologica Neerlandica*, **14**, 175-204.

References

Wise, M.E. (1971). Skew probability curves with negative powers of time and related to random walks in series. *Statistica Neerlandica*, **25**, 159-80.

Wise, M.E. (1974). Interpreting both short and long-term power laws in physiological clearance curves. *Mathematical Biosciences*, **20**, 327-37.

Wise, M.E. (1975). Skew distributions in biomedicine including some with negative powers of time in *Statistical distributions in scientific work*, Vol. 2, (G.P. Patil et al., eds). D. Reidel, Dordrecht, Holland, 241-62.

Wise, M.E., Osborn, S.B., Anderson, J., and Tomlinson, R.W.S. (1968). A stochastic model for turnover of radiocalcium based on the observed laws. *Mathematical Biosciences*, **2**, 199-224.

Worsley, K.J. (1987). Inverse distributions, censored data and GLIM. *GLIM Newsletter*, No.15, 32-4.

Yamamoto, E. and Yanagimoto, T.(1993). The use of the inverse Gaussian model for analyzing the lognormal data. *Statistical Science & Data Analysis*, K.Matusita et al. (eds),489-99.

Yanagimoto, T. (1991). Estimating a model through the conditional MLE. *Annals of the Institute of Statistical Mathematics*, **43**(4), 735-46.

Yang, G.L. and Chen, T.C. (1978). On statistical methods in neuronal spike - train analysis. *Mathematical Biosciences*, **38**, 1-34.

Yule, G.U. (1939). On sentence-length as a statistical characteristic of style in prose; with applications to two cases of disputed authorship. *Biometrika*, **30**, 363-90.

Zigangirov, K.Sh. (1962a). Expression for the Wald distribution in terms of normal distribution. *Radio Engineering and Electronic Physics*, **7**, 145-8.

Zigangirov, K.Sh. (1962b). Representation of Wald's distribution by means of a normal distribution. *Radiotekhnika Electronika*, **7**, 164-6, (in Russian).

AUTHOR INDEX

A

Abouammoh, A. M 317
Abd-El-Hakim, N. S 317
Agrafiotis, G. K 322
Agrawal, R 337
Ahmad, K. E 317
Ahmad, M 317
Ahmed, A.N 317
Ahsanullah, M 317,328
Aitkin,M 317
Akman,O 325
Al-Hussaini,E.K 317
Amin,N.A.K 27,32,33,285
Aminzadeh,M.S 317
Amoh,R.K 317
Ang,A.H.S 61,317
Anderson,J 339
Anscombe,F.J 136,317,318
Ashford,J.R 298,304,318
Ashkar,F 203,320
Athreya,K.B 143,318
Atkinson,A.C 123,124,125

B

Babu,G.J 270,277,280–2,318
Bachelier,L 1,318
Badhwar,G 324
Bahadur,R.R 38,70–72,318
Bain,L.J 324
Baker,R 318
Balakrishnan, N 23,34,318
Banerjee,A.K 20,21, 60,61,137
Barbour,C.D 162,318
Bardsley,W.E 230,231,318
Bar-Lev,S.K 42,318,319,332
Barndorff-Nielsen,O.E 125,146,
 198,201,250,319
Barr,A 319
Bartlett,M.S 174,224,319
Basu,A.K 319
Beekman,J.A 319
Berg,P.T 167,171,319
Berge,E 326
Bernicot,M 284,323
Betró,B 21,319
Bhattacharyya,G.K 20,21,60,
 61,136,139,155,172,183,
 185,188,207,210,212,214,
 213,269,270,271,275,276,
 282,318,319,320,324

Bhattacharya,S.K 14,142,143,
 150,320
Bickel,P.J 98,320
Birnbaum,A 276,320
Birnbaum,Z.W 144,154,155,
 156,157,158,320
Blaesild,P 318,320
Bobeé,B 203,320,332
Box,G.E.P 48,320
Bravo,G 320
Brown,J.H 162,318
Bshouty,D 319
Buckholtz,P.G 337
Bulmer,M.G 320
Burbeck,S.L 320

C

Capocelli,R.M 320
Chan,M.Y 33,35,320,321
Chandra,N.K 321
Chang,D.S 61,96,97,99,106,
 108,110,111,156,158,321,336
Chapman,G.R 309,312,323
Chatfield,C 136,321
Chaturvedi,A 73,87,89,321,327
Chaubey,Y.P 270,277,280,
 281,282,305,307,308,318,321
Chen,P 321
Chen,T.C 240,242,339
Chen,W.S 23,34,318
Cheng,R.C.H 27,30,31.32,
 285,321,322,327
Chernoff,H 166,322
Chhikara,R.S 1,9,10,17,39,
 40,45,47,49,52,53,54,
 56,59–61,94,103,206,224,
 322,324,334
Clayton,D 317
Cohen,A.C 33,34,35,320,321,
 322
Cohen,J.E 286,322
Corbet,A.S 324
Cox,D.R 38,65,147,250,260,322
Cramèr,H 69,155,322
Crepeau,H 324
Cressie,N.A.C 333
Currie,I.D 322

D

Dagpunar,J.S 322
Damaniou,C 322
Daniel,C 276,322

Author index 341

Davis,A.S 47,323
Dean,C 163,323
De Groot,M.H 214,323
Deheuvels,P 2284,323
Dennis,B 290,292,294,295,298,
 301,304,323,330
Desmond,A.F 61,144,154,155,
 156,309,312,323
Dey,D.K 327
Dhulesia,H 284,285,323
Diaconis,P 323
Diekmann,A 323
Doksum,K.A 98,214,218,320,
 323
Dugue,D 323
Durairajan,T.M 70,71,323

E

Eaton,W.W 234,323
Edgeman,R.L 73,76,90,91,114,
 115,116,323,324,332
Ehrenberg,A.S.C 136,321,324
Eisenstat,S 276,324
Embrechts,P 324
Engelhardt,M 156,324
Enis,P 318,332
Essenwanger,O.M 230,231,324,
 336

F

Ferdous,J 332
Ferguson,M 257,324
Fisher,R.A 26,48,94,123,324
Fletcher,H 154,155,324
Folks,J.L 1.9,10,13,17,39,40,
 56,59,60,61,103,206,
 224,322,324
Fortin,V 332
Foster,H.A 203,324
Fries,A 61,155,172,183,185,
 188,207,210,212,213,214,
 269–271,275,276,282,319,
 320,324

G

Gacula,M.C.Jr 206,324
Gani,J 204,324
Gendron,M 324
Gerber,H.U 324
Gerstein,G.L 333
Ghosh,B.K 86.324
Goel,P.K 214,323
Goh,C.J 61,96.324,336

Good,I.J 324
Goodhardt,G.J 321,332
Gossiaux,A 134,325
Govindarajulu,Z 276,325
Gradshteyn,I.S 13,325
Green,J 333
Griffith,W.S 327
Grimmet,G 198,325
Gumbel,E.J 203,325
Gupta,M 321
Gupta,R.C 317,325
Gupta,R.P 257,324
Guttman,I 52,53,54,322
Gydesen,H 175,176,178,181,326

H

Hadwiger,H 1,191,193,325
Halgreen,C 319
Halphen,E 1,203,213
Harrington,L 252,253,255,
 257,330
Hartley,H.O 276,332
Harvey,A.C 171,325
Hashimoto,K 336
Hasofer,A.M 204,325
Hass,R.W 330
Hassairi,A 204.325
Hawkins,D.M 309,310,312,325
Hazen,A 203,325
Hendrick,D.J 333
Hiridoglou,M.A 307,333
Hiramatsu,K 336
Hirano,K 326
Hoem.J.M 326
Holden,A.V 326
Holgate,P 121,326
Holla,M.S 121,123-4,153,326
Homer,L.D 239,326
Horton,R.E 203,326
Hougaard,P 326
Howlader,H.A 326
Hoýland,A 214,218,323
Hsieh,H.K 13,50,51,61,
 104–106,108,110,326
Huff,B.V 326

I

Ikeda,N 337
Iles,T.C 30,31,322
Iliescu,D.V 326
Isogai,T 326
Iwase,K 23,327

Iyengar,S 327

J

Jain,G.C 204,327,328
Jaisingh,L.R 327
Jensen,J.L 320
Johnson,L.G 157,327
Johnson,E.V 294,335
Jones,G 32,327
Jørgensen,B 76,115,143,145,
   147,149,154,327
Joshi,S 73,84,86,87
Juritz,J 159,161,162,336

K

Kabaila,P 314,316,339
Kabe,D.G 337,337
Kalbfleish,J.D 298,299,300,330
Kappenman,R.F 91,327
Karan Singh,R 327
Karlin,S 291,314,327
Kemp,A.W 133,134,328
Kemp,C.D 133,134,328
Kempton.R.A 298,299,328
Kendall,D.G 204,327
Keyfitz,N 191,193,328
Khan,M.S.H 204,327,328
Khatree,R 328
Khatri,C.G 328
Kiang,N.Y.S. 333
Kirkeby,S 337
Kirmani,S.N.U.A 318,328
Kmenta,J 167,331
Kobayashi,Y 328
Kocherlakota,S 317,328
Kokonendji,C.C 328
Konstantinowsky,D 154,328
Korwar,R.M 23,326,328
Koudou,A.E 198,319
Kourouklis,S 328
Kovner,J.L 14,44,58,59,332
Koziol,J.A 328
Krapivin,V.F 328
Kronberg,D 337
Kubala,J.J 206,324
Kumar,S 150,320
Kurze,U.J 259,328

L

Laha,R.G 329
Lancaster,T 223,224,328
Lande,R 286,287,289,290,328
Lánský,P 242,329,335

Laurent,A.G 327
Lawless,J.F 323
Le Cam,L 329
Lee,M.T 152,154,338
Lehmann,E.L 7,12,39,42,96,329
Lemaire,J 134,324
Leslie,P.H 286,287,329
Letac,G 319,329
Levine,M.W 242,243,39
Lewis,P.W.A 147,322
Liao,Q 67,68,69.243,250,329
Lieblein,J 118,329
Lim,S.C 324,336
Lingappiah,G.G 329
Littel,R.C 48,332
Lombard,F 78,79,81,83,
   330
Lotka,A.J 330
Luce,R.D 320
Lukacs,E 49,329

M

MacGibbon,K.B 320,330
Mahmoud,M 330
Malik,H.J 332
Mann,N.R 102,330
Marcus,A.H 259,260,261,330
Maxim,L.D 252,253,255,257,330
McCool,J.I 35,330
McCullagh,P 330
Mehta,J.S 330
Michael,J.R 330
Miller,H.D 260,322
Mitter,P 323
Miura,C.K 40,45.182,183,
   185,330,335
Mora,M 329,330
Morimoto,H 337
Morlat,G 203,329,330
Munholland,P.L 290,298,
   299-301,304,323,330,331

N

Nâdas,A 43,59,60,330,331
Nagi,S 317
Nath,G.B 331
Nebebe,F 321
Nelder,J.A 167,330,331
Nelson,W 206,212,214,218,276,
   282,331
Neufeldt,A.H 232,338
Newby,M 331

O

Oberhofer,W 169,331
O'Brien,C.M 331
Olwell,D.H 309,310,312,325
Ord,J.K 123,124,331
O'Reilly,F.J 114,116,118,120, 331
Orzack,S.H 286,287,289,290, 328,337
Osborn,S.B 339
Ostle,B 331

P

Padgett,W.J 26,27,30,52–57,59, 94–6,99,331,332
Padmanabhan,P 332
Pakes,A.G 5,332
Pal,N 332
Palmer,J 332
Pandey,B.N 332
Pandey,M 332
Pandey,S.K 321
Panjer,H.H 339
Park,E 23,339
Parsian,A 332
Patel,J.K 332
Patel,R.C 37,332
Patil,S.A 14,44,58,59,332
Patwardhan,G 327
Pavur,R,J 116,118,323,332
Pearson,E.S 224,276,332
Pericchi,L.R 333
Perng,S.K 48,333
Perrault,L 320,332
Pierce,D.A 324
Pitts,D 324
Podwysocki,M.H 255,333
Prabhu,N.U 204,324
Proschan,F 147,333

Q

Qasrawi,H 304,333

R

Radil,T 329
Ragab,A 333
Ramachandran,B 333
Rao,C.R 28,69,116,333
Rasmussen,P 320
Ratani,R.T 336
Read,K.L.Q 298,301,318,333
Reiser,B 42,318

Rhyzhik,I.M 13,325
Ricciardi,L.M 320
Rodieck,R.W 333
Rodriguez-Iturbe 332
Rotondi,R 21,319
Roy,L.K 333,337
Rubinstein,G.Z 333
Ruchti,W 325
Rueda,R 114,116,118,120,331
Rukhin,A.L 326

S

Sahin,I 333
Saleh,E.J 329
Salzburg,P.M 73,75,323
Samanta,M 48,49,333,334
Sanjarifarsipour,N 332
Sankaran,M 121,124,133,141,334
Sarndal,C 305,307,334
Saunders,S.C 144,154,155, 156,157,158,320,334
Savage,L.J 235
Schafer,R.E 330
Scheffe,H 7,12
Schrödinger,E 1,154,334
Schuchany,W.R 330
Schwarz,C.J 334
Scott,R.C 290,323,332
Scott,M.J 323
Seal,H.L 167,334
Seshadri,V 1,9,12,42,43,144,154 327,328,329,333,334,328
Setô,N 23,326
Shaban,S.A 334
Shah,M 73,84,86,87,327
Shapiro,S.S 157,158,334
Sheppard,C.W 235,334,335
Sherif,Y.S 33
Shorrock,G.E 330
Shoukri,M.M 335
Shuster,J.J 43,183,188,334,335
Sichel,H.S 123,124,125,127, 133,135,153,335
Silcock,H 220,222,335
Singh,U 337
Singpurwalla,N.D 330
Sinha,B.K 35,317,332
Sinha,S.K 335
Small,A 239,326
Smith,C.E 242,328,335
Smith,M.L 335
Smoluchowsky,M.V 1,335

Snyder,N.F.R 294,335
Soejeti,Z 214,320
Stark,G.E 335
Stein,G.Z 123-5,127-8,153,
 159,161,162,336
Stephens,M.A 322
Stewart,C 323
Stewart,D.A 230,231,324,336
Sundarier,V.H 336

T

Tadikamala,P.R 336
Takagi,R 259,336
Takhashi,R 336
Tang,L.C 61,96-7,99,106,108,
 110-1,156,158,324,336
Tang,W.H 61,317
Tapiero,C.S 336
Taylor,H.M 291,314,327
Teugels,J 121,123,336
Thompson,J.R 57,336
Tomlinson,R.W.S 339
Toren,N 336
Travadi,R.J 336
Tremblay,L 336
Tsoi,S.K 52,54,331
Tuckwell,H 242,337
Tuljapurkar,S.D 286,293,337
Tweedie,M.C.K 1,2,172,173,
 182,273,274,337

U

Uchida,H 326
Uddin,Md.B 332
Uffer,M.B 235,335
Upadhyay,S.K 53,57,337

V

Villmann,H 194,195,337
Vickers,G.G 318
Voda,V.G 326,337
Von Alven,W.H 56,58,59,96,
 101,106,141,337

W

Wald,A 2,73,75-6,84,195,337
Wani,J.K 337
Wasan,M.T 73,75,205,319,
 333,337,338
Washio,Y 338
Watson,G.N 338
Wedderburn,R.W.M 167,331
Wei,L.J 26,27,30,331,97,99,
 104,344
Weiss,G.H. 259,260,338
Weiss,M 237,238,239,338
Whitmore,G.A 6,123,124,139,
 141,152,154,220,222-3,
 232,234,262-4,269,277,
 323,327,329,331,338
Whitten,B.J 33-5,320-2,333
Wicksell,S.D 191,338
Wilbur,S.R 294-338
Wilk,M.B 157,158,334
Williams,C.B 324
Willmot,G.E 121-5,129,134,
 323,336,338,339
Winterton,J 224,331
Wise,K 3314,316,338
Wise,M.E 235,237,239,339
Worsley,K.J 339
Wright,F.T 324

Y

Yalovsky,M 6,338
Yamamoto,E 339
Yamamoto,T 336
Yanagimoto,T 339
Yang,G.L 240,242,339
Yarnold,P 329
Yeh,L 123,124,125,318
Ylvisaker,D 323
Yule,G.U 135,339

Z

Zelen,M 118,329
Zigangirov,K.Sh 2.339
Zucchini,W 336

SUBJECT INDEX

A

Analysis of reciprocals
    172,173,175
    182,183,186
    187-9,273,275-7
Analysis of variance 172
    273,274
Anderson Darling test
    116,118
Average run length 310-3
Average sample number 78
    86,87

B

Bahadur efficiency 70–72
Bayes estimat(or)e 94-6
    142-3
Bessel function 12,67
    69,142,260
Beta 10,11,17,40,183
Bonferroni inequality 65
    158
Bootstrap 277,280-3
Brownian motion 1,38,43
    59,61,78,140
    154-5,175,284

C

Censored data 141,218,269,286
Chisquared(χ^2) 8,9,13,18,41,42
    48,50.53,54,70,71
    79,80,88,91,97,108
    109,111,117,118,133
    135,136,145,148-150
    161,17-5,182,186,188
    190,213-4,221-2,225-9
    300,311-2
Coefficient of variation 26,50
    61,84.133,136,238,261
Compound law 121-3,136
Confidence bounds 96
    1101,107,210
Confidence interval 51,59,60
    61,91,97,106,108
    111,155,158,165,292
Control chart 90,91
Cramer-von Mises test
    116,118
Critical time 103-6,108
    110-1

Cumulant 23,27,236,315
Cumulant transform 237
Cusum (charts) 309-13

D

Deviance 162,186
Dystroph(ic)y 194-7

E

Erlang 136,261
Exponential dispersion
    model 145

F

F distribution 11,40,43,47
    48,49,53-54,174-5
    183,189,262-4,275,277
First passage time 1,58,60
    121,140-1,150,152,154-5
    219,240,285,290

G

Gamma law 3,20,27,30,94-5
    234-6,238,241,284
    311
Gaussian process 117,217
Generalized IG 7,258
Goodness-of-fit 23,114
    135-6,155-7,203

H

Hazard rate 102-3,106,150-3
    241
Hyperbolic normal 242-51

I

Information matrix 94,117
    147,159,165,169,269

K

Kolmogorov-Smirnov 114-5
Kuhn-Tucker 98,11,206
Kurtosis 26,50,122

L

Laplace transform 2,3,5,8
    122,151,193,204
    235,301,315
Least squares 158,188,190
    210-1,219,276
Likelihood 20,27,29,30
    34,38,57,147,165
    184,208,216-8

266,268-9,274,291
300,304
Likelihood ratio 38-40,42,45-51
    61,79,81,94,141,146-7
    168,172-4,182,274-5
Linear models 159
Log likelihood 7,27-31,50
    128-9,141-2,148-50
    159,168,185-6,207
    216,265,300,306,315-6
Lognormal 27,30,37,64-9
    114-5,206,234-5,241-51
    261-2,284,291

M

Maximum likelihood 7,13,23,27-9
    32-3,40,45,50,65-6,70,95
    105,108-11,116,125,127
    129,133,136,141,146,149
    155,159,161,172,176-7
    185-6,208-10,216,251
    254-7,265-6,269,277,283
    285,291,300,315
Minimum variance 7,23
    92,95

N

Natural conjugate 20
    21,137
Nested hypothesis 147,162,185
Newton-Raphson 105,127
    132,161,165
Normal law 1,3,6,9,10,21
    31,33,39,40-2,46
    82,93,117,144,155
    176,242,265,282.292
    299

O

One-way layout 172
Operating characteristic
    74-7,84

P

Poisson(law)process 121-3,153-4
    126,161,163,165,259
    300,315-6
Posterior density 19-21
    60,94,142-3
Power 38,44,45,51,59,79
    82,83,116
Prediction interval 38,52-3
    56,57,264

Prior 19,20,57,94-5,142-3
Pseudo- mle 269,281-2,307
    323

R

Rao-Blackwell 92,116,118
Reciprocal gamma 242-251
Reciprocal IG 7,198-9,201-2,257-8
Regression 262,265,274-5
    277,305,307
Reliability 51,60,73
    92,95,104,111,150

S

Separate families 65,2506
Sequential analysis 2,73,84
    86
Sequential estimation 87
Sequential procedure 78-9,82-3
    87-8
Sequential test 73,76,84
Shot noise process 260
Skewness 26,30-1,33-5,50,122
Spike train 239-242
Student's law 11,20,25-6,40-41
    47,49-50,52,58,79,82
    93-5,97-8

T

Tolerance limits,38 61-65,96,101
Tree 198-201
Two-way classification
    183

U

Uniformly most powerful
unbiased test 39-43,45,47
    70,71
Umvue 12,23-6,90,110
U-statistic 12,13

V

VALT 214

W

Wald's identity 76
Weibull law 27,114-5
    230-1,234,241
Wiener process 96,139,150
    206,215-6,219,220
    224,232,241-2,284
    288,290

GLOSSARY

$\mathcal{L}(X) = IG(\mu, \lambda)$	=	law of the random variable X is inverse Gaussian with parameters μ and λ
$1_{R^+}(x)$	=	the indicator random variable X
$\mathbb{E}(X)$	=	expected value of the random variable X
$\Gamma(\alpha, \beta)$	=	a gamma law with shape parameter α and scale parameter β
$Be(\alpha.\beta)$	=	the Beta function with two arguments α, β
$\Rightarrow$	=	tends to
$\approx$	=	approximates to
$GIG(\alpha, \chi, \psi)$	=	generalized inverse Gaussian law (Halphen's law), where α is a shape parameter, χ is the coefficient of the term in x^{-1} and ψ the coefficient of the term in x
$RIG(\mu, \lambda)$	=	the reciprocal inverse Gaussian law obtained from the generalized inverse Gaussian by letting $\alpha = 1/2, \chi = \lambda$ and $\psi = \lambda/\mu^2$
$M - IG(\mu, \lambda, p)$	=	a mixture inverse Gaussian law with parameters μ, λ, p
$P - IG(\mu, \lambda)$	=	a compound Poisson-inverse Gaussaian law with parametes μ, λ
$\propto$	=	is proportional to
χ_n^2	=	the Chi-squared law (n degrees of freedom)
t_n	=	Student's t (n degrees of freedom)
$F_{a,b}$	=	Snedecor's F (with parameters a,b)
$N_p(0, \Lambda)$	=	p-variate Gaussian with zero mean and covariance matrix Λ
$K_\lambda(.)$	=	Bessel function of the third kind(indexλ)
$\otimes$	=	a symbol for product measure
$\perp\!\!\!\perp$	=	is independent of
tr A	=	trace of the matrix A
X^t	=	transpose of vector(matrix) X
X^{-1}	=	inverse of the matrix X
e	=	identity vector
VALT	=	(variable)accelerated life test
ARL	=	average run length
ASN	=	average sample number
Cusum	=	cumulative sums
LCL(UCL)	=	lower(upper) control limits
LN	=	lognormal
MLE	=	maximum likelihood
(U)MVUE	=	(uniformly)minimum variance unbiased estimator
$\sim$	=	is distributed as
$\in$	=	belongs to
$\xrightarrow{D}$	=	tends in law to
UMPU	=	uniformly most powerful unbiased
SPRT	=	sequential probability ratio test

Lecture Notes in Statistics

For information about Volumes 1 to 63 please contact Springer-Verlag

Vol. 64: S. Gabler, Minimax Solutions in Sampling from Finite Populations. v, 132 pages, 1990.

Vol. 65: A. Janssen, D.M. Mason, Non-Standard Rank Tests. vi, 252 pages, 1990.

Vol 66: T. Wright, Exact Confidence Bounds when Sampling from Small Finite Universes. xvi, 431 pages, 1991.

Vol. 67: M.A. Tanner, Tools for Statistical Inference: Observed Data and Data Augmentation Methods. vi, 110 pages, 1991.

Vol. 68: M. Taniguchi, Higher Order Asymptotic Theory for Time Series Analysis. viii, 160 pages, 1991.

Vol. 69: N.J.D. Nagelkerke, Maximum Likelihood Estimation of Functional Relationships. V, 110 pages, 1992.

Vol. 70: K. Iida, Studies on the Optimal Search Plan. viii, 130 pages, 1992.

Vol. 71: E.M.R.A. Engel, A Road to Randomness in Physical Systems. ix, 155 pages, 1992.

Vol. 72: J.K. Lindsey, The Analysis of Stochastic Processes using GLIM. vi, 294 pages, 1992.

Vol. 73: B.C. Arnold, E. Castillo, J.-M. Sarabia, Conditionally Specified Distributions. xiii, 151 pages, 1992.

Vol. 74: P. Barone, A. Frigessi, M. Piccioni, Stochastic Models, Statistical Methods, and Algorithms in Image Analysis. vi, 258 pages, 1992.

Vol. 75: P.K. Goel, N.S. Iyengar (Eds.), Bayesian Analysis in Statistics and Econometrics. xi, 410 pages, 1992.

Vol. 76: L. Bondesson, Generalized Gamma Convolutions and Related Classes of Distributions and Densities. viii, 173 pages, 1992.

Vol. 77: E. Mammen, When Does Bootstrap Work? Asymptotic Results and Simulations. vi, 196 pages, 1992.

Vol. 78: L. Fahrmeir, B. Francis, R. Gilchrist, G. Tutz (Eds.), Advances in GLIM and Statistical Modelling: Proceedings of the GLIM92 Conference and the 7th International Workshop on Statistical Modelling, Munich, 13-17 July 1992. ix, 225 pages, 1992.

Vol. 79: N. Schmitz, Optimal Sequentially Planned Decision Procedures. xii, 209 pages, 1992.

Vol. 80: M. Fligner, J. Verducci (Eds.), Probability Models and Statistical Analyses for Ranking Data. xxii, 306 pages, 1992.

Vol. 81: P. Spirtes, C. Glymour, R. Scheines, Causation, Prediction, and Search. xxiii, 526 pages, 1993.

Vol. 82: A. Korostelev and A. Tsybakov, Minimax Theory of Image Reconstruction. xii, 268 pages, 1993.

Vol. 83: C. Gatsonis, J. Hodges, R. Kass, N. Singpurwalla (Editors), Case Studies in Bayesian Statistics. xii, 437 pages, 1993.

Vol. 84: S. Yamada, Pivotal Measures in Statistical Experiments and Sufficiency. vii, 129 pages, 1994.

Vol. 85: P. Doukhan, Mixing: Properties and Examples. xi, 142 pages, 1994.

Vol. 86: W. Vach, Logistic Regression with Missing Values in the Covariates. xi, 139 pages, 1994.

Vol. 87: J. Müller, Lectures on Random Voronoi Tessellations.vii, 134 pages, 1994.

Vol. 88: J. E. Kolassa, Series Approximation Methods in Statistics. Second Edition, ix, 183 pages, 1997.

Vol. 89: P. Cheeseman, R.W. Oldford (Editors), Selecting Models From Data: AI and Statistics IV. xii, 487 pages, 1994.

Vol. 90: A. Csenki, Dependability for Systems with a Partitioned State Space: Markov and Semi-Markov Theory and Computational Implementation. x, 241 pages, 1994.

Vol. 91: J.D. Malley, Statistical Applications of Jordan Algebras. viii, 101 pages, 1994.

Vol. 92: M. Eerola, Probabilistic Causality in Longitudinal Studies. vii, 133 pages, 1994.

Vol. 93: Bernard Van Cutsem (Editor), Classification and Dissimilarity Analysis. xiv, 238 pages, 1994.

Vol. 94: Jane F. Gentleman and G.A. Whitmore (Editors), Case Studies in Data Analysis. viii, 262 pages, 1994.

Vol. 95: Shelemyahu Zacks, Stochastic Visibility in Random Fields. x, 175 pages, 1994.

Vol. 96: Ibrahim Rahimov, Random Sums and Branching Stochastic Processes. viii, 195 pages, 1995.

Vol. 97: R. Szekli, Stochastic Ordering and Dependence in Applied Probability. viii, 194 pages, 1995.

Vol. 98: Philippe Barbe and Patrice Bertail, The Weighted Bootstrap. viii, 230 pages, 1995.

Vol. 99: C.C. Heyde (Editor), Branching Processes: Proceedings of the First World Congress. viii, 185 pages, 1995.

Vol. 100: Wlodzimierz Bryc, The Normal Distribution: Characterizations with Applications. viii, 139 pages, 1995.

Vol. 101: H.H. Andersen, M.Højbjerre, D. Sørensen, P.S.Eriksen, Linear and Graphical Models: for the Multivariate Complex Normal Distribution. x, 184 pages, 1995.

Vol. 102: A.M. Mathai, Serge B. Provost, Takesi Hayakawa, Bilinear Forms and Zonal Polynomials. x, 378 pages, 1995.

Vol. 103: Anestis Antoniadis and Georges Oppenheim (Editors), Wavelets and Statistics. vi, 411 pages, 1995.

Vol. 104: Gilg U.H. Seeber, Brian J. Francis, Reinhold Hatzinger, Gabriele Steckel-Berger (Editors), Statistical Modelling: 10th International Workshop, Innsbruck, July 10-14th 1995. x, 327 pages, 1995.

Vol. 105: Constantine Gatsonis, James S. Hodges, Robert E. Kass, Nozer D. Singpurwalla(Editors), Case Studies in Bayesian Statistics, Volume II. x, 354 pages, 1995.

Vol. 106: Harald Niederreiter, Peter Jau-Shyong Shiue (Editors), Monte Carlo and Quasi-Monte Carlo Methods in Scientific Computing. xiv, 372 pages, 1995.

Vol. 107: Masafumi Akahira, Kei Takeuchi, Non-Regular Statistical Estimation. vii, 183 pages, 1995.

Vol. 108: Wesley L. Schaible (Editor), Indirect Estimators in U.S. Federal Programs. viii, 195 pages, 1995.

Vol. 109: Helmut Rieder (Editor), Robust Statistics, Data Analysis, and Computer Intensive Methods. xiv, 427 pages, 1996.

Vol. 110: D. Bosq, Nonparametric Statistics for Stochastic Processes. xii, 169 pages, 1996.

Vol. 111: Leon Willenborg, Ton de Waal, Statistical Disclosure Control in Practice. xiv, 152 pages, 1996.

Vol. 112: Doug Fischer, Hans-J. Lenz (Editors), Learning from Data. xii, 450 pages, 1996.

Vol. 113: Rainer Schwabe, Optimum Designs for Multi-Factor Models. viii, 124 pages, 1996.

Vol. 114: C.C. Heyde, Yu. V. Prohorov, R. Pyke, and S. T. Rachev (Editors), Athens Conference on Applied Probability and Time Series Analysis Volume I: Applied Probability In Honor of J.M. Gani. viii, 424 pages, 1996.

Vol. 115: P.M. Robinson, M. Rosenblatt (Editors), Athens Conference on Applied Probability and Time Series Analysis Volume II: Time Series Analysis In Memory of E.J. Hannan. viii, 448 pages, 1996.

Vol. 116: Genshiro Kitagawa and Will Gersch, Smoothness Priors Analysis of Time Series. x, 261 pages, 1996.

Vol. 117: Paul Glasserman, Karl Sigman, David D. Yao (Editors), Stochastic Networks. xii, 298, 1996.

Vol. 118: Radford M. Neal, Bayesian Learning for Neural Networks. xv, 183, 1996.

Vol. 119: Masanao Aoki, Arthur M. Havenner, Applications of Computer Aided Time Series Modeling. ix, 329 pages, 1997.

Vol. 120: Maia Berkane, Latent Variable Modeling and Applications to Causality. vi, 288 pages, 1997.

Vol. 121: Constantine Gatsonis, James S. Hodges, Robert E. Kass, Robert McCulloch, Peter Rossi, Nozer D. Singpurwalla (Editors), Case Studies in Bayesian Statistics, Volume III. xvi, 487 pages, 1997.

Vol. 122: Timothy G. Gregoire, David R. Brillinger, Peter J. Diggle, Estelle Russek-Cohen, William G. Warren, Russell D. Wolfinger (Editors), Modeling Longitudinal and Spatially Correlated Data. x, 402 pages, 1997.

Vol. 123: D. Y. Lin and T. R. Fleming (Editors), Proceedings of the First Seattle Symposium in Biostatistics: Survival Analysis. xiii, 308 pages, 1997.

Vol. 124: Christine H. Müller, Robust Planning and Analysis of Experiments. x, 234 pages, 1997.

Vol. 125: Valerii V. Fedorov and Peter Hackl, Model-oriented Design of Experiments. viii, 117 pages, 1997.

Vol. 126: Geert Verbeke and Geert Molenberghs, Linear Mixed Models in Practice: A SAS-Oriented Approach. xiii, 306 pages, 1997.

Vol. 127: Harald Niederreiter, Peter Hellekalek, Gerhard Larcher, and Peter Zinterhof (Editors), Monte Carlo and Quasi-Monte Carlo Methods 1996, xii, 448 pp., 1997.

Vol. 128: L. Accardi and C.C. Heyde (Editors), Probability Towards 2000, x, 356 pp., 1998.

Vol. 129: Wolfgang Härdle, Gerard Kerkyacharian, Dominique Picard, and Alexander Tsybakov, Wavelets, Approximation, and Statistical Applications, xvi, 265 pp., 1998.

Vol. 130: Bo-Cheng Wei, Exponential Family Nonlinear Models, ix, 240 pp., 1998.

Vol. 131: Joel L. Horowitz, Semiparametric Methods in Econometrics, ix, 204 pp., 1998.

Vol. 132: Douglas Nychka, Walter W. Piegorsch, and Lawrence H. Cox (Editors), Case Studies in Environmental Statistics, viii, 200 pp., 1998.

Vol. 133: Dipak Dey, Peter Müller, and Debajyoti Sinha (Editors), Practical Nonparametric and Semiparametric Bayesian Statistics, xv, 408 pp., 1998.

Vol. 134: Yu. A. Kutoyants, Statistical Inference For Spatial Poisson Processes, vii, 284 pp., 1998.

Vol. 135: Christian P. Robert, Discretization and MCMC Convergence Assessment, x, 192 pp., 1998.

Vol. 136: Gregory C. Reinsel, Raja P. Velu, Multivariate Reduced-Rank Regression, xiii, 272 pp., 1998.

Vol. 137: V. Seshadri, The Inverse Gaussian Distribution: Statistical Theory and Applications, xi, 360 pp., 1998.